Bioscience
at the
Physical Science Frontier

BIOSCIENCE
AT THE
PHYSICAL SCIENCE FRONTIER

Proceedings of a Foundation

Symposium on the 150th Anniversary

of Alfred Nobel's Birth

Edited by

CLAUDIO NICOLINI

Chair of Biophysics, School of Medicine,
University of Genova, Genova, Italia

HUMANA • CLIFTON, NEW JERSEY

Crescent Manor
PO Box 2148
Clifton, NJ 07015

Printed in the United States of America

This volume contains the proceedings of the Symposium held to celebrate the 150 birthday of Alfred Nobel that was held in San Remo, Rome.

This volume comprises volume 9 of CELL BIOPHYSICS.

Preface

Since early Greek and Roman times, atoms were assumed—after undergoing their various interactions—to take on the stable configurations of either the living or the inanimate world. This simple and unitary theory has evolved markedly, even while maintaining its validity over several centuries of vicissitudes, and in essence constitutes the first example of a synthesis between the physical and the life sciences.

In modern times, a similar relationship between the structure of various macromolecules and the function of living cells has also emerged as one of the most striking findings of those scientists active at the convergence of the physical and life sciences. This fundamental result is reported in the first two sections of the present work, namely those on "Oncogenes and Cancer" and "Lower-to-Higher-Order DNA Structure," in many cases by those recent Nobelists who themselves have been major contributors to work at the intersection of these fields.

Many other significant areas of forefront scientific inquiry today (for example, research on the brain and vision), as well as some of the most exciting technological developments (e.g., work on renewable fuels and materials, biotechnology, and NMR and X-ray tomography) and methodological advances (e.g., studies in the fields of statistical mechanics and cancer chemotherapy) depend on the complex but harmonious collaborations of physicists, chemists, mathematicians, engineers, biologists, and physicians.

This cooperation among scientists across a variety of disciplines not only opens exciting new frontiers, but also permits a fresh synthesis of both theoretical and empirical knowledge through its frequent interconnection of expertises and its encouragement of role exchange among the traditionally compartmented soft and hard sciences.

The Symposium on "New Frontiers at the Crossing of Life and Physical Sciences"—held not long ago in San Remo (a beautiful town on the Italian Riviera) to celebrate the 150th birthday of Alfred Nobel, along with four other Symposia in other scientific fields that saw the participation of over 30 Nobel Prize winners, all under the sponsorship of the Nobel Foundation and the City of San Remo with the patronage of the King of Sweden and the President of Italian Republic—contains their reports on many representative topics on today's new frontiers. With two exceptions (i.e., James Watson and Alexander Rich), all the lectures delivered within that Symposium—which I had the privilege to chair—have been included in the book. It has been a sometimes difficult but always exciting undertaking, and I hope that the good fruit of our symposium, now happily in hand, maintains the spirit of that unique occasion.

Claudio Nicolini

Contents

Preface, CLAUDIO NICOLINI v

List of Contributors ix

Introduction

LINUS PAULING
One Aspect of the Physical Sciences in Relation to Biology 3

Section I Oncogenes and Cancer

HOWARD M. TEMIN
Retroviruses and Evolution 9

DAVID BALTIMORE
The Beginning of the Molecular Description of a Cancer 17

Section II Lower-to-Higher DNA Structure

E. M. BRADBURY AND J. P. BALDWIN
Neutron Scatter and Diffraction Techniques Applied to Nucleosome and Chromatin Structure 35

C. NICOLINI
Nuclear Structure: From the Pores to the High-Order Gene Structure 67

Section III Brain and Vision

DAVID H. HUBEL
Blobs and Color Vision 91

LEON N COOPER
Neuron Learning to Brain Organization 103

DONALD GLASER
A Physicist's View of Vision 145

Section IV Technological Developments

A. M. CORMACK
Scanning in Medicine and Other Fields 151

HARLYN O. HALVORSON
Genetic Engineering: A New Biotechnology 171

MELVIN CALVIN
Renewable Fuels and Materials: Oil From Plants 189
PAUL LAUTERBUR
NMR Imaging in Biomedicine 211

Section V Methodological Developments

ILYA PRIGOGINE
Life and Physics: New Perspectives 217
LEO SACHS
Cell Differentiation and Malignancy........................... 225
GIANNI BONADONNA
New Strategies in Cancer Chemotherapy 243
WILLIAM F. RAUB
Summary and Prospects 257
AUTHOR INDEX 263
SUBJECT INDEX 265

Contributors

J. P. BALDWIN • *Department of Physics, Liverpool Polytechnic, Liverpool, United Kingdom*

DAVID BALTIMORE • *Department of Biology, Whitehead Institute for Biomedical Research, Massachusetts Institute of Technology, Cambridge, Massachusetts*

GIANNI BONADONNA • *Division of Medical Oncology, Instituto Nazionale Tumori, Milan, Italy*

E. M. BRADBURY • *Department of Biological Chemistry, School of Medicine, University of California, Davis, California*

MELVIN CALVIN • *Department of Chemistry and Lawrence Berkeley Laboratory, University of California, Berkeley, California*

LEON N COOPER • *Department of Physics and Center for Neural Science, Brown University, Providence, Rhode Island*

A. M. CORMACK • *Physics Department, Tufts University, Medford, Massachusetts*

DONALD GLASER • *Physics Department, Tufts University of California at Berkeley, Berkeley, California*

HARLYN O. HALVORSON • *Rosenstiel Basic Medical Sciences Research Center, Brandeis University, Waltham, Massachusetts*

DAVID A. HUBEL • *Department of Neurobiology, Harvard Medical School, Boston, Massachusetts*

PAUL C. LAUTERBUR • *Department of Chemistry, State University of New York at Stony Brook, Stony Brook, New York*

C. NICOLINI • *Chair of Biophysics, School of Medicine, University of Genova, Viale, Italy*

LINUS PAULING • *Linus Pauling Institute of Science and Medicine, Palo Alto, California*

ILYA PRIGOGINE • *Universitę Libre de Bruxelles, Brussels, Belgium*

WILLIAM F. RAUB • *National Institutes of Health, Bethesda, Maryland*

LEO SACHS • *Department of Genetics, The Weizmann Institute of Science, Rehovot, Isreal*

HOWARD M. TEMIN • *Department of Oncology, McArdle Laboratory, University of Wisconsin, Madison, Wisconsin*

INTRODUCTION

One Aspect of the Physical Sciences in Relation to Biology

LINUS PAULING

Linus Pauling Institute of Science and Medicine, Palo Alto, California 94306

About forty years ago I decided to write a book about "One Aspect of the Physical Sciences in Relation to Biology," which is the topic of this meeting. At the time, I went so far as to formulate the title "The Molecular Basis of Biological Specificity," later I changed the title to "The Molecular Basis of Life," and now I have even gone so far as to write a few pages of the book. This is what I shall relate here, "One Aspect of the Physical Sciences in Relation to Biology."

It represents, in part, a portion of my life for about twenty years, beginning in 1929 to 1947 or 1948, when I nearly felt satisfied; and when I say "satisfied," I mean I felt satisfied that I understood "life." It started in 1929 because it was the year when Thomas Hunt Morgan and a good number of his associates, namely, those who had been involved with him in discovering the gene (Sturtevant and Bridges), came to Pasadena to start, at the new California Institute of Technology, a division of Biological Sciences.

I became well acquainted with these biologists, including Tyler and Emerson and other younger ones who came along as graduate students, and I found very quickly that I was puzzling over the problem of biological specificity. Morgan, for example, was working on self-sterility in Ciona, which was a difficult problem to explain: How Ciona can be sterile against themselves, but fertile against nearly all other individual Ciona. Well, of course, there are many other examples of biological specificity. I do not know whether at first I had any feeling, real feeling, that I would understand biological specificity in the course of time, but then something happened in 1936.

I had been working with Charles Coryell, one of my students, on the magnetic and other properties of hemoglobin. When I gave a seminar on that subject at Rockefeller Institute for Medical Research in 1936, Karl Landsteiner asked me to come to his laboratory, which I did the next day. In those days, when it took five days to go from California to New York, one did not go for a day and return the next day; one stayed a week and went to musical comedies, museums, and such during the week. So, I was able to see Landsteiner the next day. He asked me how I would explain the specificity of interaction of antibodies and antigens, and I replied that I could not. In fact, Landsteiner later spent a week with me, in 1937, giving me a sort of private course in the field of immunology that was very valuable because I had found the literature pretty confusing. So, in 1940 I published a paper on the nature of antibodies, the structure of antibodies, and the nature of serological reactions.

A young immunochemist (Dan Campbell) came to work with me as a Rockefeller fellow, then went back to Chicago for a year, and finally returned and stayed on as a member of our department. Dan Campbell, David Pressman, and many other students and visiting scholars worked with me in this field, so that by 1948 I was convinced that many of the ideas I had published in the 1940 paper were correct, but that one of them was incorrect, although I am now trying to find some way of salvaging it. That day I was finally satisfied with my understanding of the nature of life.

Since that time, I have often thought that I ought to sit down and write the book. The one question I have tried to answer is how a human being is able to synthesize macromolecules that are characteristic of him or her. There are clearly two simple alternative ideas about this: One is that a macromolecule with a certain structure may interact with other molecules in such a way as to stabilize the complex of two identical molecules; the other is that a macromolecule strongly interacts with another molecule if it is complementary in structure to the first molecule. This idea of complement pairing of the structure is now in the elementary biology textbooks. As far as I can see, it is just assumed, with no emphasis on the idea, that the basis of biological specificity is molecular complementariness. The fact is ignored that fifty years ago this was not recognized. In 1940, after I had written my paper on molecular complementariness and the interaction of antibodies and antigens, Max Delbruck met me on the campus outside the Atheneum and asked if I had seen the paper by Pascual Jordan, who collaborated with Born in developing matrix mechanics. I had not seen it. In this note, Jordan said that because of the quantum mechanical principles of resonance, two identical molecules interact more strongly than two molecules that are not identical. I said, without having read the paper, "I am sure this is nonsense, because the difference in interaction energy of identical and nonidentical macromolecules is so small compared to the energy of thermal agitation, that this cannot possibly be correct. The gene must consist of two mutually complementary molecules that can separate, so that each of them can act as the template for the other." We published a little note to that effect in 1940, and I gave many lectures about it (Jim Watson and Francis Crick attended my lectures) and this idea was ultimately recognized as being correct.

I am pleased that I was born in 1901, two months after Planck had published his paper about quantum theory, containing the Planck constant, and that I have lived through this extraordinary period, when our understanding of the world has increased so greatly.

When I was a boy, in fact up until 1928–1929, nobody knew why argon or xenon gases condensed to liquids. It was not until 1929 that the physicists developed an understanding of the Van der Waals force of attraction. In 1918, Kesom, in Holland, attempted to discuss the Van der Waals attraction of hydrogen chloride moelcules to one another. You have electric dipole moments, and if they are oriented "head-to-tail," i.e., "plus to minus," there is stability, if the other way, instability. Kesom was able to show that this gives rise to an inverse sixth-power Van der Waals energy of attraction, but it is far too small to explain the Van der Waals forces of HCl. Debye introduced the idea that the electric dipole moment would polarize the electrons in another molecule, and this added somewhat to Van der Waals forces of attraction. This was about 1919 or 1920, but it was still not enough, by a large factor of two or three, to explain the observed Van der Waals forces of attraction.

Finally, in 1928 and 1929, Fritz London developed the quantum mechanical theory of Van der Waals attraction that applies whether a molecule has a permanent dipole moment or not. For example, the xenon atom has electrons that are described as being spherically symmetrical around the nucleus, with no electric dipole moment. The electrons are more or less independent of one another: Sometimes they pile up on one side, creating an instantaneous electric dipole moment that polarizes another xenon molecule that in turn gives rise to the principal kind of Van der Waals force of attraction.

In 1920, in Berkeley, Latimer and Rodebush wrote their paper about the hydrogen bond. This formation of a bond involving a hydrogen atom interacting with a pair of electrons on another atom to give a weak bond, which contributes to weak intramolecular interactions, is of course fundamental to biology. So, by 1940 we had a good understanding of the nature of intramolecular forces. Actually, at that time, Landsteiner asked me to write a chapter on intramolecular forces for the second edition of his book on serological reactions, which I did.

In conclusion, by 1948, when my students, my colleagues, and I had finished publishing our long series of papers on serological reactions, I felt satisfied that I knew enough physics to understand how atoms and groups of atoms interact with one other, and how these interactions can give rise to living organisms. It is detailed molecular complementariness that is responsible for life.

Section I ONCOGENES AND CANCER

Retroviruses and Evolution

HOWARD M. TEMIN

McArdle Laboratory, University of Wisconsin, Madison, Wisconsin 53706

INTRODUCTION

This discussion will deal with one of the exceptions to the grand ideas that laid the basis of all modern work in molecular biology. This will also serve as an introduction to the next two articles. Together, the three articles will cover an entire area of research.

Figure 1 is a reminder of the general scheme of information transfer in living cells—DNA transfers information to RNA, RNA transfers information to proteins, and a cell divides, giving rise to two daughter cells.

Some modification of this scheme is seen after infection by the usual kind of viruses. DNA viruses take control of the infected cell—the viral DNA becomes the master of the cell and serves as a template for formation of viral RNA. The viral RNA is translated into viral proteins. The RNA viruses short circuit this entire scheme of information transfer as a result of the viral RNA acting as the master template for further RNA synthesis. In addition, the cells infected with these DNA and RNA viruses usually die.

The viruses I am going to deal with are called retroviruses. Retroviruses are viruses that have RNA as their genetic material, but, upon infection of cells, transfer their information to DNA (*1*). Thus, they reverse the usual flow of information in cells. The newly synthesized viral DNA does not supplant the action of the cell DNA, but is inserted into it to make a provirus. Both the proviral DNA and the cellular DNA act as templates for the formation of messenger RNA, which translated into proteins. However, the infected cell now makes virus as well as progeny cells.

Retroviruses can also infect germ-line cells and become part of the germ-line. Then the virus genome is transmitted as a normal cellular gene. Such viruses are known as endogenous viruses, in distinction to those infecting somatic cells, which are known as exogenous viruses.

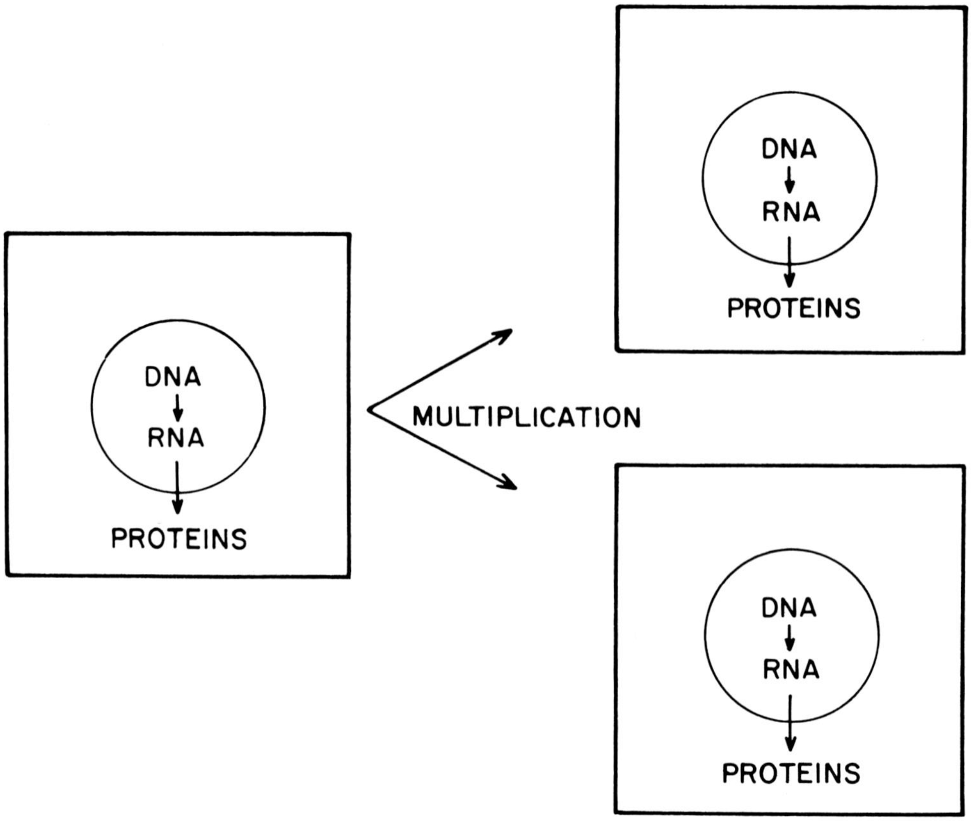

Fig. 1. Information transfer in cells.

INFORMATION TRANSFER IN RETROVIRUS REPLICATION

The retrovirus particle contains two identical molecules of RNA, called plus strands because they have the polarity of the viral messenger RNAs. (The reasons for the existence of two identical RNA molecules are not entirely clear. They may relate to the details of viral DNA synthesis. However, in any event, the virus is not effectively diploid.)

In the infected cell the viral RNA acts as a template for the synthesis of a double-stranded linear molecule of DNA. This DNA is transferred to the nucleus, where it is ligated to form circular molecules and then integrated with cell DNA. The integrated cell DNA acts as a template for viral RNA synthesis.

Figure 2 shows these steps with more details of the structures. Only one of the molecules of viral RNA is shown. It has some special features: At the 5′-end there is a small sequence that is also found at the 3′-end,

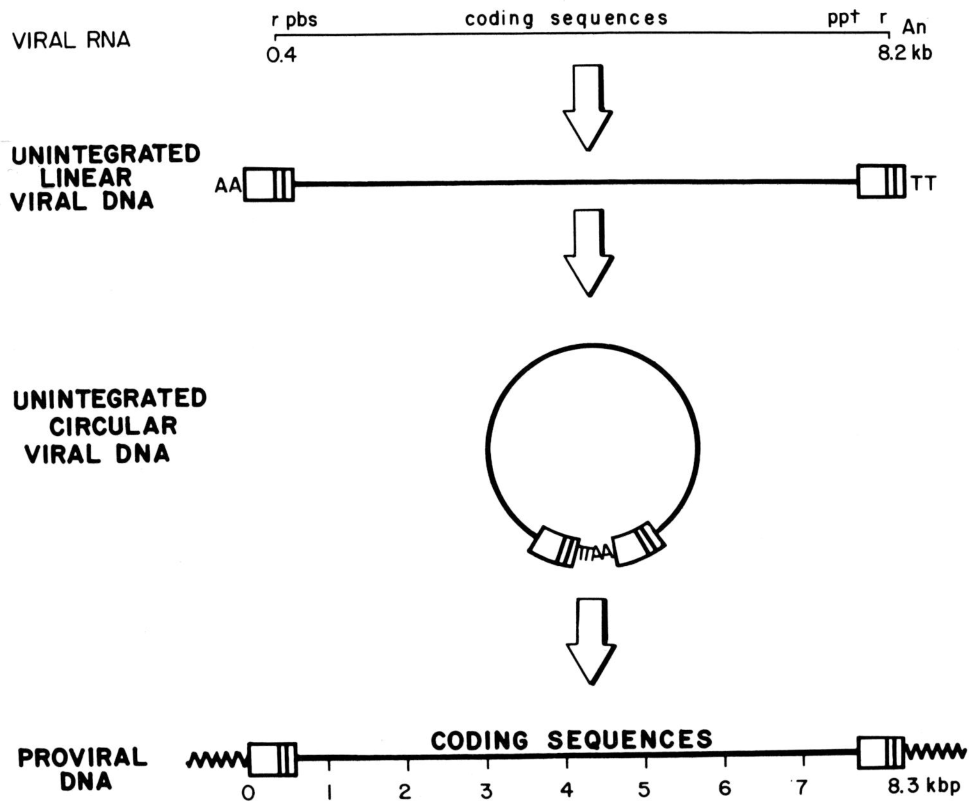

Fig. 2. Information transfer in retrovirus infection. Abbreviations are defined in the text (from ref. *16*)

the repeat (r); near the 5′-repeat is the primer binding site (pbs); and near the 3′-repeat is the polypurine tract (ppt). The latter two are important in viral DNA synthesis. [There is also a 3′-poly(A) sequence (An).] During information transfer from RNA to DNA, sequences from near the repeat up to the pbs or ppt are repeated, such that viral DNA contains a larger terminal repeat, called the LTR (box in Fig. 2).

At the end of the LTR in unintegrated viral DNA, there is an inverted repeat of 5–15 base pair (bp). In the virus we study, spleen necrosis virus (SNV), this inverted repeat is 5 bp—AATGT . . . . . ACATT (*2*). Upon integration, 2 bp are cut off the inverted repeat and a direct repeat of 5 bp of cell DNA is formed.

Thus, there are many different inverted and direct repeats in this series of structures. A similar pattern of inverted and direct repeats is also found in other genetic structures—movable genetic elements and bacterial tranposable elements (*3*).

SIMILARITIES OF RETROVIRUSES AND MOVABLE GENETIC ELEMENTS

Movable genetic elements (in eukaryotes) and transposons (in prokaryotes) have the characteristic of not being essential to cells (that is, a cell may have none of a particular element), of being in different locations in different cells, and of transposing from one location to another.

In particular, as shown in Fig. 3, a bacterial transposon carrying antibiotic resistance and a movable genetic element from *Drosophila* have the same pattern of direct and inverted repeats of cell and element DNA found in retroviruses. Furthermore, when the ends of these elements were sequenced, it was found that all the eukaryotic cellular movable genetic elements of this type had the same nucleotides at their ends—TG . . . . . . CA (*4,4a*). In some cases there are further nucleic acid homologies in the terminal sequences (*5*).

This similarity suggests that there may be evolutionary homology among these elements. This hypothesis was strengthened by a recent report of retrovirus-like particles and circular DNA in *Drosophila* cells carrying *copia* elements and a reverse transcription step in transposition of Tyl in yeast (*6,6a*).

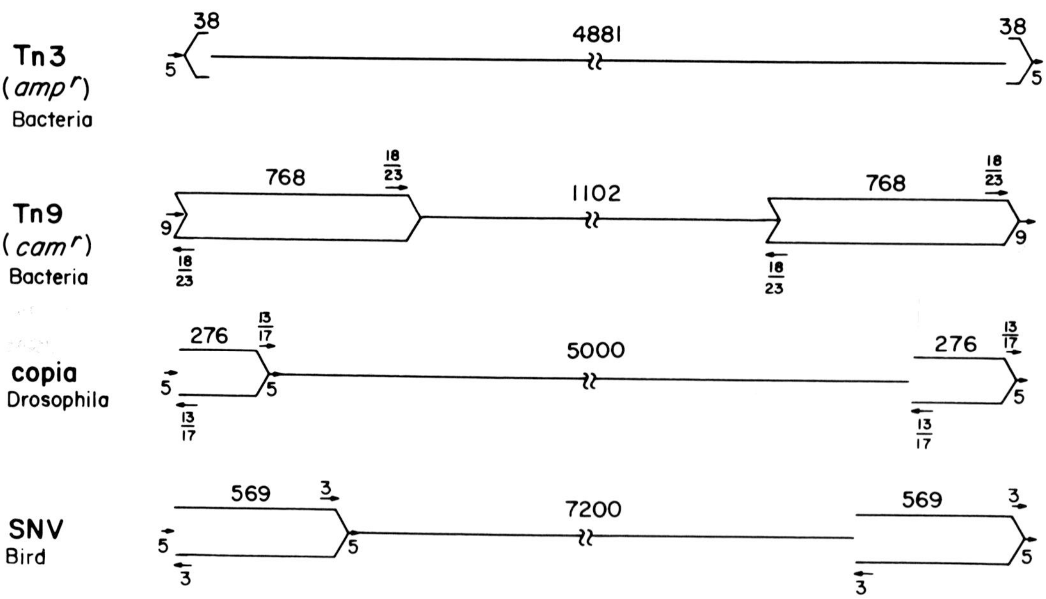

Fig. 3. Structure of bacterial transposons, *Drosophila* movable genetic element, and avian retrovirus DNAs. *amp*ʳ Is ampicillin resistance; *cam*ʳ is chloramphenicol resistance; SNV is spleen necrosis virus

ROLE OF INVERTED REPEATS

We wanted experimentally to ask: What is the significance of the inverted repeats for retroviruses? To do this, we made small deletions at the ends of a retrovirus LTR from a nearby restriction enzyme cleavage site (*7*). We then transfected cells with virus DNA containing these deletions in both LTRs and measured three aspects of viral replication: production of infectious virus, formation of unintegrated viral DNA, and formation of integrated viral DNA.

When we made deletions starting 8 bp from the 5′-end of the LTR, going into the LTR, we found normal amounts of virus and of unintegrated and integrated DNA. When we made deletions that removed all of the inverted repeat and ppt, we got no virus, no unintegrated DNA and no integrated DNA.

When we made deletions going just up to the inverted repeat or just into it, we found a new phenotype. There is no viral DNA integration, but there is formation of reduced amounts of unintegrated viral DNA and infectious virus.

Analogous results were obtained with deletions at the other end of the virus LTR, i.e., deletions of the inverted repeat and PBS abolished all relication, whereas deletions going up to the inverted repeat abolished only integration.

Thus, we can conclude that the conservation of the terminal inverted-repeat sequences in cellular movable genetic elements relates to the essential role these play in integration (or transposition).

OTHER ELEMENTS USING REVERSE TRANSCRIPTION

Another relationship of retroviruses and other genetic systems has recently been suggested. Summers and Mason have demonstrated that hepatitis B-like viruses replicate through an RNA intermediate, although previously these viruses appeared to be orthodox DNA viruses, and they do not normally integrate in replication (*8*). Furthermore, recently, the same suggestion has been made about the plant DNA virus, cauliflower mosaic virus (*9*). These two viruses would then be similar to the nonintegrating retroviruses, except that the DNA rather than the RNA stage would be packaged.

We can then propose that all of the genetic elements using reverse transcription are genetically related, as shown in Fig. 4. Unfortunately, except for the experimental formation of nonintegrating retroviruses from integrating retroviruses and the introduction of retroviruses into the germ-line, we cannot determine the direction of these evolutionary

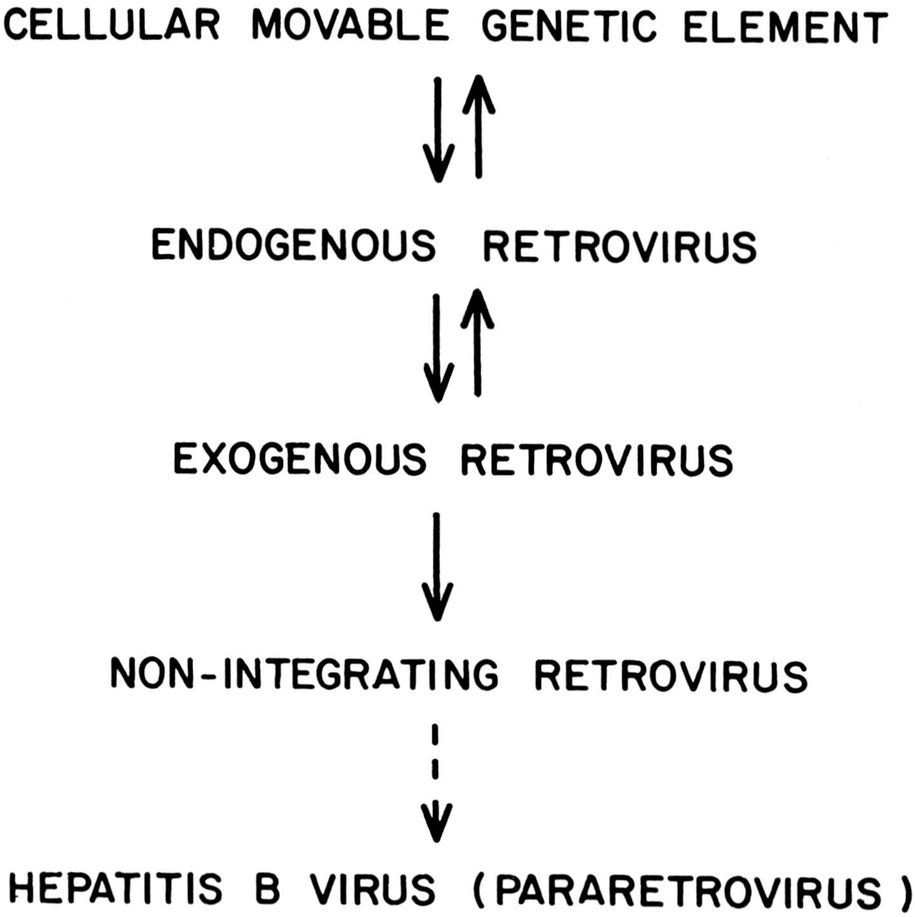

Fig. 4. Evolution of genetic elements using reverse transcription.

relationships (*1,7*). I, however, believe that at some point retroviruses did evolve from cellular, movable genetic elements (*4,4a*).

In addition, there are other DNA sequences in the eukaryotic genome that may be related to this evolutionary pathway. These are the elements that appear to have used reverse transcription in their evolution. They include small nuclear RNA genes and Alu family sequences, cDNA or processed genes, and repeated DNA sequences with similar structures, for example, the *Drosophila* F and G elements (*4,4a,10–12*).

All of these have DNA sequences that suggest they were transposed through an RNA intermediate. The hallmarks are a poly d(A) tract at one end, a direct repeat around the element, and, in the case of cDNA genes, loss of intervening sequences.

Genes with this structure have been described for globulin, globin, tubulin, and others. The frequency of this kind of transposition has not been high in the germ-line, but the products have accumulated in the genome. At present, we cannot say whether these extra sequences have any functional role. They may be duplicated as a result of nucleotide sequence relationships to cellular movable genetic elements or there may be a cellular process for their transposition.

This kind of genetic transposition, if repeated, may explain the origin of retroviruses from small genetic elements. By a process of transposition and deletion, the original element might enlarge and include sequences that code for retrovirus proteins (*13*). Such a structure would then become a virus when it gained sequences for packaging the RNA in the viral proteins.

We have performed deletion experiments that localize the encapsidation sequences between the primer binding site and the start of the *gag* gene (*14*). Thus, almost all of the sequences required for a retrovirus to function, except for the coding sequences, are found at the ends of the virus genome.

We can experimentally substitute other DNAs into the middle of the retrovirus and form virus vectors to introduce exogenous DNA into cells (*15*). In addition, it has been found that highly oncogenic retroviruses have usually deleted many of their coding sequences and substituted modified cellular sequences to make an active oncogene (*1*).

Thus, the processes involved in retrovirus evolution are still occurring and can be employed in useful ways.

SUMMARY

A wide variety of different genetic elements and DNA sequences have used reverse transcription in their replication or transposition. Some of these elements may be evolutionarily related and show a continued evolution to autonomy from the cell genome. The evolution of a cancer reflects a similar drive for autonomy from the organism.

ACKNOWLEDGMENTS

The work from my laboratory is supported by PHS grants CA-07175 and CA-22443 from the National Cancer Institute. I am an American Cancer Society Research Professor.

REFERENCES

1. Weiss, R., Teich, N., Varmus, H., and Coffin, J. (eds.) (1982), *RNA Tumor Viruses*, 2nd ed., Cold Spring Harbor Laboratory, Cold Spring Harbor, NY.
2. Shimotohno, K., Mizutani, S., and Temin, H. M. (1980) *Nature* **285,** 550.
3. Shapiro, J. A. (ed.) (1983) *Mobile Genetic Elements,* Academic, NY.
4. Temin, H. M. (1980), *Cell* **21,** 599.
4a. Temin, H. M. (1985), *Mol. Biol. Evol.,* **2,** 455.

5. Kugimiya, W., Ikenaga, H., and Saigo, K. (1983) *Proc. Natl. Acad. Sci. USA* **80,** 3193.
6. Shiba, T., and Saigo, K. (1983), *Nature* **302,** 119.
6a. Boeke, J. D., Garfinkel, D. J., Styles, C. A., and Fink, G. R. (1985), *Cell* **40,** 491.
7. Panganiban, A. T., and Temin, H. M. (1983), *Nature* **306,** 155.
8. Summers, J. and Mason, W. S. (1982), *Cell* **29,** 403.
9. Pfeiffer, P., and Hohn, T. (1983), *Cell* **33,** 781.
10. Temin, H. M. (1982), *J. Cell. Biochem.* **19,** 105.
11. Sharp, P. A. (1983), *Nature* **301,** 471.
12. DiNocera, P. P., and David, I. B. (1983), *Nucl. Acids Res.* **16,** 5475.
13. Shimotohno, K., and Temin, H. M. (1981), *Cold Spring Harbor Symp. Quant. Biol.* **45,** 719.
14. Watanabe, S., and Temin, H. M. (1982) *Proc. Natl. Acad. Sci. USA* **79,** 5986.
15. Temin, H. M. (1984), *Recombinant DNA Research and Virus,* (Becker, Y., ed.) Martinus Nijhoff, B. V., The Hague, Netherlands, pp. 3–14.
16. Panganiban, A. T., and Temin, H. M. (1984) *Cell* **36,** 673.

Oncogenes

The Beginning of the Molecular Description of a Cancer

DAVID BALTIMORE

Whitehead Institute for Biomedical Research, Massachusetts Institute of Technology, Cambridge, MA 02139

Let me continue from where Howard Temin left off. That is, to deal with the genes that can become incorporated into retroviruses and enable a retrovirus to change from a relatively innocuous virus to one that can cause cancer with great efficiency and speed.

Cancer itself is an alteration of the normal pattern of differentiation during the growth and migration of cells. It entails an unbelievably fundamental alteration in many other properties because a cell changes from one that participates in the normal life of the body to one that grows in an anomalous fashion. How can such a fundamental alteration in behavior be induced? The best way to answer this is to study the viruses that cause cancer.

That viruses do in fact cause cancer was first pointed out to me by James Watson in 1958. Cancer-causing viruses have associated with them the minimal amount of genetic material needed to induce the transformation of normal growth into tumorous cells. Thus, we study viruses, at least partly, because of the expectation of finding packed into them and their genetic material all the information needed to develop a minimal understanding of normal cell transformation.

The viruses we study come to us from natural historians, i.e., people who have looked at the induction of cancer in domestic or wild animals. It has turned out that most of these viruses are retroviruses having the capability of replicating detailed structures and behavior, as described earlier by Howard Temin; but retroviruses do not normally transform cells. The life cycle of a retrovirus, in which it uses its own genes to enable its own reproduction, is one that has no effect on, or is at all necessary for, cell growth. How can it be that viruses that grow perfectly well without changing cells, can also become viruses that transform a wide range of cells, causing tumorous growths throughout the entire body of experimental animals. There are really only two hypotheses about this transformation that are seriously entertained: In the first case, the virus acquires some new genetic capability; and in the second case it works via interaction with normal cellular genetic information. Since viruses integrate themselves (or at least a copy of themselves) into normal cells, one can imagine that the integration process itself is somehow able to induce the transformation of normal cells into tumor cells. The discoveries of the last few years have shown that both of these notions are correct, that viruses can acquire new genetic notions and cause tumors, or can cause tumors either at the site of integration or by the integration process itself, in a sense. I am mainly going to discuss the first form of induction of tumors, only briefly mentioning the second form, that of the induction of tumors by acquisition, in a retrovirus structure, of new genetic information. There are several essential questions to be answered here: From where does the genetic information come? How many genes does it take? Can a single gene cause a cancer? Does it take multiple genes to produce oncogenic information? The answers to these questions are not so clear. It is clear, however, that viruses can cause cancer by acquisition of a single gene. Thus, a single gene is capable of causing cancer, that is, a single gene added to a single cell.

A tumor virus, the Rous sarcoma virus, known to cause cancer was in fact the first retrovirus ever isolated, although it was not known in those days how to subclassify viruses. Rous first identified this virus in 1911, and was subsequently awarded a Nobel prize 55 yr later because it took that long to understand how fundamental his discovery had been.

Work in just the last 10 yr has shown that the Rous sarcoma virus is different from the basic retrovirus structure in that it has acquired a single new gene from the cell in which it grew, but, as has been shown recently, it critically modified it so that the gene found in the Rous sarcoma virus can no longer be identified as a normal cellular gene: it is a slightly derivatized form, although there is no formal proof that that derivatization is what changed it from a normal gene into a tumor-inducing gene. To distinguish a normal gene from a tumor-producing one, we call the normal gene a protooncogene and the tumor-producing gene an oncogene, from oncogenesis, i.e., the process of producing cancer.

The virus that I am going to describe is similar to the Rous sarcoma virus in that it has acquired a new cellular gene; but because it is the one we studied and with which we are most familiar, I will use it as an illustration, it is called the Ableson murine leukemia virus (Fig. 1). This virus has the very handy property of having been used in a controlled experiment, whereas most viruses people have worked with came from a natural population, and their history is not as well understood.

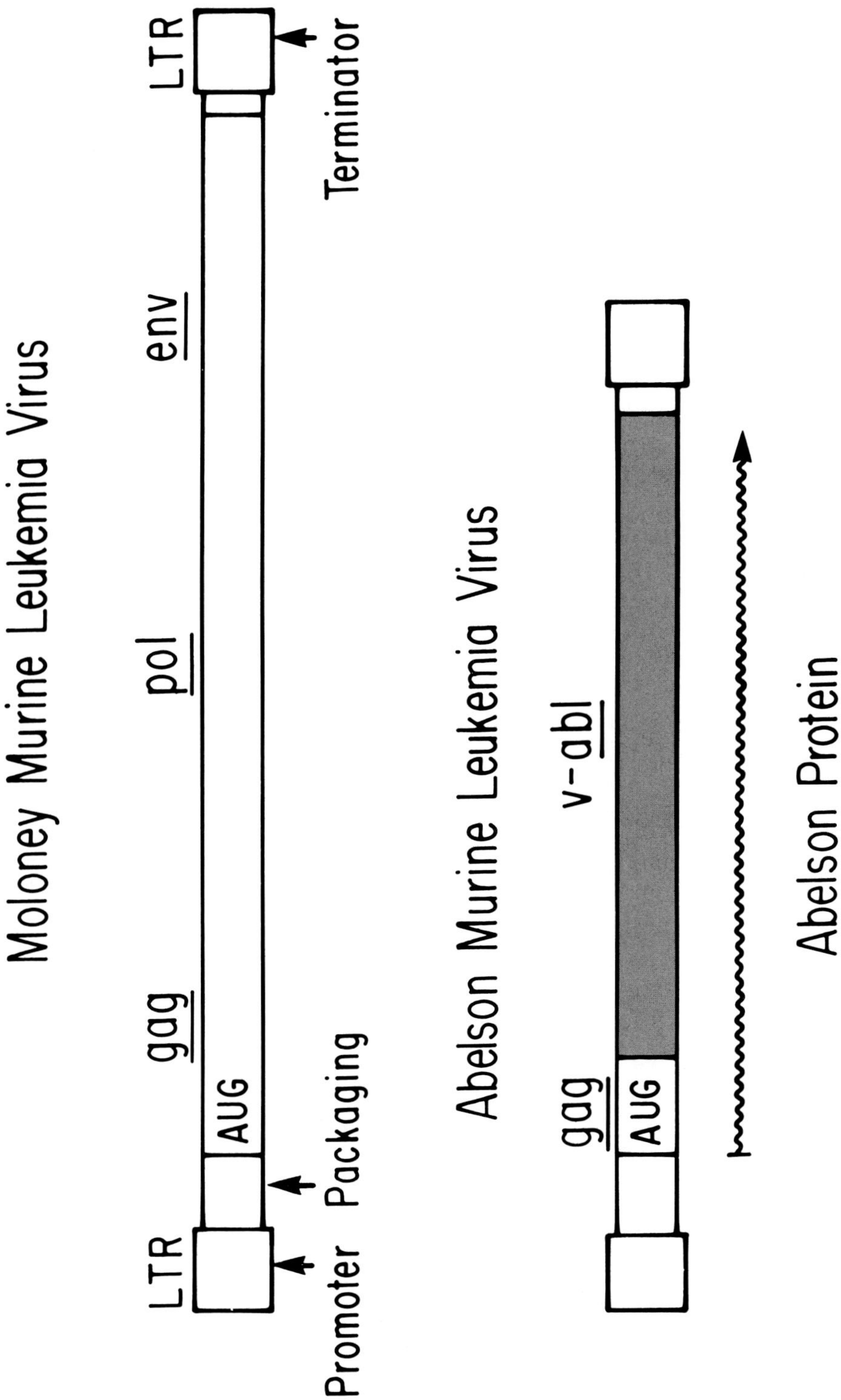

Fig. 1. Structure of A-MuLV genome.

In an experiment in which Dr. Abelson infected a steroid-treated mouse with a relatively innocuous virus—for these purposes we can consider it an innocuous virus, although it actually can cause a tumor—the animal came down with an unknown disease, quite different from any yet identified, that could be passed from animal to animal. This virus was named the Abelson murine leukemia virus.

This disease is one in which large numbers of tumor cells appear in the spleen and bone marrow of the infected animal, and the animal dies within 3–5 wk after innoculation with the virus. The cells found growing were early cells of what is called a B-lymphoid series of cells, which make antibodies. At the end of the paper I will make a slight digression to discuss antibody synthesis, as a footnote to Dr. Pauling's opening remarks. This virus can cause a tumor to form in blood cells because the cells that are involved in making antibodies are ordinarily circulated for part of their life in the blood. This virus can also transform fibroblast cells growing in monolayers and thus has two kinds of activities. I will later return to the genetics of these two kinds of transforming activities, i.e., the ability to transform adherrent cells and the ability to transform blood cells.

The power of this virus is shown in a very simple experiment. We can make the tumor cells simply in culture (althouth animals injected with the virus will grow these tumor cells) by taking normal bone marrow out of a normal mouse, exposing it to the virus, and then growing it in a semisolid medium. What happens is visually obvious; colonies of cells grow in the agar under the influence of the virus, whereas if we use any other kind of virus that we know about or no virus at all, we find absolutely no growth in mice. These growing cells can be transferred from the agar, grown into a continuous population of cells, and innoculated back into the animals from which they came, causing tumors to grow. Thus, the process of tumor induction has occurred entirely in culture. Repeat experiments done in many different systems over the years quite clearly show that the virus has acquired the ability to be a tumor-producing virus.

Actually, we can study these cells and determine what they are by looking at their gene organization, because they happen to be cells of the antibody-forming system, and this system is one in which the differentiation of cells and the acquisition of specific functions by cells is molded by changes at the level of DNA organization. It is one of those situations that was described earlier, one in which DNA is fundamental to the acquisition of a specific cellular process. Thus, we can state quite clearly that in the differentiation of cells of the immune system, different tumor cells fall somewhere between stem cells of the immune-system cells that are just beginning to rearrange their heavy-chain genes and cells that have acquired immunoglobulin (acquired antibody molecules at their cell surface).

This process, which takes a few days in animals, first involves the rearrangement of genes on one chromosome and then the rearrangement of genes on another one or two chromosomes, followed by the acquisition of antibody specificity. Is it the tumor virus that is causing these cells, which would ordinarily differentiate, to stop their differentiation and then grow as continuous lines of cells? The virus is able to carry out this quite specific process because of the acquisition of cellular genes. Let me next tell you how we found that cellular gene and were able to study it.

The methods of recombinant DNA analyses allow us to pull out from the cell intermediates in retrovirus growth and clone them into bacteria, where they can grow indefinitely. The reason for this, as Howard Temin remarked, is that there are a variety of DNA intermediates in retrovirus growth—circular molecules, linear molecules that are integrated into the cell, unintegrated linear molecules, and the like—and using the techniques of recombinant DNA methodology, they can be cloned into bacterial plasma or viruses. Thus, we can pull out from the cell those molecules that are very low in copy number and grow them to indefinite quantities. Such a molecule is shown in Fig. 2, inserted into a bacterial virus.

By analyzing those molecules, sequencing them, and doing experiments of the form that Howard Temin showed earlier, one knows that the parental virus has, as he described, long terminal repeats at either end, the coding sequence for a series of virus proteins, and a number of signals, i.e., a packaging signal, an initiation signal for the initiation of proteins synthesis, a promotor in LTR that allows this to be expressed as an RNA molecule, and a terminator to terminate RNA synthesis (Fig. 1). This is a genetic package designed to express a specific set of genes, that is, the parental virus.

The Abelson virus was derived by deleting almost all of the genetic information necessasry to make it a virus and replacing it with some new genetic information, as shown in Fig. 2. This means that the virus is not able to grow by itself, but only in the presence of a helper virus; however, this is easily arranged and thus presents no experimental problem.

The coding information that Abelson's virus has is partly derived from its retrovirus parents—a protein of about thirty thousand molecular weight is made by coding from its retrovirus parent—and the rest of the information comes from a normal cellular gene, a gene called the *abl* gene. We can find this gene by cutting out a probe from cloned DNA, a piece that is specific to the insert that gets rid of all the retrovirus DNA, leaving only the DNA inserted in the Abelson virus. With this as a probe, we can search the normal cell genome for DNA that is complementary to the DNA and identical in sequence to the DNA incorporated into the Abelson virus. In this way (Fig. 3) a segment of cell DNA called the c-*abl* gene (cellular *abl* gene), was found. It is about 35,000–40,000 nucleotides

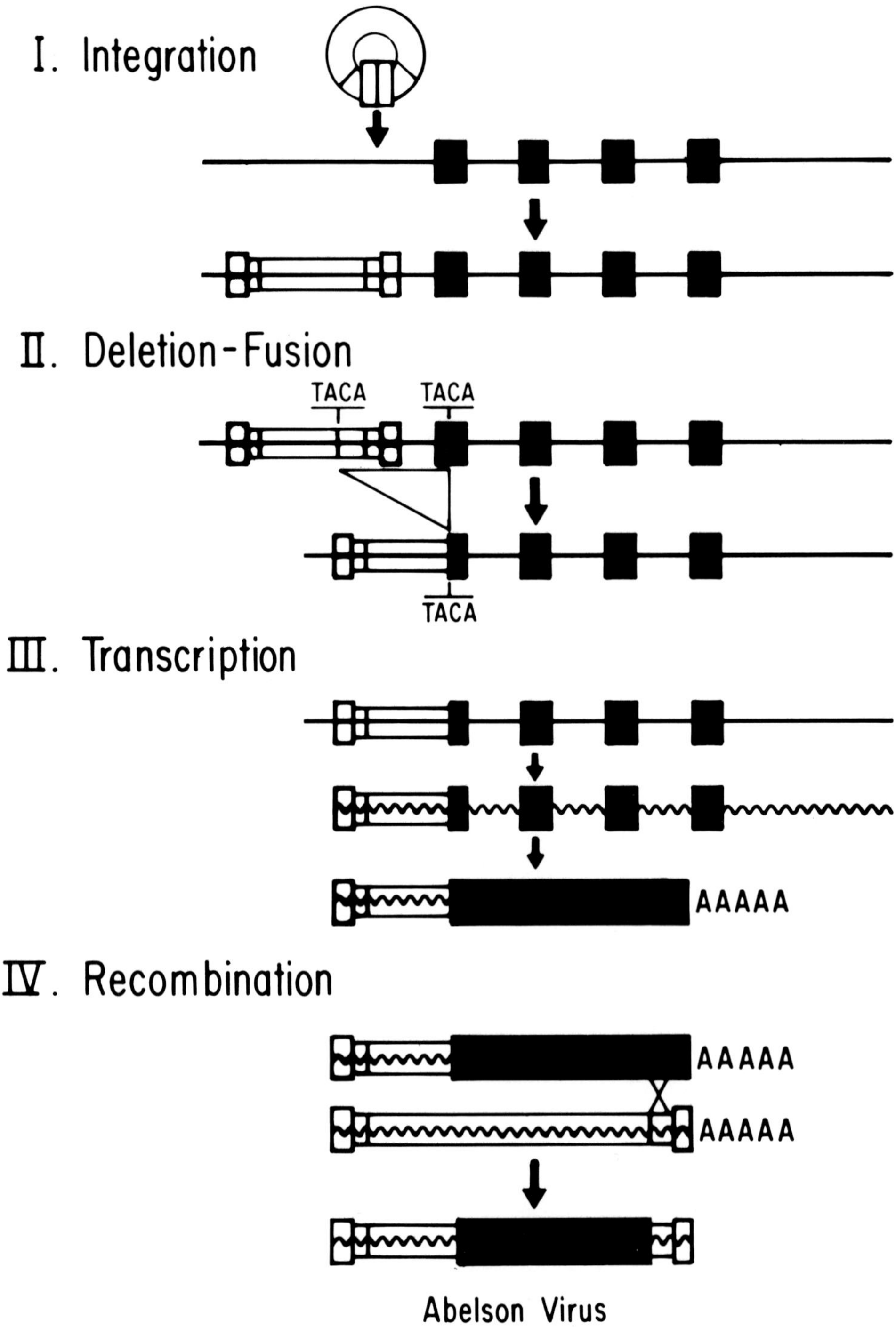

Fig. 2. Model of A-MuLV formation.

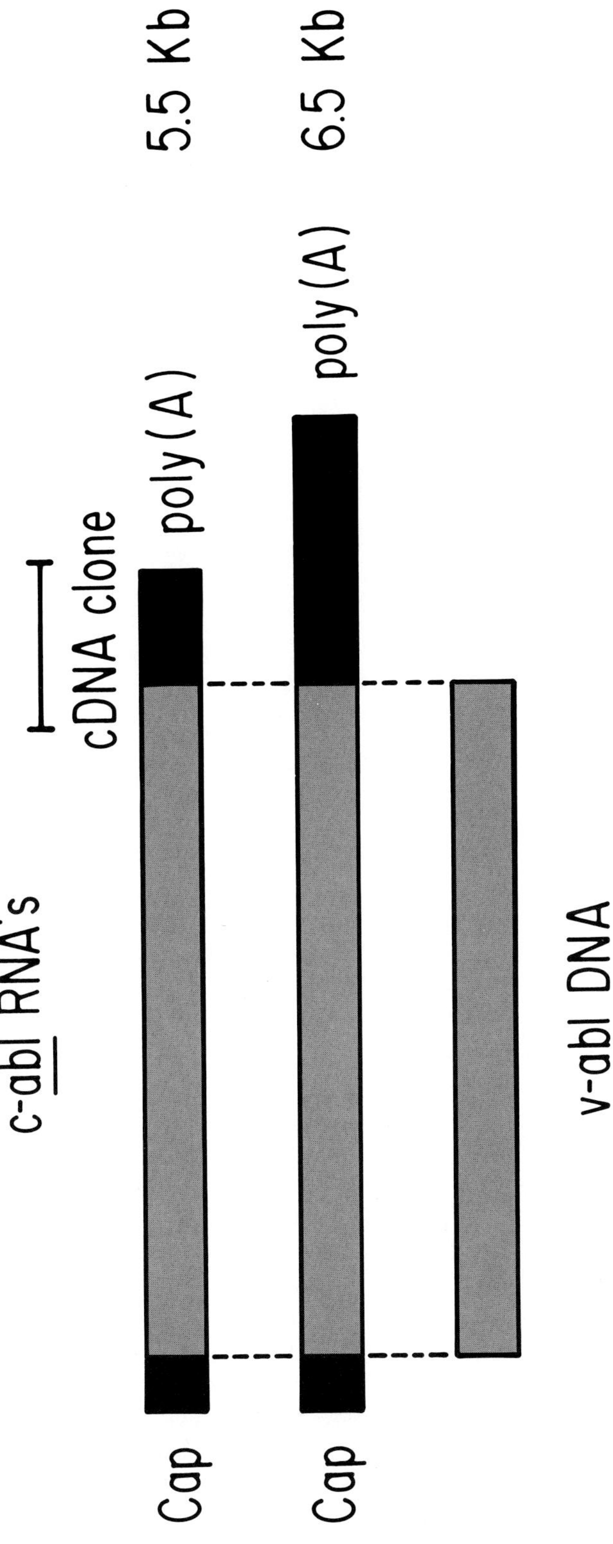

Fig. 3. Relationship between v-*abl* and c-*abl* RNA.

in length and has strewn through it many regions that together make up the v-*abl* insert, the insert that transforms Abelson's virus from an innocuous virus to a tumor-inducing one (Fig. 4).

Now, as Temin pointed out, this organization of little bits of coding sequences with lots of intervening DNA is the standard organization of almost all known genes in invertebrates. It involves what are called the intervening sequences, whose functions are not generally known—they may not have functions—but that split genes. In this case, the organization is interesting from an evolutionary point of view because it involves a very large number of small coding regions, small meaning on the order of a hundred or a couple of hundred nucleotides, separated by regions that can even be thousands of bases in length plus one block of sequences, which altogether involves about two-thirds or three-quarters of the v-*abl* insert.

As I will show later, the "business" part of this gene is split into tiny pieces and a larger part that has no known function from the point of view of cancer induction, consisting of a large block of sequences. There is reason to believe that the gene evolved from a set of little pieces that later had appended to it a large exon. Since this is a normal gene, it should have normal functions. It should make an RNA molecule, and that RNA molecule should in turn make a protein molecule, both of which have been found. We have found, in fact, that all cells that we have examined—from mice, humans, and whatever—made at least some RNA that comes from this gene; thus, although we isolated it as a tumor-inducing gene, it must possess a normal function that is probably very important in all cells.

Once more, because it is common for genes of this sort to be tumor-inducing genes, it is a highly conserved gene. It is one that can be found in humans, mice, and chickens, and, most impressively, can be found in basically the same sequence, although not in exactly the same organization, in a fruit fly. Thus, it is not only important to the functioning of all kinds of cells, it is probably important to the functions of cells from very primitive to very complex animals.

Basically, this gene has not changed very much over half a billion years of evolution. The RNA molecules that we found in mouse cells (Fig. 3) have v-*abl* coding sequences in them, but also have some extra sequences at both ends. All this indicates is that there are two kinds of RNA with a difference between them at their ends, but a difference that is probably not significant. This, in turn, means that when the normal gene is expressed, it is expressed from an RNA molecule that is initiated a little before the v-*abl* sequence, terminates within it, and has extra coding sequences at both ends. We thus expect to find in the DNA this coding sequence and, in fact, we do. At one end, it represents an extension of a long exon, whereas at the other end it amounts to the extension of yet another little exon upstream. Transcription probably initiates

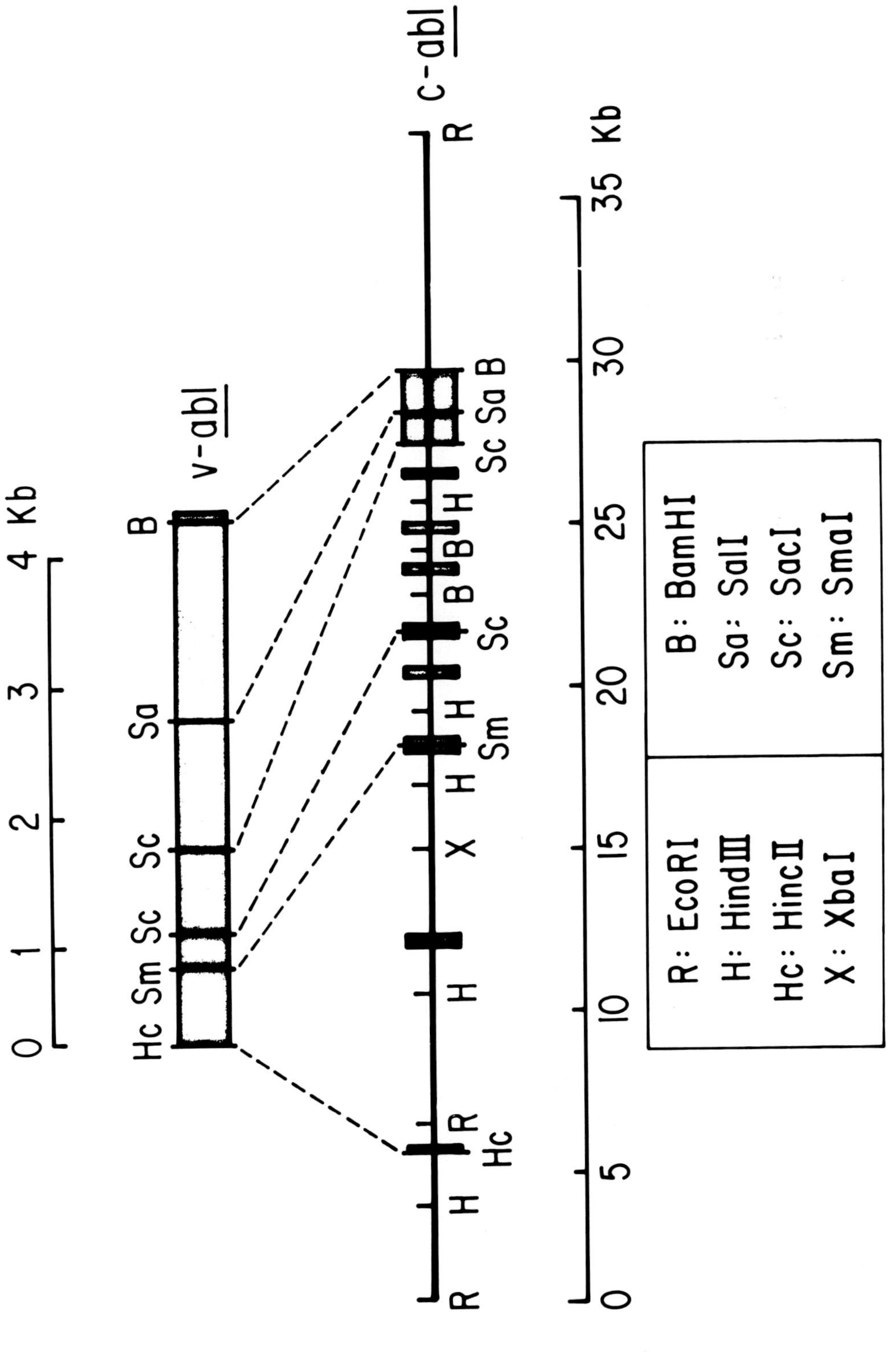

Fig. 4. Relationship between v-*abl* and c-*abl*.

around these coding sequences, making a long transcript that goes all the way through, and terminates on the sequence.

Now, how does this gene get picked up in a retrovirus? We find it in a retrovirus, but we also find it in the DNA of the normal cellular genes. How did it get picked up? We suggested a number of years ago—and it is from a variety of systems that it has been proven quite conclusively—that events of the sort mentioned above occurred in some form. That is, in one cell in the animal that Dr. Abelson infected, the innocuous virus integrated itself near the Abelson gene. It made itself part of the chromosome that the gene is on, and then by a series of events (Fig. 2) that involved deletion of DNA, fusion of the retroviral or proviral DNA to the v-*abl* sequences, transcription, and running through the new construct, would move the intervening sequences at the level of RNA; and that further recombination event generated what we know today as the Abelson virus. The probabilities of most of these events are all very low. The probability of integration is probably one in ten to the fifth or sixth power. The probability of the deletion event we do not know, but it is probably on the same order; and the probability of the second recombination is somewhat higher than that, but the overall probability in an animal of one of these events, or of one such series of events of this type, occurring is probability on the order of ten to the minus twelve.

Roughly, the immune system has at least that number of cells in it; over a reasonable period of time, there are about ten to the eighth power cells made per day. And so, there are enough orders of magnitude to make this a consistent explanation of how the Abelson virus arose. It is also then not surprising that it has never arisen again.

Seeing that it was a product of a series of rare events, we would expect that the virus is not a minimal virus because it has not been subjected to much further selection after this series of more or less random events occurred. And so, it is an experimental question: How much of this structure is really necessary to induce tumors, and how much of it is merely a consequence of the history of how the virus happened to form? To answer these questions, we have undertaken a series of experiments to delete various pieces of information from the virus and determine which ones are needed and which ones are not.

We found that the minimal transforming region (Fig. 5) involves very few sequences, although the virus picked up a very large gene. It is only a small domain of that gene's function that is involved, and that domain, in fact, represents the part that is encoded by all those exons. The part that was appended to it is not involved in transformation in any direct manner, although it may slightly modulate it.

I earlier stated that the Rous sarcoma virus has a transforming gene similar to the one I have just described for the Abelson virus. As I will discuss in a moment, many other viruses also have it. There are at least seventeen known genes about which I could tell the same story as I told about the Abelson gene. We then looked at its structure as compared to

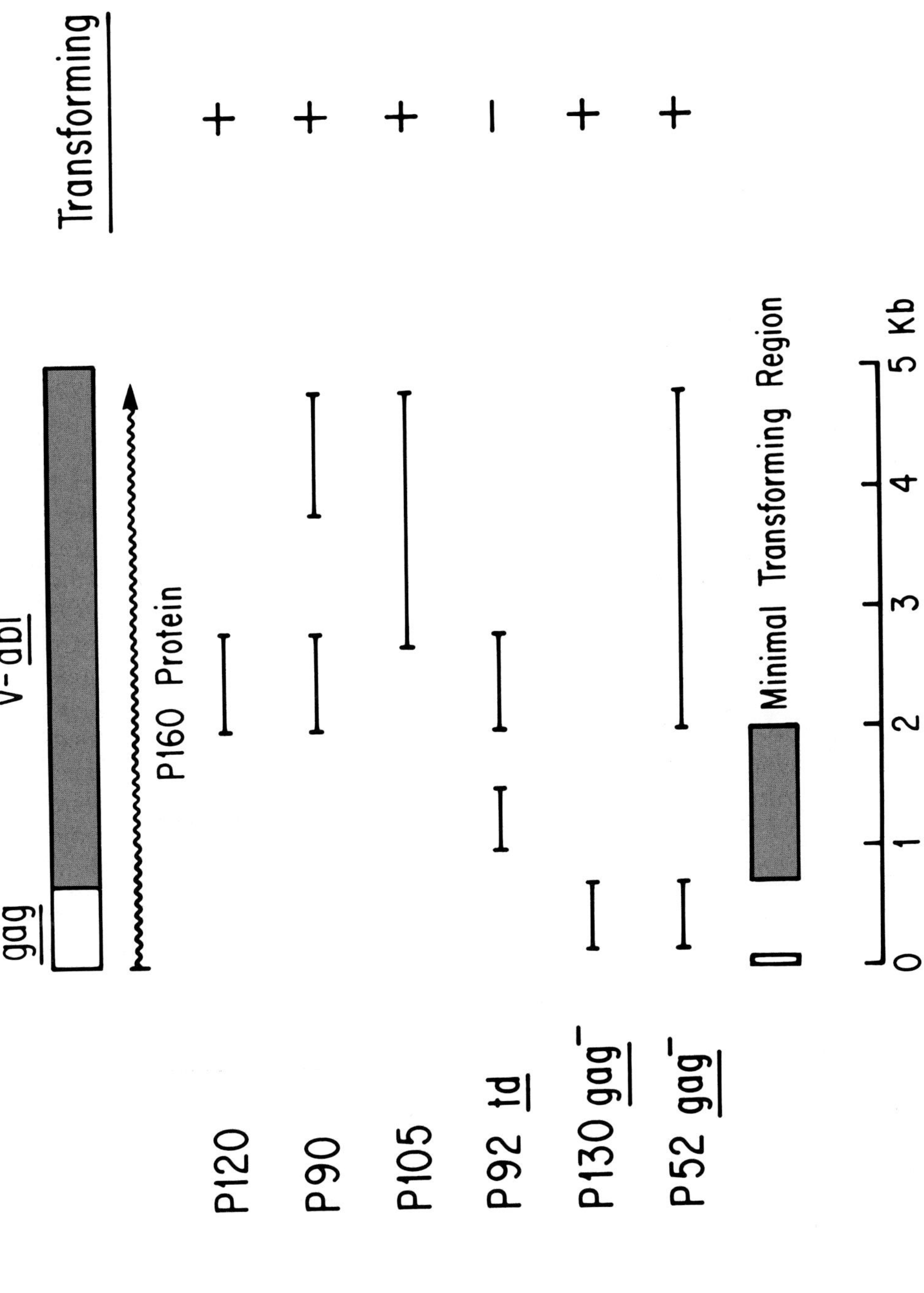

Fig. 5. Region need for transformation.

the Rous sarcoma and other genes. And much to our surprise, this is now the v-*abl* gene, described on the top line running through Fig. 5, which is described by the amino acid encoded, using a single letter code for the amino acid. We can see in comparing the top line with the second line, which is the *src* sequence, that there are many regions of identical sequences. The identities are sufficient that there is absolutely no question that these are closely related genes from an evolutionary point of view. One of these genes is cloned from chickens, the other from mice. Both of them have homologies in *Drosophila,* for which both sequences, done at Harvard, are again very closely related and show the same kinds of identities, especially through the region of greatest homology.

The amusing thing is that for the *abl* gene, we are reading in Fig. 4 from the N-terminal end, and have stopped writing the sequence roughly at the point at which the region necessary for the transformation runs out. Thus, there is much more coding sequence.

Although the Rous sarcoma virus gene has its necessary portion at its C-terminal end, the *abl* virus gene has its necessary portion at its N-terminal end, namely at the two opposite ends of the protein. This shows quite clearly that the evolution of these two genes has been modified by appending sequences at either one end or the other to create two independent genes.

I have said that there are seventeen genes known to transform cells in retroviruses: these are retroviral oncogenes. Do they all fall into the same sequence class or not? When 15 are compared, it appears that there is a set falling into the same sequence class; that is, if placed one under another, I could have drawn red boxes showing close evolutionary relationships. These are the *src* gene and a few others.

There is one kind of intermediate gene, the murine sarcoma virus gene, which shown some, but not a great, homology. There is a second gene of that class, called the *raf* gene, and there are many others, about two-thirds of the total, which are absolutely unrelated to any of the sequences found in this first group. They form other sequence classes: two of them are related to one another (i.e., the *raf* gene) and others are not related at all (The so-called *myb* and *erb A* genes, for example, are totally unrelated by sequence). Some of the gene sequences have not been determined yet, and so are not clear. But a couple have recently appeared that are in new classes.

What this means is that, although most genes able to cause cancer are related, there are in fact also a whole series of unrelated protooncogenes, indicating that there are lots of types of genes in the normal genome of the animal that when picked up and modified in an appropriate way can become oncogenes, i.e., cancer-inducing genes. And, in fact, the products of these are very different—some are in the nucleus, some in the cytoplasm, some have specific biochemical activities, and others lack those activities—we know from this that they are unrelated. All of the genes in the sequence Class 1 have one common biochemical activity.

They are all enzymes able to transfer a phosphate residue onto a tyrosine in target proteins, either in themselves or in other proteins that they can affect. These are the so-called protein kinases. We can demonstrate that the Abelson virus proten is a protein kinase that is specific for tyrosine; even when we delete the region described earlier, and put it into *Escherichia coli,* allowing it to make protein in *E. coli.* So this insert, the *abl* insert, will make a tryosine kinase in *E. coli* that then phosphorylates many *E. coli* proteins (which is not very useful because it ultimately will kill the bacteria).

Finally, it is possible to determine, by deletion mapping, the exact region of the protein in *E. coli* that is necessary for this biochemical activity. The borderline turns out to be almost exactly the borderline that we were able to define by transformation activity in cells or by their ability to cause tumors. Thus, it seems reasonably clear, although it is still a correlative observation, that the activity of this protein kinase domain is what causes the cells to become tumor cells, and thus focuses our attention on this specific biochemical event as one that is central to growth control in the cells: There is indeed an increasing literature consistent with this finding.

I want to add a footnote that continues the discussion Dr. Pauling started. Because the cells transformed by Abelson's virus have a uniquely useful property whereby they represent cells of the early stages of lymphoid cell differentiation (i.e., the differentiation of cells of the antibody-forming system), they will allow us to look in detail at how antibody genes are made. Antibodies have two chains—heavy chains and light chains—and the regions of those chains involved in "seeing" the antigen are encoded at the N-terminal end. Within that N-terminal end there are three critical regions called the hypervariable regions. Two of them come about through evolutionary processes and are represented in a diversity of genes encoded in the animals. I will not discuss those, although their evolution is quite specific, interesting, and important. The region of interest here is the third, which makes protein that is actually involved in binding to antigen and is called the third hypovariable region. It is formed by a combinational process in which bits of DNA are brought together from different places on a chromosome to form a single continuous region.

We have recently found that there is a development of diversity there that goes beyond anything that we had imagined or that has been previously seen in biology. The actual creation of random DNA sequences. I want to present a little evidence for that and for where these sequences occur.

To put together an antibody gene involves putting together bits of DNA from three different libraries of DNA sequences. One library is called the variable region proper or VH segment (VH1, VH2, and others). There is certainly on the order of hundreds of VH regions, all encoded one after another along a stretch of DNA millions of bases in length.

There is another library of sequence, the D library, that involves maybe 20 different elements. And there is a third library of sequence, the J sequence, that involves a library of four elements. The way to compose a complete coding sequence for the variable region of a heavy-chain gene is by joining together one element of the D library with one from each of the V and J libraries, as we have recently proven. These are put together because there are signal elements at the borderlines that tell enzymes which DNA to join. You might therefore expect to find that every time one element from the V library joins with one from the D library, the junction is a fixed junction, and that, similarly, when a D element joins with a J element, it is a fixed junction. But, in fact, this is not true because the junctions are highly variable. The surprising thing is that there are sequences in each of these junctions that are not encoded in either the D element, J element, or V element. In fact, as far as we can tell, although it is a hard thing to prove rigorously, they are not encoded anywhere in the DNA.

We can take these Ableson transformed cells out of agar, as I described earlier, and start growing them in culture. They grow indefinitely and they can be cloned, so we can take an individual cell and grow a new population. We can, in each of these populations, look at the structure of the DNA and, in fact, clone out the DNA and read its nucleotides sequence. So we can watch the evolution of this problem, when first the D joined to a J, which turns out to be the first event, and then a V joins to a preformed DJ.

Now, if we look at the parental configuration, that is the configuration along the DNA as it is seen in the germ line, we can recognize J elements, D elements, and regions representing signal sequences. The coding sequences for D and J occur together on one chromosome, which we have isolated in sequence. You can then read the J element i.e., GGGGTC, GGGGTC, right up to a point where it suddenly stops. We can read the D element coming in the other way, i.e., CCTA, CCTA, and so on, but, suddenly, in between these two bits of clearly encoded sequences, there is a sequence that is not found at all in either the J or D elements.

We can look at the other chromosome in this cell, because all cells have two chromosomes of any given kind, and see the same thing again. It happens to recombine to J3, so we can read the J3 sequence; it happens to be at the same point on the J3 sequence, but it could have been anywhere (we have seen others that are anywhere). Upon reading the D sequence, the D-coding sequence, we suddenly find a new sequence not found in D, or J. Through the addition of a V, you can again get cutting away of D sequences and replacement with new nucleotides that came in without being encoded within either V or D. We believe we know what enzyme makes this. It is an enzyme called terminal deoxynucleside transferase, an enzyme that makes random DNA polymers and that had been isolated from lymphoid cells many years ago; incidently, it has never been given a function because no one imagined that synthesis of random DNA might be useful. But I think synthesis of random DNA is useful and it contributes to the development of the remarkable specificity that antibodies have. So antibodies come about as a consequence of both coded DNA sequences and little bits of random DNA sequences, as they appear in the animal.

What does the future hold? Well, I think first of all we have got to put the genes, as well as the differentiation processes, that I have been talking about into the context of human cancer. I think the first steps along those lines have been taken very eloquently, as shown by experiments of the last couple of years. But I also think that we have got to go beyond genetics because, although we can find the genes involved in cancer now and can make a good case for their involvement in human cancer, what goes on inside the cell is important.

If Dr. Nicolini's projection that basic research will give us a clue to handling human cancer in new ways is to come about, and I certainly believe it will, we have to find out what the biochemistry of cancer is. Because I think the only rational projection for how we are going to turn this kind of information into new treatments for cancer is through intervention in biochemical events—that is, ultimately, the synthesis of drugs—just as we have synthesized them before, but now with an understanding of the biochemistry of transformation rather than simply continuing to use a random panoply of cytotoxic drugs that have had only limited success.

I think, also, that we have to be able, in many different systems, to relate the tumor cells to the progression of events in normal cells because the tumor cells represent the fine stages in differentiation. There may be targets that one can find that relate to those stages of differentiation. I would not be surprised at all if Dr. Sachs in the later session would pick up on that notion. We have to understand differentiation better, and we have to understand, finally, one thing that none of us will talk about, I think, because none of us understand it very well, the migration of cells.

A frank tumor cell is one that migrates to inappropriate places in the body, and we have no idea what are the genetic bases of those kinds of cell recognition phenomena. But I think we can project that progress in this area will occur. And that a molecular understanding of cancer—something no one would have projected as a possibility ten or fifteen years ago—almost certainly lies in our future. That measure of progress to date excites all of us, and keeps us all moving.

Section II LOWER-TO-HIGHER ORDER DNA STRUCTURE

Neutron Scatter and Diffraction Techniques Applied to Nucleosome and Chromatin Structure

E. M. BRADBURY*† AND J. P. BALDWIN‡

†Department of Biological Chemistry, School of Medicine, University of California, Davis, California 95616 and ‡Department of Physics, Liverpool Polytechnic, Liverpool, United Kingdom

INTRODUCTION

The application of neutron scatter and diffraction techniques to structural studies of the nucleosome and of chromatin provides a striking example of the interface of modern physics and the life sciences. The structure and function of the eukaryotic chromosome is a major problem in molecular cell biology. One approach toward an understanding of the functions of such a complex system is to determine (i) the structure of the repeating chromatin subunit, the nucleosome, (ii) the arrangements of nucleosomes in the different orders of chromatin structure, and (iii) how these structural parameters change in response to the functional requirements of cells, i.e., to approach function through an understanding of structure.

The nucleosome contains one tetramer of histones $(H3, H4)_2$, two dimers of histones (H2A, H2B) and one molecule of very lysine-rich histone, H1, associated with about 200 base pairs (bp) of DNA. The DNA content of the nucleosome is 195 ± 5 bp DNA for most cell types in higher eukaryotes, with the exception of certain specialized cells (*1*). Also, the DNA content of the nucleosomes from lower eukaryotes are lower than those found for higher eukaryotes (*1*). Specialized cells contain different subtypes of histones H1, H2A, and H2B. Very recently, Stein's laboratory showed (*2*) that the nucleosome DNA repeats of HeLa cells (198 bp), chicken erythrocytes (212 bp), and sea urchin sperm (241 bp) can be generated from poly d(AT) and the purified histones from

*Author to whom all correspondence and reprint requests should be addressed.

each of these cell types. It follows that variations in nucleosome DNA repeats are most probably determined by their histone contents, particularly the subtypes of H1, H2A, and H2B. Although histones are very highly conserved proteins, particularly histones H3 and H4, they are subjected to reversible chemical modifications (*3,4*) that change the character of the modified amino acid and, in effect, relax the rigid sequence conservation in response to cell requirements. There are three major reversible chemical modifications; (i) acetylation of lysines in the core histones H2A, H2B, H3, and H4 that convert a basic charged lysine to a neutral residue (*3,4*), (ii) ubiquitination of lysines in histones H2A and H2B to give uH2A and uH2B in which the small globular protein ubiquitin is covalently linked to the amino group of a lysine side chain (*5–7*), and (iii) phosphorylations of serines and threonines in histones H1, H2A, H2B, and H3 that convert a neutral residue to a negatively charged residue (*3,4,8,9*). These modifications have been associated with major chromosome functions: histone acetylation with chromatin replication and transcriptional activity (*3,4*); ubiquitination of H2A and H2B with the labeling of potentially active chromatin (*5,10,11*); and H1 and H3 phosphorylation with chromosome condensation (*3,4,8,9*). Also associated with active chromatin is a well-defined subgroup of nonhistone chromosomal proteins called high-mobility group or HMG proteins 14 and 17 (*12,13*). Major questions, therefore, concern the structures of the nucleosome and the different organization of chromatin and in the long term how these structures change in response to the reversible chemical modifications of histones and the binding of nonhistone chromosomal proteins associated with different functions.

NEUTRON TECHNIQUES

Neutron scatter and diffraction techniques are particularly powerful in structural studies of two-component biological systems, such as chromatin (*14*). With this technique it is possible, with minimum perturbation to the biological system, to obtain the spatial arrangements of either the DNA or the protein components of chromatin. This ability to separate the neutron scattering or diffraction effects of histone from those of the DNA comes from the quite different neutron scattering-lengths of proton $H(^{1}H)$ compared to the scattering lengths of deuterium $D(^{2}H)$ and the elements found in biological macromolecules, carbon, oxygen, nitrogen, and phosphorus. The neutron scattering length is analogous to the X-ray scattering factor, and both of these parameters are given for the different elements in Table 1. X-rays are scattered by electrons and scattering factors of different elements, i.e., the ability to scatter X-rays increases with increasing atomic weights. There is no similar simple relationship for neutron scattering lengths. Neutrons are scattered by atomic nuclei that act as true point scatters and the neutron scattering length is a complex

TABLE 1

Element	Neutrons: Scattering length, 10^{-12} cm	X-ray: Scatter factor
H (^{1}H)	−0.38	0.28
D (^{2}H)	0.65	0.28
C	0.66	1.7
N	0.94	2.0
O	0.58	2.3
P	0.51	4.2

TABLE 2

	Scattering length densities, 10^{11} cm^{-2}	Mixture of D_2O:H_2O to give same scattering length density, %
D_2O	0.63	100
DNA	0.38	63
Histone	0.20	37
H_2O	−0.06	0

property of the atomic nucleus. As can be seen in Table 1, the neutron scattering length for hydrogen is negative, whereas the other elements have positive scattering lengths and, except for nitrogen, similar values. The difference in sign can be regarded as a difference in the phase of the scattered neutron wave for the heavier nuclei compared to hydrogen. A major consequence of this difference in signs between the scattering length of ^{1}H and those of C, O, N, and P is that the scattering length density (i.e., the sum of the scattering length of the atoms in a macromolecule divided by its volume) of a biological macromolecule will be strongly dependent on the proportion of hydrogen in the molecule. Thus, the scattering length density of histone is 0.20×10^{11} cm^{-2}, whereas for DNA it is 0.35×10^{11} cm^{-2}. This very large difference can be exploited by studying the neutron scatter and diffraction of nucleosomes or chromatin in mixtures of light water, H_2O, and heavy water, D_2O. Table 2 shows the scattering length densities of H_2O, D_2O, histone, and DNA. Because of the difference in scattering lengths of H and D, the scattering length density of H_2O is -0.06×10^{11} cm^{-2}, whereas that of D_2O is $+0.63 \times 10^{11}$ cm^{-2}. Thus, by taking simple mixtures of H_2O and D_2O, all values of scattering length density between these extremes can be obtained. The range of contrast variation for neutrons is so large that it includes the neutron scattering-length densities of all the classes of biological molecules, nucleic acids, proteins, lipids, and so on. The water mixtures that match those of the histones and of DNA are given in Table

2. This technique, called neutron contrast matching, is illustrated in Fig. 1 for the nucleosome core particle. The neutron scattering-length density of 37% D_2O:63% H_2O matches that of the histones. Thus, in this water mixture the histones will be largely invisible and the DNA component will dominate the neutron scatter. In 63% D_2O:37% H_2O, the DNA component will be matched, and the histones will dominate the scatter. In H_2O, both the histone and the DNA have larger scattering-length densities than for H_2O and in effect are in positive relief, whereas in D_2O, both components have smaller scattering-length densities than for D_2O. In D_2O, therefore, high reverse contrast can be obtained, a situation not achievable for X-ray scatter studies. This neuron contrast variation technique has been developed largely by Stuhrmann and colleagues (*15–17*). Because of its minimum structural perturbation, D_2O is the ideal small molecule to use in neutron contrast variation studies of the spatial distribution of different components of a macromolecular assembly. Furthermore, because deuterium scatters neutrons strongly, neutron scattering techniques can be considerably extended by the isomorphous substitution of deuterium in nonlabile sites in the macromolecule. Engelmann and colleagues (*18,19*) have shown that analysis of neutron scatter curves from reconstituted ribosomes containing deuterated components give the distances between the centers of any selected pair of protein subunits in the assembly. The process involves the deuteration of nonlabile sites in the pair of proteins. A sufficient number of distances between pairs of

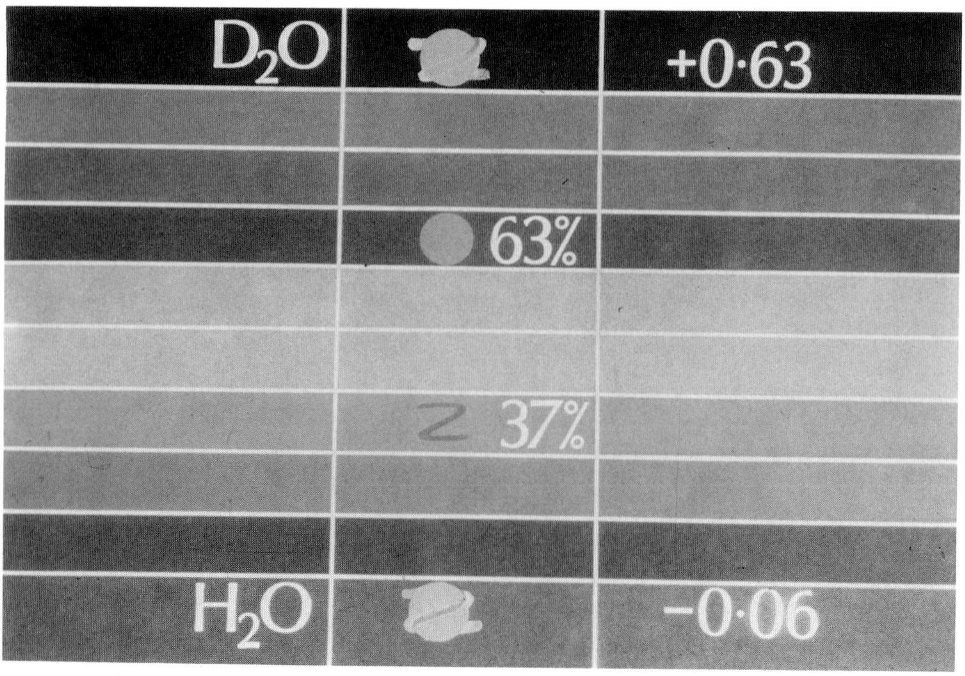

Fig. 1. Neutron contrast matching experiment.

proteins have now been determined so that the process of triangulation will allow the topology of all the proteins in the 30S ribosomal subunit to be determined.

NUCLEOSOME AND SUBNUCLEOSOME PARTICLES

Micrococal nuclease digestion of chromatin, either in vitro or in isolated nuclei, cleaves largely in the linker DNA between adjacent nucleosomes to give mononucleosomes and oligonucleosomes. With longer times of nuclear digestion, the DNA within a nucleosome is digested and well-defined subnucleosome particles can be obtained; in particular, the chromatosome and the nucleosome core particles. The DNA and histone components of these related particles are as follows:

Nucleosome = ~200 bp DNA + $[(H3_2H4_2)(H2A,H2B)_2]$ + H1
Chromatosome = 168 bp DNA + $[(H3_2H4_2)(H2A,H2B)_2]$ + H1
Core particle = 146 bp DNA + $[(H3_2H4_2)(H2A,H2B)_2]$

The core particle is a very well-defined particle and has been subjected to intense physical studies. Two major approaches have led to the low-resolution structure of the core particle. First, neutron scatter studies of the core particle in solution that led to the initial determination of its major structural features (*20–23*) and second, X-ray and neutron diffraction studies of single crystals of core particles that are now giving higher-resolution structures (*24–27*).

NEUTRON SCATTER STUDIES OF NUCLEOSOMES AND CORE PARTICLES

Examples of neutron scatter curves from chromatin subunits (22) in a range of mixtures of H_2O and D_2O are given in Fig. 2. They show how these scatter curves change markedly as a function of contrast, i.e., $D_2O:H_2O$ mixtures. Neutron scatter curves have been obtained for chicken erythrocyte subunits (*22*) and chicken-erythrocyte and calf-thymus core particles (*23*) and each scatter curve contains data on the structural parameters of these particles. From the low-angle regions of the scatter curves, Guinier plots (*28*) give the radius of gyration of the particle at each $D_2O:H_2O$ mixture. Figure 3 gives these plots of Ln $I(\kappa)$ vs κ^2 for chicken-erythrocyte and calf-thymus core particles; $I(\kappa^2)$ is the intensity of the neutrons scattered at an angle θ and $\kappa = 4\pi \sin \theta/\lambda$ ($=2\pi s$), where λ is the wavelength of the neutrons. A linear Guinier plot is expected for a monodisperse solution of core particles of slope $\frac{1}{3}R_g^2$ where R_g is the radius of gyration of the particle. As can be seen in Fig. 3, the slope of the Guinier plot close to the contrast matched $H_2O:D_2O$ mixture for proteins (35% D_2O) is very much larger than that close to the DNA contrast matched point (60% D_2O). For such plots it was determined that

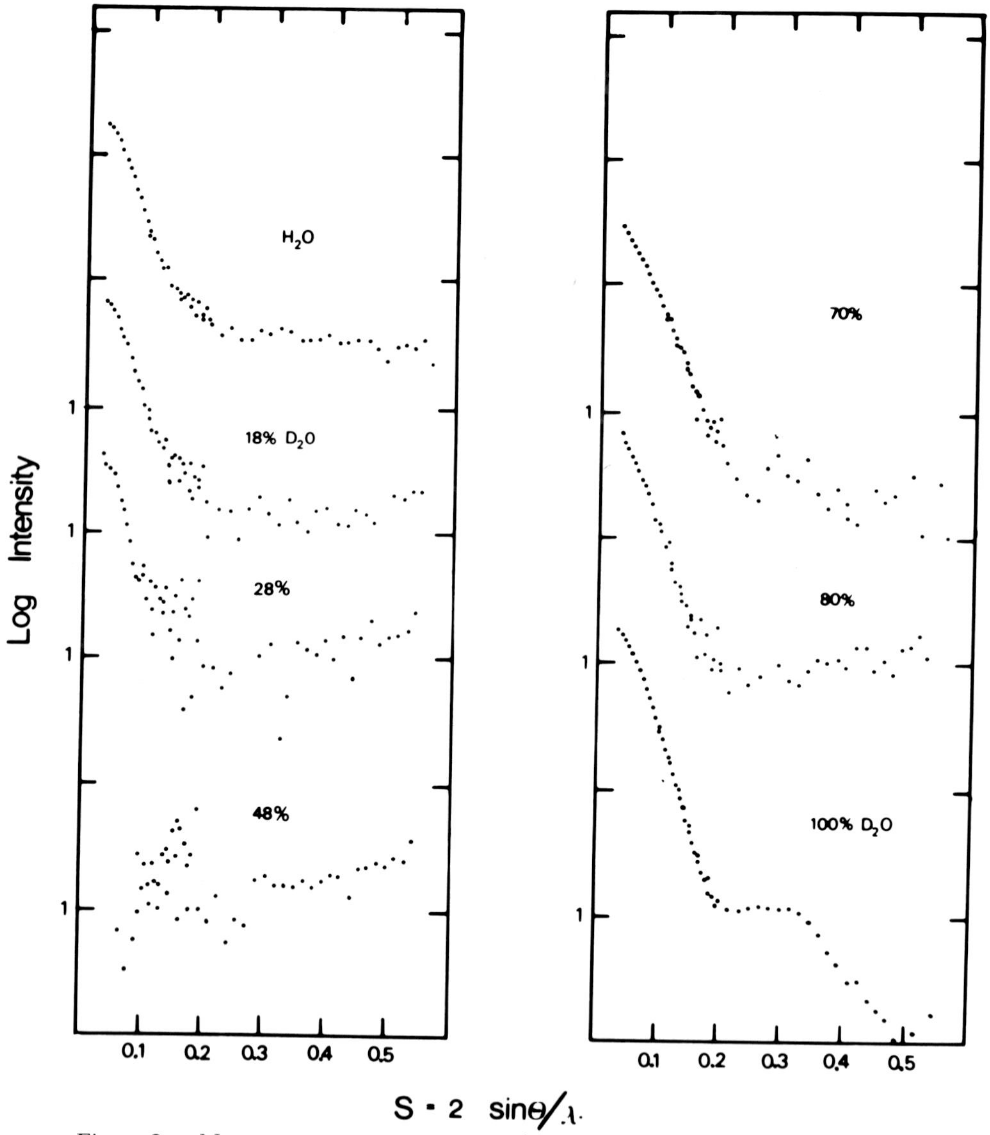

Fig. 2. Neutron scatter curves from chicken-erythrocyte chromatin subunits in different mixtures of $D_2O:H_2O$.

R_{gDNA} = 4.8 nm and $R_{gprotein}$ = 3.3 nm, demonstrating conclusively that DNA was external to a histone core in the nucleosome and in the core particle (*20–22*).

In addition to the particle R_g at different contrasts, the Guinier plots of Fig. 3 also give the zero-angle scatter intensities I_o by extrapolation to zero angle. $\sqrt{I_o}$ is proportional to the contrast $\bar{\rho} = \rho_{particle} - \rho_{solvent}$ and varies linearly with $\rho_{solvent}$. The intercept on the abscissa at $\sqrt{I_o} = 0$ occurs when $\rho_{particle} = \rho_{solvent}$ and gives a measure of the mean scattering-length density of the particle, ρ_{mean} (*22,23*). Knowing the labile protons in the particle that exchange with deuterium allows ρ_{mean} to be used to ob-

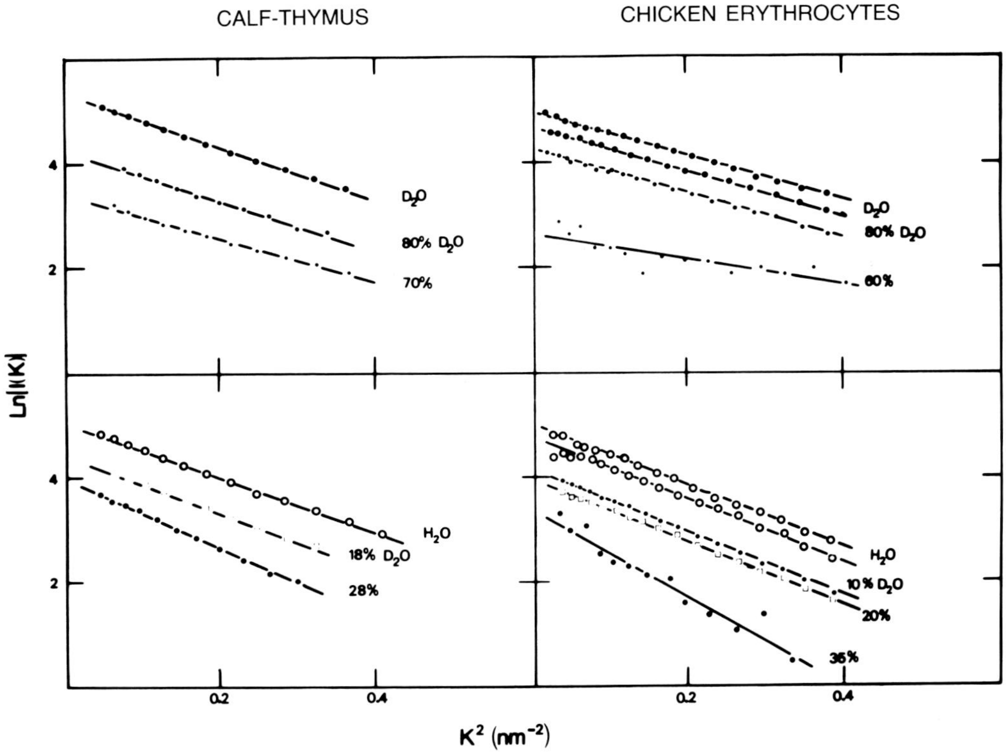

Fig. 3. Guinier plots from the low-angle regions of the neutron scatter curves for core particles from calf thymus (left) and chicken erythrocytes (right).

tain the volume of the "dry" particle V_f. The V_f value and other structural parameters obtained from the neutron-scatter curves are given in Table 3. The value of $V_f = 233$ nm^3 is in very close agreement with the volume of the dry particle obtained from the partial specific volumes of the DNA and histones of 218 nm^3. As can be seen, the quality of the data obtained from the Guinier plots of Fig. 3 is very high.

From Stuhrmann's analysis (*15–17*) of the variations of R_g with contrast $\bar{\rho}$ is given by the equation:

$$R_g^2 = R_c^2 + \alpha/\bar{\rho} - \beta^2/\bar{\rho}^2 \qquad [1]$$

From a plot of $R_g{}^2$ vs $1/\bar{\rho}$, the parameters R_c, α, and β can be obtained. At infinite contrast ($1/\bar{\rho} = 0$) fluctuations in internal structure are too small to contribute to the scattered intensity and R_c gives the radius of gyration of the volume occupied by the particle in solution, i.e., of the shape envelope of the particle, as though it was filled with a uniformly dense scattering material. The coefficients α and β give information on the internal structure of the particle. The relationship of these coefficients to particle structure have been derived by Ibel and Stuhrmann (*17*). The values of α and β are given in Table 3. The positive value of the coefficient α shows that the scattering length density increases with increasing radical vector from the center of the particle, i.e., DNA must be located on the outside

TABLE 3
Parameters of the Chicken Erythrocyte Nucleosomes and Core Particles

	Core particle,[a] 137 bp DNA	Nucleosomes,[b] 195 ± 20 bp DNA
ρMean (i.e., average neutron scattering density of particles	2.79×10^{10} cm^{-2}	2.97 ± 10^{10} cm^{-2}
α [From Stuhrmann (1974) analysis]	$4.13 \pm 0.12 \times 10^{-4}$	$4.5 \pm 0.2 \times 10^{-4}$
β [From Stuhrman (1976) analysis]	$8.7 \times 10^{-10} \pm 1.1 \times 10^{-9}$ nm^{-2}	
R_g (DNA dominating the scatter)	4.8 nm	4.9 nm
R_g (Protein dominating the scatter)	3.3 nm	3.0 nm
R_c [Equivalent to radius of gyration (R_g) of uniformly dense particle]	3.92 ± 0.05 nm	4.05 ± 0.05 nm
Radius R of equivalent uniform sphere	5.06 nm	5.23 nm
Volume of equivalent sphere	543 nm^3	599 nm^3
Volume of histones (2H2A, 2H2B, 2H3, 2H4) from particle specific volumes	139 nm^3	139 nm^3
Volume of DNA from partial specific volumes	79.1 nm^3	110.2 nm^3
Total estimated "dry" volumes	218.1 nm^3	249.2 nm^3
V_f neutron determined "dry" volume	223 nm^3	249 nm^3

[a]From Suau et al., 1977.
[b]From Hjelm et al., 1977.

of the particle. The coefficient β is very small and close to zero, from which it can be estimated that the separation of the centers of scattering length densities of the protein and DNA is less than 10Å.

Basic Scattering Functions

The above analysis comes from Guinier plots of the low-angle region of the scatter curves of Fig. 2. The full scatter curves, however, contain all of the structural information that can be obtained from a neutron scatter study of monodisperse particles in solution. The basic scattering functions are obtained from the experimental scatter curves of Fig. 2 by parabolic regression, using the scatter equation for the intensity of coherent neutron scatter into solid angle $\Delta\Omega$.

$$I_{(s)} = \phi_s(d\delta/d\Omega)_s\Delta\Omega n_t T_s \qquad [2]$$

Where $s = 2 \sin \theta/\lambda$, ϕ is the incident flux, T_s the transmission, and n_t the number of particles contributing to the scatter. The analysis of the scattering curves into basic scattering functions allows the contributions of shape $I_C(s)$ and internal structure $I_S(s)$ to be separated. It uses all available scattered neutrons and considerably reduces statistical error. Figure 4 gives the three basic scatter functions $I_S(s)$, $I_C(s)$, and $I_{CS}(s)$ (22,23).

The Shape Function $I_C(s)$

This function corresponds to the scattering that would be observed at infinite contrast, i.e., of the shape envelope of the particle filled with a uniformly dense material. Note that the shape function for the core particles does not show maxima at 5.5 and 11.0 nm, which are found in the fiber neutron and X-ray diffraction pattern and will be discussed later. After correcting for exchangeable protons, the small angle region of $I_C(s)$ gives the "dry" volume of the core particle of 223 nm^3 and the radius of gyration of the shape envelope of the particle, $R_c = 3.94 \pm 0.05$ nm (Table 3). An oblate ellipsoid of axial ration 0.5 has been found to give the best fit to $I_C(s)$ for the nucleosome (22) and the core particle (23).

The Internal-Structure Function $I_S(s)$

The internal-structure function results from the internal fluctuations in scattering length density within the core particles. This function corresponds also to the neutron scatter curve from core particles in a solvent of contrast equal to the mean scattering-length density ρ_{mean} of the particle, i.e., the contrast matched position in 48% D_2O:52% H_2O (see scatter curve, Fig. 2). The maxima in $I_S(s)$ at $s \simeq 0.1$ - 0.15 nm^{-1} and $s \simeq 0.25$ - 0.30 nm^{-1} are characteristic of a particle with domains of different scattering-length densities separated along the radius of the particle.

Scattering Function $I_{CS}(s)$

$I_{CS}(s)$ results from the cross-product of shape and internal structure. The minimum in $I_{CS}(s)$ at $s \simeq 0.07$ nm^{-1} results from the clear separation of protein and DNA domains in the particle. The negative trough in $I_{CS}(s)$ indicates that the stronger scattering component, i.e., the DNA, is located on the periphery of the particle. The reverse of this situation, i.e., DNA on the inside and protein on the outside, would invert this curve to give a position peak in $I_{CS}(s)$.

The Radial-Distribution Function

Fourier transformation of each of the fundamental scatter functions gives their corresponding Patterson function. The core particles in solu-

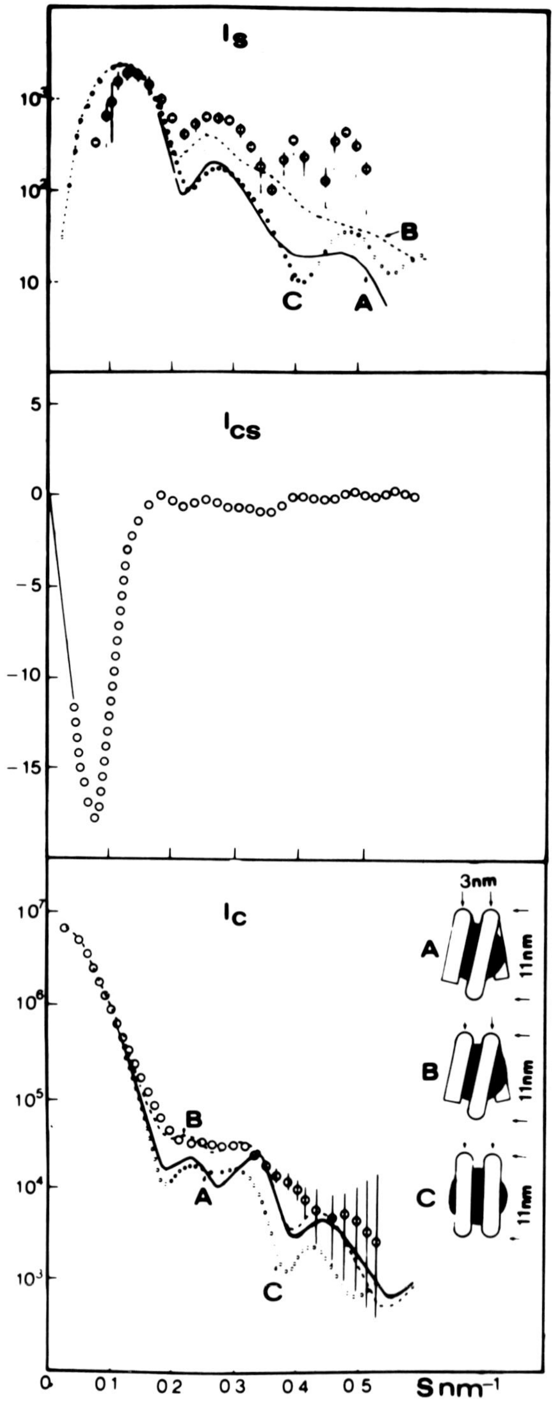

Fig. 4. Basic scatter functions $I_S(s)$, $I_C(s)$, and $I_{CS}(s)$ for chicken-erythrocyte core particles. (From ref. *23*, with permission.)

tion assume all orientations with respect to the neutron beam, and when the resulting Patterson is spherically averaged it gives the distance distribution function $D(r)$, where r is a chord length of the shape of the particle. Fourier transformation of $I_C(s)$ give the distance-distribution function $D_C(r)$ for the particle shape.

$$D_C(r) = 8\pi r s I_C(s) \sin(2\pi r s) ds \quad [3]$$

The $D_C(r)$ function goes to zero at the maximum chord length r_{max} in the particle that corresponds to the maximum dimension of the particle. The functions $D_C(r)$ for native and reconstituted core particles (*29*) are given in Fig. 5. Note that $D_C(r)$ for the reconstituted-core particles is identical to that of the native-core particles, showing faithful reconstitution. In all cases the maximum value of r is 11.5 nm, which is the maximum distance in the particle, and the most probable distance is 4.3 nm. The maximum distance in the particle is an important parameter and provides constraints to model building.

Low-Resolution Structure of the Core Particles

The shape that gives the best fit to the low-angle region of $I_C(s)$ is an oblate spheroid with an axial ratio of 0.5 (*22,23*). The maximum dimension of the particle from the radial-distribution functions $D_C(r)$ in Fig. 5 is 11.0–11.5 nm (*22,23*). Thus, the overall shape of the core particle in solution must be a flat disk 11–11.5 nm in diameter by 5.5–6.0 nm in thickness. When we consider the spatial arrangements of the histones and the DNA, it is clearly demonstrated by the values of 4.8 nm for R_{gDNA} and 3.3 nm for $R_{gprotein}$ (Table 2) that the DNA must be external to the protein. Within the confines of a flat disk 11–11.5 by 5.5–6.0 nm, the DNA can be located only on the periphery of the disk to give the measured R_{gDNA} of 4.8 nm. The $R_{gprotein}$ of 3.0–3.3 nm shows that the core particle contains a core of the histone octamer [$(H2A\ H2B)_2(H3_2\ H4_2)$]. The DNA length is 146 bp DNA, i.e., about 48 nm of B form DNA. To give the observed R_{gDNA}, this length of DNA would have to be coiled around the circumference of the disk shaped particle with a mean radius of 4.5 nm. This radius would require 1.7 ± 0.2 turns of DNA.

Model Calculations

The total structural information from neutron scatter of core particles in aqueous solution is given by the scatter curves of Fig. 2, from which the fundamental scatter functions $I_C(s)$, $I_S(s)$, and $I_{CS}(s)$ have been derived (Fig. 4). The correct model for the core particle in solution must therefore give calculated fundamental scatter functions that agree closely with the observed functions. In making these model calculations it was found that there was only one type of model that fitted all of the neutron scatter data (*30*). The structure was constrained by the following limits: *DNA*—the DNA component is a flat helix with 1.6–1.8 turns of pitch between 3.0 and 3.5 nm; histone—the measured value of 3.4 ± 0.2 nm for $R_{gprotein}$ is too large for all of the histone octamer to be confined within the DNA superhelix. It is estimated that about 20–25% of the histone is external to a protein core consisting of the remaining 75–80% of the apo-

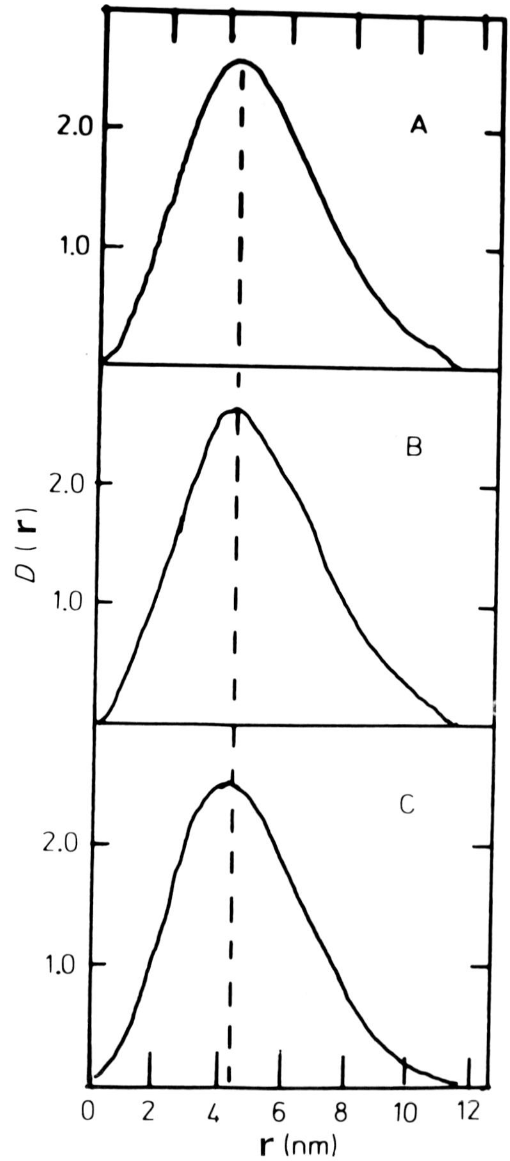

Fig. 5. Distance distribution functions in D_2O, $D(r)$ vs r, for (A) control core particles, (B) core particles at 20 mM NaCl, pH 7, and (C) reconstituted core particles. In all cases the largest distance r(max) = 11.5 nm and the most probable distance is r_{mp} = 4.3 nm. (Taken from ref. 29, with permission.)

lar globular regions of the core histones. This hydrophobic core of the apolar globular regions has a thickness of 4 nm and a diameter of 6.4–7.5 nm. For a range of shapes for the histone core, the fits to the fundamental scatter functions were improved by including in the histone core a hole along the helix axis of the core particle with a diameter of less than 1 nm. Thus, a hole is included along the helix axis of the core particle.

When one or more of these structural parameters deviated from the above limits, poorer fits to the fundamental scatter functions were obtained. Taking all of the neutron scatter data, the best low-resolution structure for the core particle in solution is given in Fig. 6. This is very close to the low-resolution structure obtained for the core particle from X-ray and neutron diffraction (*24–27*) of crystals, showing that there is little effect of crystal packing forces on the core particle structure in crystals. Nuclear magnetic resonance studies (*31*) of core particles show that the N-terminal regions of H2A and H2B are not bound within the core particle, and the N-terminal regions of H3 and H4 are released between 0.3 and 0.6*M* NaCl. These regions are located on the outside of the core particle.

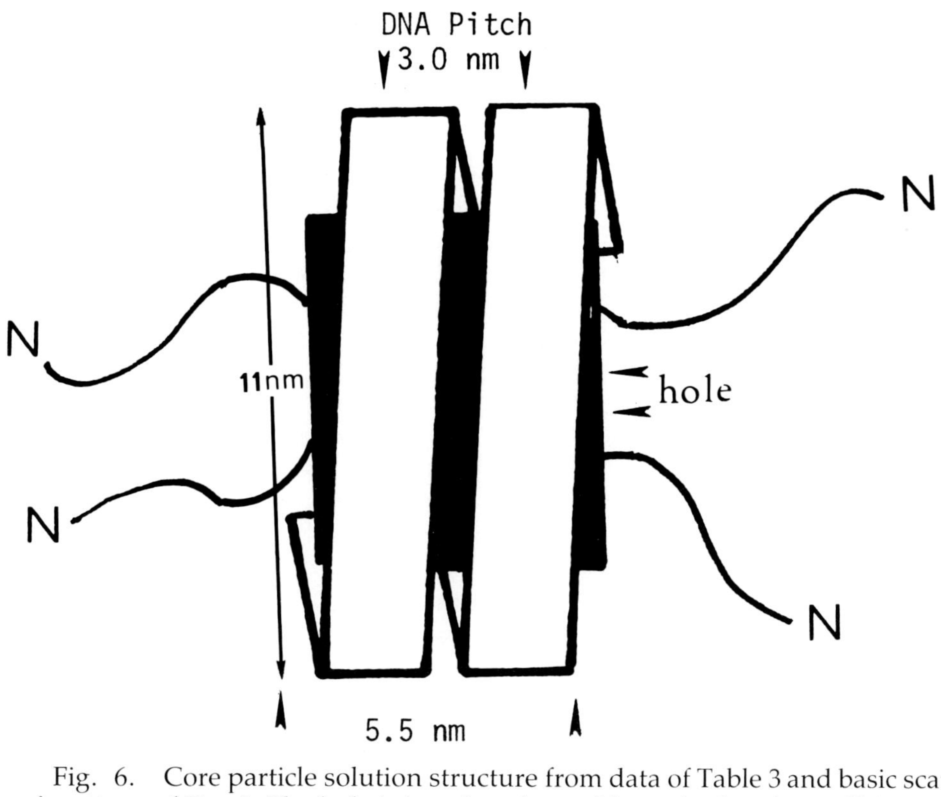

Fig. 6. Core particle solution structure from data of Table 3 and basic scatter functions of Fig. 4. The hole is less than 1 nm. N-terminal regions of histones H2A and H2B are mobile in core particles (*31*).

Chromatosome

The chromatosome contains 168 bp DNA, the histone octamer, and histone H1. It was proposed that histone H1 sealed off two turns of the DNA by binding to one side of the nucleosome at the entry and exit points of the DNA (*32,33*). We have shown that histone H1 contains three well-defined domains; (i) a basic flexible, N-terminal region from residues 1–40, (ii) an apolar globular central region from residues 40–120 of about 3.0 nm in diameter, and (iii) a very basic, flexible C-terminal half of the molecule (*34–36*). Other classes of very lysine-rich histones, H5 from avian erythrocytes and mammalian H1°, have similar three-domain structures (*37*). Based on the properties of the very lysine-rich histones, it was proposed (*38*) that the apolar glubular central region bound to the DNA at the exit and entry regions of the chromatosome to seal off two turns of DNA. This model for the chromatosome containing the different classes of the very lysine-rich histones is given in Fig. 7 (*37*) and is supported by nuclease digestion studies of H1 depleted chromatin to which peptides corresponding to the different H1 domains have been added back. Only the central globular region was able to partially protect the 168 bp DNA length from nuclease digestion (*39*).

HIGHER-ORDER CHROMATIN STUDIES

There is much evidence to suggest that DNA in metaphase chromosomes and interphase nucleus is organized into discrete loops containing 30,000–90,000 bp of DNA (*40*) or in chromatin of average size of 34,000 bp DNA (*41,42*). When the DNA of these loops or domains is packaged into nucleosomes with a packing ratio of 6–7:1, their lengths will be reduced from 10–30 to 1.5–4.5 μm, showing that orders of chromatin structure above the linear array of nucleosomes are required in interphase and metaphase chromosomes.

In the electron microscope, chromatin fibrils of different diameters have been observed under different ionic conditions. At low ionic strengths these chromatin fibrils have a diameter of 10 nm (*43,45*). Increase of ionic strength, either monovalent or divalent cations, results in a transition from the 10 nm fibril to a fibril of diameter 25–30 nm (*43,45*). From neutron fiber diffraction (*46*) and electron microscopy (*47*) it has been proposed that the 25–30 nm fibril consists of the 10 nm fibril coiled at an outside diameter of 30 nm with a pitch of 11 nm. We now discuss the neutron scatter results of the 10–30 nm fibril transition and the neutron diffraction results from fibers of chromatin.

NEUTRON SCATTER STUDIES OF HIGHER-ORDER CHROMATIN STRUCTURE

Neutron scattering techniques have been applied to large oligomers of nucleosomes under a variety of pH and ionic strength conditions to

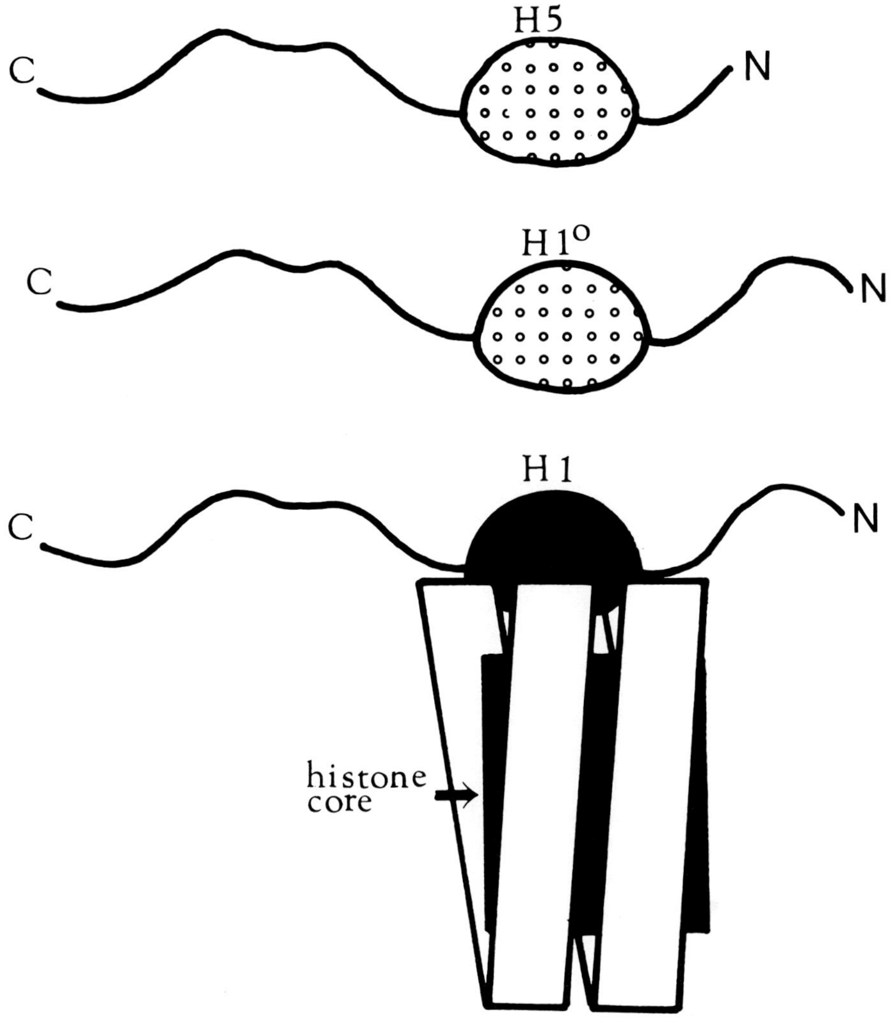

Fig. 7. Model for chromatosome with two turns of DNA sealed off by the globular central regions of the very lysine-rich histones. The globular region of H1° is homologous with that of H5. (From ref. *37*, with permission.)

obtain an undertanding on the transition from the 10 → 30 nm filament and the arrangement of nucleosomes in both of these chromatin structures (*48*). Because the nucleosome is a flat disk 11–11.5 by 5.5–6 nm, it can assume quite different orientations with respect to the axis of the chromatin fibril.

As with the core particles, the neutron scatter curve of chromatin in solution is a measure of the intensity of neutrons $I(\kappa)$ scattered from solution as a function of the scattering angle 2θ and plotted against the parameter $\kappa = 4\pi \sin \theta/\lambda$, where λ is the neutron wavelength. For extended filaments, the Guinier analysis (*28*) is modified and $\kappa I(\kappa)$ is plotted against κ^2. From the slope of the linear region of this plot, the cross-section radius of gyration R_{gx} is obtained, and its intercept on the ln $[\kappa I(\kappa)]$ axis gives the mass/unit length of the filament.

The cross-section radius of gyration of the chromatin filaments in solution will change with contrast $\bar{\rho}$. Similar to the above discussion for the core particle when the scattering-length density of the water mixture is equal to that of the protein (37% D_2O:63% H_2O), (Fig. 1) then the measured transverse R_{gx} will correspond to that of the DNA component and at the contrast matched position for DNA R_{gx} for the protein will be obtained. The dependence of R_{gx} on contrast $\bar{\rho}$ is given by the same form of the equation as used previously for the core particle, i.e.

$$R_{gx}^2 = R_{cx}^2 + \alpha/\bar{\rho} - \beta/\bar{\rho}^2 \qquad [4]$$

A plot of R_{gx}^2 vs $1/\rho$ allows R_{cx}, α, and β to be obtained; R_{cx} is the cross-section R_g of the filament at infinite contrast, α is the first moment of the internal-structure scattering-length density projected on the filament axis, and a nonzero β implies an asymmetric distribution about the filament axis of the scattering-length density of the protein compared to the DNA.

• The experimental scatter curves of $\kappa I(\kappa)$ vs κ can be integrated to give a function

$$D(r)/r = 2\pi P(r)$$

where $D(r)$ is the chord distribution function and has the property of going to zero when r is equal to the largest chord that can be drawn in the cross-section of the fibril. $P(r)$ is the circular average of the Patterson function of the cross-section of the filament.

Extended Chromatin—the 10 nm Filament

The modified Guinier plots for rods (*28*), ln $[\kappa I(\kappa)]$ vs κ^2, obtained from the scatter curves of the nucleofilaments in different mixtures of D_2O:H_2O at low-ionic strength are given in Fig. 8. From these modified plots two parameters are obtained; (i) the cross-section R_{gx} from the slopes of the linear proteins of these plots and (ii) extrapolation of the linear portion of the plot back to zero angle gives $[\kappa I(\kappa)]_o$. A plot of $[\kappa I(\kappa)]_o$ against $\rho_{solvent}$, the solvent-scattering density, is linear and intercepts the abscissa at the mean scattering-length density for the chromatin filament ρ_{mean}.

This was found to be the same as the scattering-length density of a mixture 52% D_2O:48% H_2O corresponding to a protein:DNA ratio of 1·1 (*48*). From the zero angle $[\kappa I(\kappa))^*_o$, the mass/unit length of the nucleofilament could be estimated to be 26,000 ± 6000 g/mol/nm, which corresponds to 1 nucleosome/10 ± 2 nm. This gives a DNA packing ration in the extended chromatin structure of 7.5 ± 1.5.

From the values of R_{gx} at different contrasts, $\bar{\rho}$ and Eq. [4] R_{gx}^2 can be plotted against $1/\rho$ to give the parameters R_{cx}, α, and β. R_{gxDNA} and $R_{gxprotein}$ can be obtained from this plot at the respective contrast matched positions for protein and DNA. A Linear plot of R_{gx}^2 vs $1/\rho$ was obtained showing that the parameter β is close to zero (*48*). These parameters are given in Table 4. Models for extended chromatin at low-ionic strengths must accord with these parameters.

Various models that incorporate the known features of the structure of the nucleosome have been explored. In these models, nucleosomes are

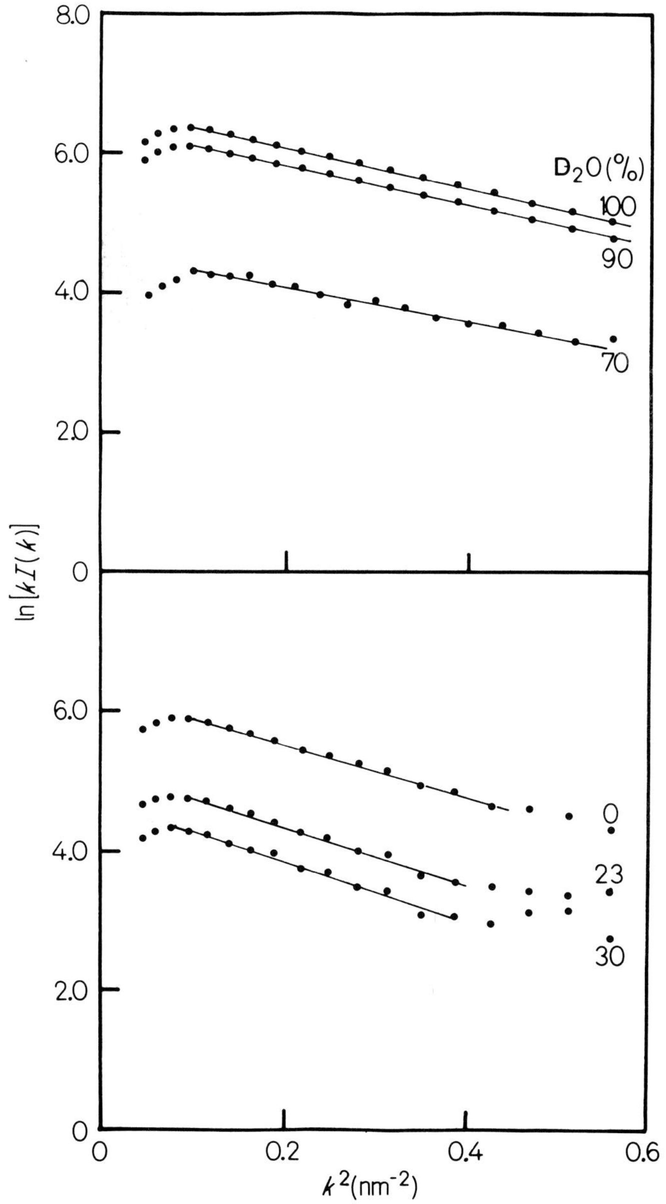

Fig. 8. Modified Guinier plots for the 10 nm nucleofilament in low-ionic strength solutions of different mixtures of D_2O:H_2O. (Taken from ref. *48*, with permission.)

TABLE 4
Neutron Scatter Parameters for Extended Chromatin

Parameter	Extended chromatin
M_L	26,000 $\pm$ 6000 g/mol/nm
or	
	1 nucleosome/10 $\pm$ 2 nm
R_{cx}	2.60 $\pm$ 0.04 nm
α_x	2.7 $\pm$ 0.3 $\times$ 10^{-4} nm
$R_{g \times \mathrm{DNA}}$	3.4 nm
$R_{g \times \mathrm{protein}}$	2.1 nm

separated by 11 nm. If the nucleosomes were arranged face-to-face with the linker DNA following a similar helical path as in the nucleosome, then $R_{gx\mathrm{DNA}}$ would be expected to be that of the nucleosome, i.e., 4.9 nm (Table 3). Putting all of the linker DNA along the axis of the filament would reduce $R_{gx\mathrm{DNA}}$ to 3.7 nm, but this would then give a mass/unit length much lower than the observed value. It follows that the nucleosomes cannot be arranged face-to-face. To reduce $R_{gx\mathrm{DNA}}$ toward the observed value, the faces of the nucleosome disks have to be rotated toward the axis of the filament. With the nucleosome disks arranged edge-to-edge, i.e., with the faces of the disks parallel to the filament axis, the calculated $R_{gx\mathrm{DNA}}$ is 3.2 nm, which is a little lower than the observed value of 3.4 nm. Thus, an arrangement of nucleosomes consistent with the measured $R_{gx\mathrm{DNA}}$ is with the nucleosome disks arranged edge-to-edge, with the faces inclined within an angle of 20° to the filament axis. In model calculations for $R_{gx\mathrm{protein}}$, in a model with the disks face-to-face, $R_{gx\mathrm{protein}}$ is calculated to be 3.0 nm, substantially larger than the observed 2.1 nm. However, with the nucleosome disks arranged edge-to-edge, $R_{gx\mathrm{protein}}$ is calculated to be 2.1 nm, which is in full agreement with the observed value. A similar agreement is also found in model calculations for R_{cx}. It follows that if the model calculations for $R_{gx\mathrm{DNA}}$, $R_{gx\mathrm{protein}}$, and R_{cx} are all close to the observed value, then the calculated parameter α must agree with its observed value. The model that gives all these parameters in close agreement with observed values is with the nucleosome disks close to edge-to-edge, with the faces of the disks inclined to the filament axis at less than 20°. This is shown in Fig. 9. There is no evidence to suggest that at low-ionic strengths the disks are in a very regular array, and some disorder is included along the filament axis. In electron microscope studies it might be expected that the disks would tend to lie flat on the grid, which would give the "10 nm" fibril. At low salt, edge-to-edge zig-zag arrangements of nucleosomes have been reported (*32*), the regularity of the zig-zag depending on the presence of histone H1.

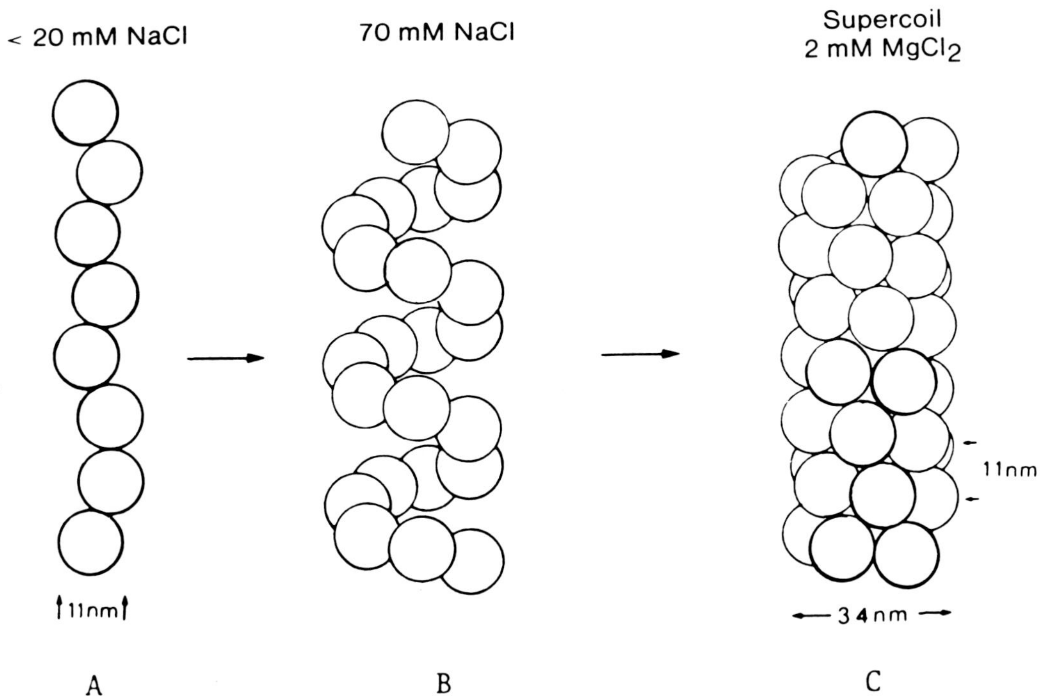

Fig. 9. Models for (A) the "10-nm" nucleofilament in which the nucleosome disks are edge-to-edge, (B) an intermediate transition point in the "10-nm" → "30-nm" transition, and (C) the 34-nm supercoil of nucleosomes of pitch 11.0 nm, which does not indicate the orientation of the nucleosomes. This is given in Fig. 12.

The 30-nm Chromatin Filament

The 10 → 30 nm nucleofilament transition is observed upon increasing the ionic strength of monovalent or divalent cations. As the ionic strength is increased from the low-ionic strength, giving the 10 nm filament, there is a sudden shift of the Guinier region to lower angles at about 20 m*M* NaCl, and an apparent R_{gx} of about 9 nm was observed. The mass/unit lengths and tR_g values have been determined over a range of ionic through the 10 → 30 nm transition (*48*). It has been assumed that the neutron rigid-rod analysis can be applied in an approximate manner, although at intermediate points of the transition, partially folded structures may not be well-defined rods. The modified Guinier plots of chromatin in D_2O at different ionic strengths are shown in Fig. 10. With increase of ionic strength, the linear region of the Guinier plots extends to increasingly higher angles, probably showing that the regions of higher-order structure become more regular. The most notable feature of the 10 → 30 nm salt-induced transition is that the slopes of the linear portions of

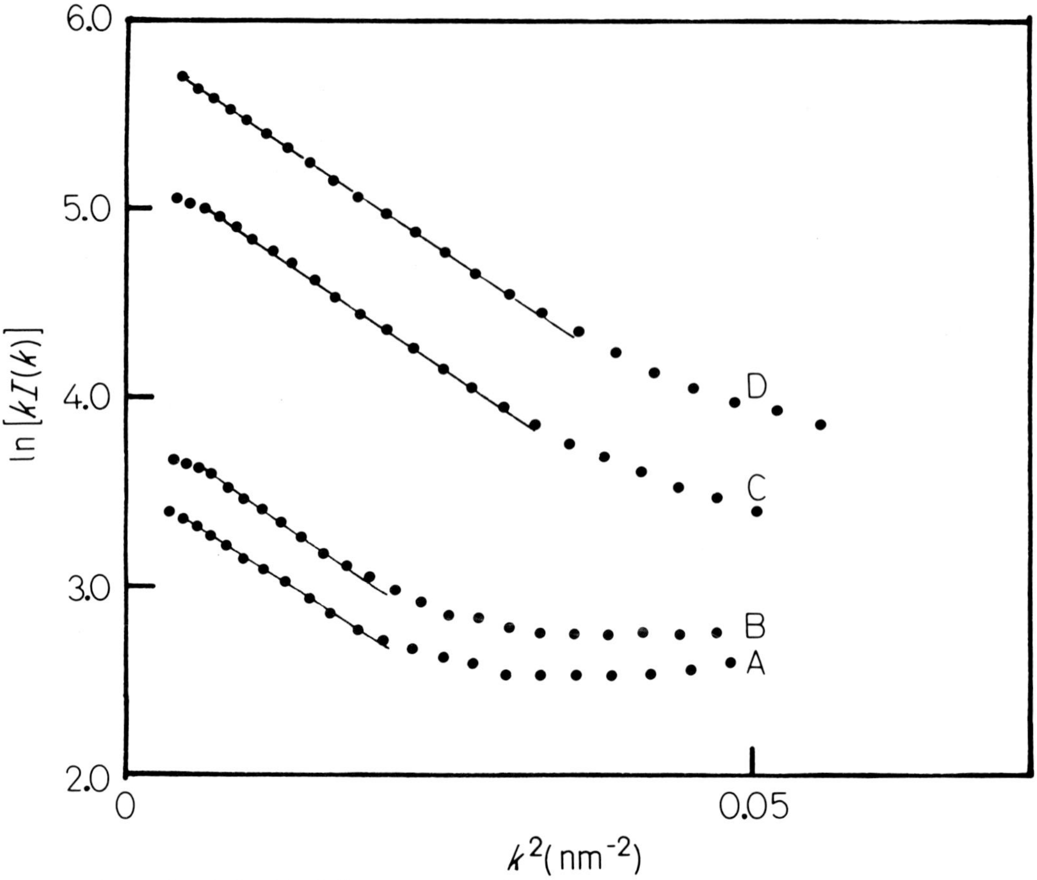

Fig. 10. Modified Guinier plots for chromatin at different-ionic strengths in D_2O; (A) 0.4 m*M* $MgCl_2$, (B) 0.8 m*M* MgCl, (C) 70 m*M* NaCl, (D) 2.0 m*M* $MgCl_2$. (Taken from ref. *48,* with permission.)

the Guinier plots, and thus the R_{gx}, remain almost constant at 9.50 nm, whereas the mass/unit length M_L, given by the intercept on the ordinate, increases markedly. These experiments have been extended by Baldwin and coworkers (*49*) for both chicken erythrocytes and calf-thymus chromatins. Although they report a similar behavior in the salt-induced transition, the values they obtained are more accurate than earlier values (*48*) and are 20% larger. These more recent values for chicken erythrocyte oligonucleosomes in different ionic strengths are as follows: 40 m*M* NaCl, R_{gx} = 10.2 nm and M_L is 0.35 nucleosomes/nm; 80 m*M* NaCl, R_{gx} = 11.3 nm and M_L is 0.54 nucleosome/nm; and 1.2 m*M* $MgCl_2$, R_{gx} = 12.5 nm and M_L is 0.8 nucleosome/nm. For calf-thymus oligonucleosomes the following parameters were obtained: 1.0 m*M* $NaCl_2$, R_{gx} = 10.8 nm and M_L is 0.42 nucleosomes/nm; and 1.2 m*M* $MgCl_2$, R_{gx} = 11.3 nm and M_L is 0.62 nucleosomes/nm.

All of the above data suggests that at 20 m*M* NaCl there is a sudden transition from extended chromatin to a loose supercoiled form that coils more tightly with increasing ionic strength until the most contracted supercoil is obtained with the parameters of Table 5. It is of interest to note that the most compact form of chicken erythrocyte chromatin that contains histone H5 has 0.8 nucleosomes/nm, whereas for calf-thymus chromatin it is 0.62 nucleosomes/nm. In parallel with the increasing compaction of the chromatin fibrils, an inflection is observed in the neutron-scatter curve that moves from 20–25 nm at 20 m*M* NaCl to 10–11 nm for the most contracted form of the supercoil. This 10–11 nm inflection probably has the same origin as the 10–11 nm semimeridional arc observed in both X-ray and neutron diffraction of chromatin fibers and attributed to the pitch of the supercoil of nucleosomes.

Transverse-Radial Distrubution Function

The transverse-radial distribution function for the most contracted form of the nucleofilament supercoil at high salt is given in Fig. 11. This function $D(r)$ goes to zero at the maximum transverse dimension of this second-order chromatin structure, which is at 34–35 nm. This observation has been repeated with the same result and it is concluded that the maximum transverse dimension, i.e., the diameter of the *hydrated* supercoiled chromatin structure is 34–35 nm. This value for hydrated chromatin is to be compared with the value of 25–30 nm obtained from electron microscopy studies of *dehydrated* chromatin.

Model for the Supercoil of Nucleosomes

The above neutron-scatter data from chicken erythrocyte chromatin of the salt-induced contracted second-order chromatin structure give the following parameters (*49*); diameter of the supercoil form 34–35 nm, mass/unit length, M_L = 0.8 nucleosomes/nm, transverse radius of gyration R_{gx} = 12.5 nm, and a pitch of 10–11 nm. It appears that these param-

TABLE 5
Neutron Scattering Parameters of the 34-nm Supercoil of Nucleosome[a]

Parameter	Supercoil
Calf thymus M_L	0.62 nucleosomes/nm
	6.82 nucleosomes/turn
$R_{g\times}$	11.3 nm
Bragg spacing of peak in scatter curve	11.0 nm
Chicken erythrocyte M_L	0.8 nucleosomes/nm
	8.8 nucleosomes/turn
$R_{g\times}$	12.5 nm
Bragg spacing of peak in scatter curve	10.11 nm

[a]Recent data from Baldwin et al. (*49*).

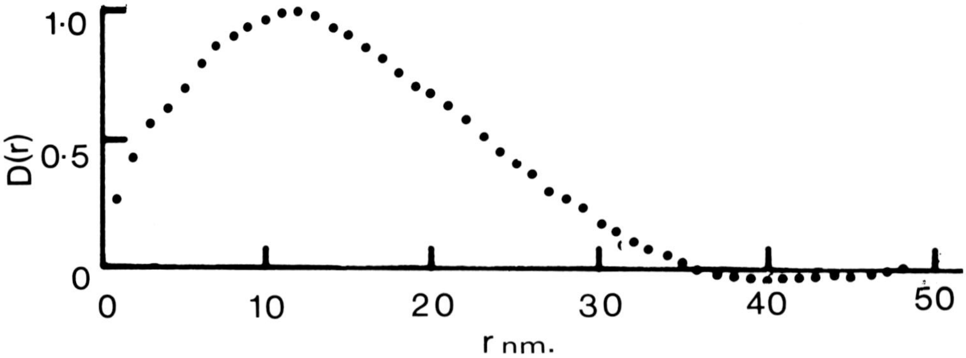

Fig. 11. The cross-section radial-distribution function *D(r)* for the 34-nm supercoil of nucleosomes in H_2O. (Taken from ref. *48,* with permission.)

eters are not constant and for calf-thymus chromatin Baldwin and coworkers (*49*) obtained parameters that correspond to 6–7 nucleosomes/turn of pitch 10–11 nm. The values of R_{gx} of 11.3 nm for calf-thymus chromatin and 12.5 nm for chicken-erythrocyte chromatin rule out models in which the faces of the nucleosome disks lie around the circumference of the supercoil (Fig. 12A). Such an arrangement would give an R_{gx} value of about 14.5 nm, which is larger than the observed values. Models with arrangements of nucleosome disks, as given in Figs. 12B and C, would give R_{gx} values of about 11.5–12.0 nm, consistent with the observed values. Neutron scatter data is unable to distinguish between arrangements B or C unless it is assumed that in the most contracted form of chromatin adjacent gyres of the supercoil are in contact. In this situation, only arrangement C would give the observed M_L, whereas arrangement B wold give a value of M_L twice as large as that observed. The model of the supercoil of nucleosomes that is fully consistent with all of the neutron-scatter data is with the nucleosome disks arranged radially. A similar arrangement with the faces of the disks inclined at about 30° to the supercoil axis has been proposed from linear dichroism of polynucleosomes orientated in electron fields (*50,51*).

A feature of all of these outline models for the 35 nm supercoil of nucleosomes is the existence of a well-defined hole down the center of the supercoil or solenoid. Such a feature would show up as an inflection in the radial-distribution function *D(r)* (Fig. 11) and no inflection is observed. Several reasons have been given as to why it may be difficult to detect a hole along the axis of the supercoil; (i) histone H1 and nonhistone proteins may be located on the inside of the supercoil, and they partially fill up the hole, as proposed for the model for the solenoid (*47*) and/or (ii) the most contracted form of the supercoil is not uniform over extended regions, and this nonuniformity smears the radial distribution function (*48*).

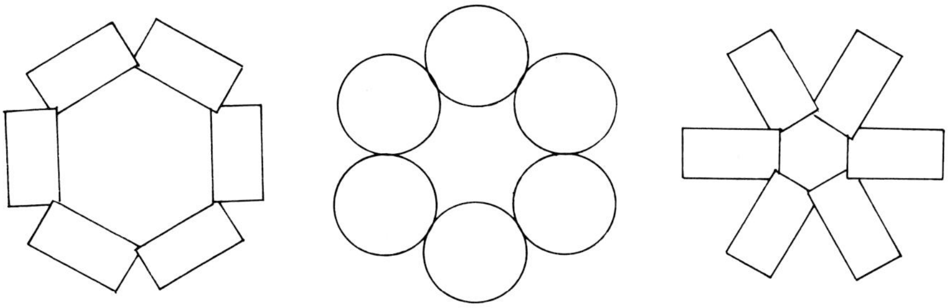

Fig. 12. Possible arrangements of nucleosome disks looking down the axis of the 34-nm supercoil of nucleosomes.

NEUTRON DIFFRACTION OF CHROMATIN

Prior to our current knowledge of the structure of the nucleosome and of the arrangements of nucleosomes in the different orders of chromatin structure, extensive X-ray and neutron diffraction studies of chromatin fibers had been made at different contrasts and water contents. Equilibration at 98% relative humidity (*r.h.*) gives a chromatin concentration of approximately 35% w/w, which is not very different from the estimated 30% concentration of chromatin in the cell nucleus. These fiber X-ray diffraction patterns gave a characteristic series of rings at 11.0, 5.5, 3.7, 2.7, and 2.2 nm that were thought initially to arise from the 11.0 nm pitch of the coil of DNA covered with histone (*52*). This 11.0-nm semimeridional arc is now attributed to the pitch of the supercoil of nucleosomes, as shown in Fig. 9C. The neutron-fiber diffraction patterns of total chromatin at 98% r.h. in different mixtures of H_2O and D_2O are shown in Fig. 13. Because the state of hydration is constant, these changes in the intensities of the diffraction intensities must be the result of changes in contrast and not structure (*53*). In D_2O and 55% D_2O the histone component would be expected to dominate the diffraction and, as can be seen, contributes strongly to the 11.0- and 3.7-nm peaks, whereas in H_2O, the DNA would dominate and contributes to the 11.0-, 5.5-, and 2.7-nm peaks. The 2.7-nm peak probably results from the pitch of the DNA around the nucleosome.

The effects of changing hydration on neutron diffraction of chromatin in H_2O and D_2O are given in Figs. 14A and B. It must be assumed that the change of hydration of either H_2O or D_2O has the same effect on chromatin structure. Thus, these marked differences between neutron diffraction in H_2O compared to D_2O must be attributed to different contrast effects caused by different states of hydration of DNA by H_2O or D_2O. We have shown (*54*) from small molecule contrast, using glycerol/D_2O and glycerol/H_2O mixtures, that the DNA regions of the nucleosome core particles is highly hydrated compared to the histone core. The volume of the water of hydration was determined to be 60 nm^3 and located in the DNA region of the structure for the core particle (Fig. 6). Consider first the changes in the neutron diffraction with D_2O hydration. It is clear that the characteristic 10.2-nm peak moves to smaller spacings on dehydration and in the dry state at 87% w/w concentration it is at 8.4 nm. The broad peak at 3.6–3.7 nm attributed to a structure feature involving the apolar protein core does not change with loss of D_2O. With

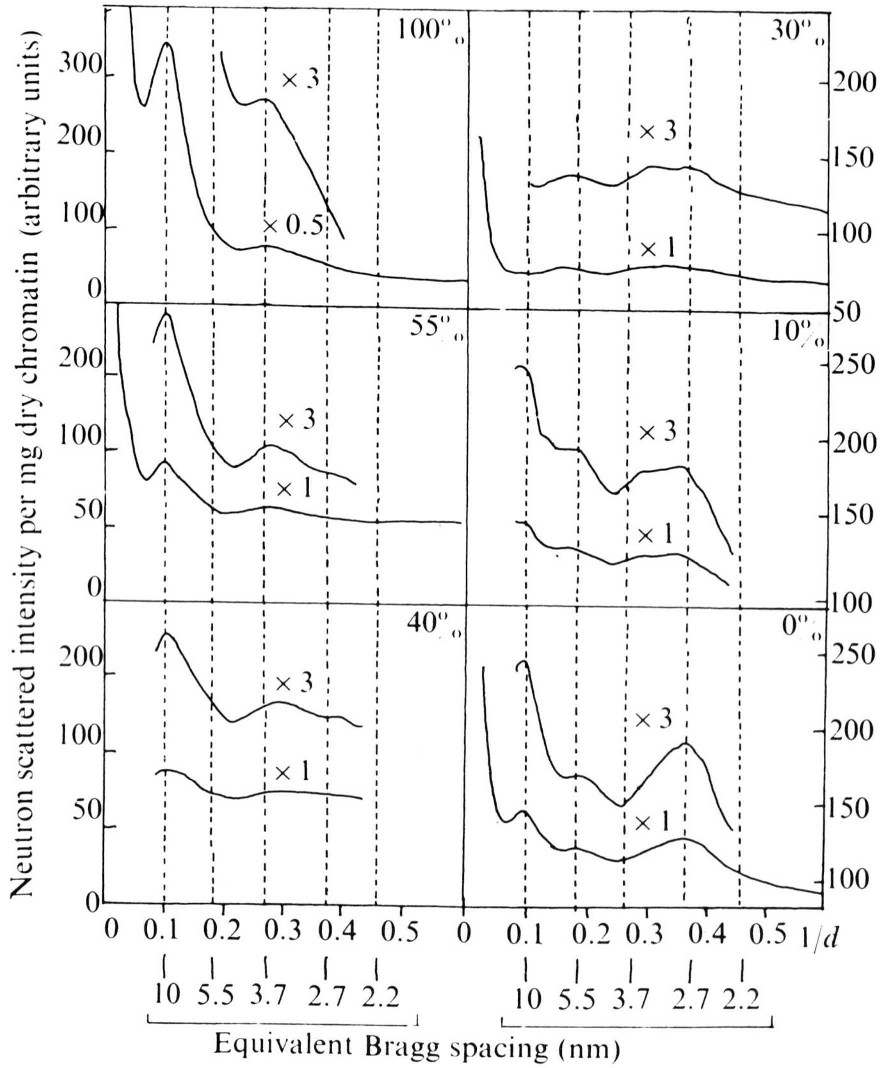

Fig. 13. Effect of contrast changes on the neutron diffraction of films of chromatin hydrated at 98% r.h. (35–60% w/w) in different mixtures of $D_2O:H_2O$. The percentage of D_2O is indicated. (Taken from ref. 53, with permission.)

H_2O dehydration, the 10.2-nm peak undergoes two effects: (i) a change in the spacing from 10.2 to 8.4 nm, as found for D_2O dehydration and (ii) a remarkable change in intensity that goes through zero at 77% w/w concentration. The change in the spacing is a structural effect, whereas the

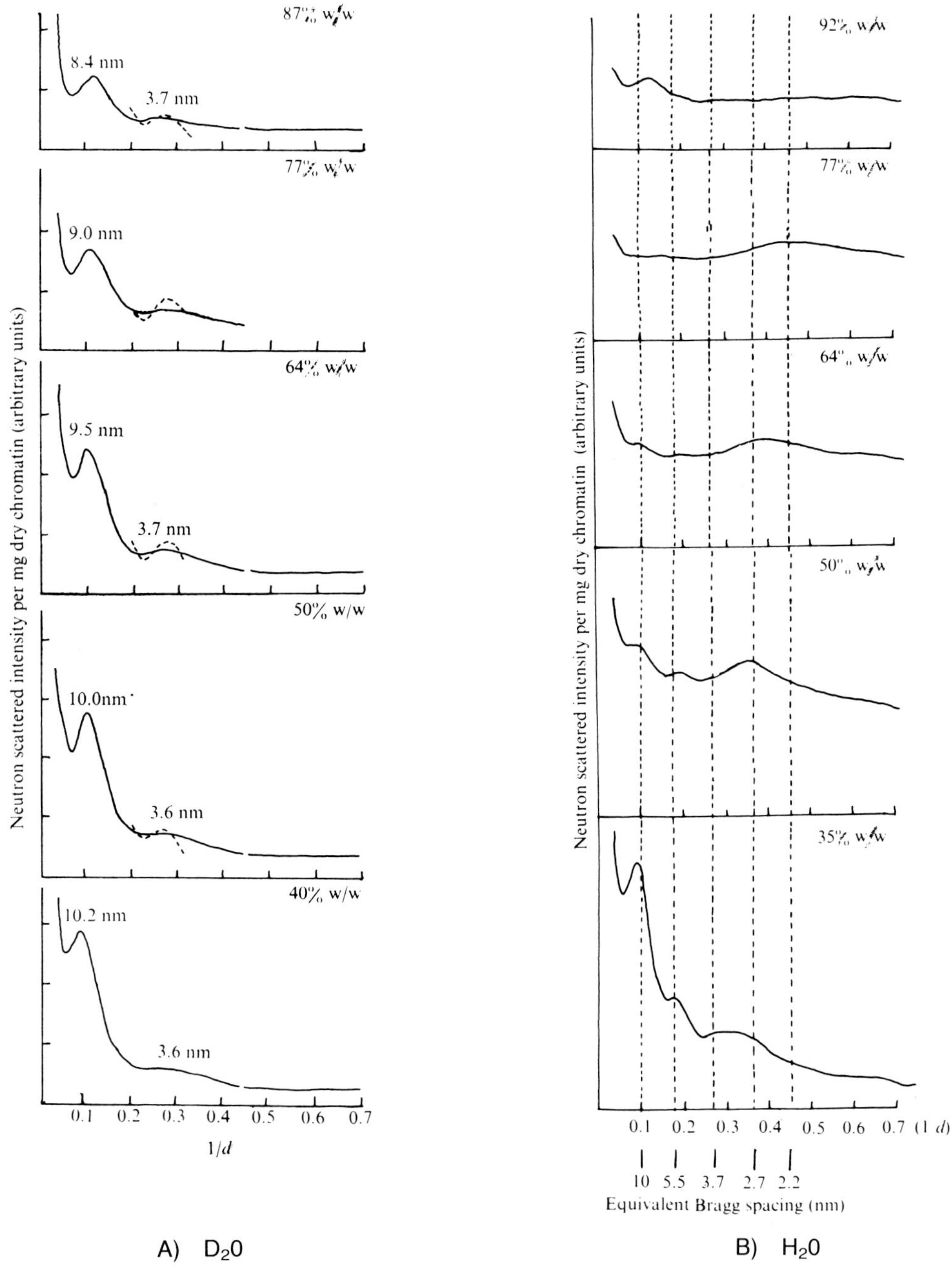

Fig. 14. Effect of concentration on neutron fiber diffractions of chromatin fibers in H_2O (right) and D_2O (left). (Taken from ref. 53, with permission.)

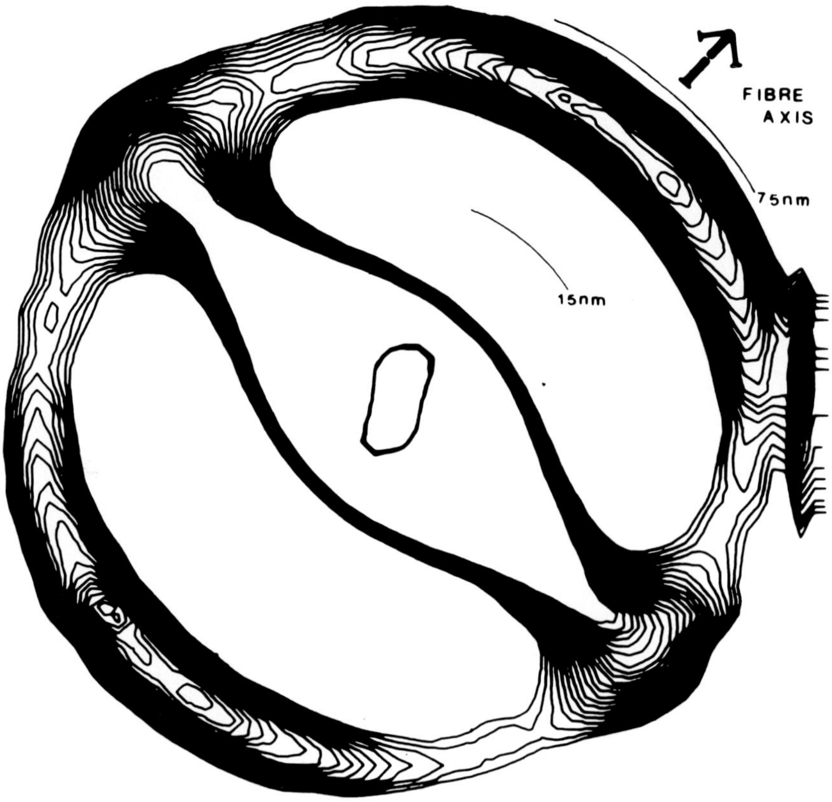

(A)

Fig. 15. Analysis of neutron intensities in the semimeridional arc from fiber neutron diffraction of (A) total chromatin at 90% r.h. D_2O (40% w/w) and (B) H1 depleted chromatin at 32% r.h. D_2O (77% w/w). This quadrant was obtained by folding the 4 quadrants of the fiber diffraction pattern. [(A) Taken from ref. *54*, with permission; (B) from ref. *46*, with permission.] (*On facing page.*)

intensity change must result largely from a contrast effect when an intermediate state of hydration of a DNA rich region has the same scattering-length density as protein-rich regions. Such a situation could not occur with D_2O because both DNA and D_2O have much higher scattering-length densities than proteins (Table 2). With H_2O, the scattering-length density of protein ($+0.22 \times 10^{11}$ cm^{-2}) is intermediate between that of H_2O (-0.06×10^{11} cm^{-2}) and DNA ($+0.37 \times 10^{11}$ cm^{-2}). This intensity behavior of the 10.2-nm diffraction is consistent with the model for the 34-nm supercoil of radial nucleosomes of Figs. 9C and 12; there are DNA-rich regions, i.e., the DNA at the edges of the nucleosomes that touch in adjacent gyres and protein-rich regions that include the core histone octamer. The progressive loss of the 5.5- and 7.7-nm rings is attributed to increasing disorder of the DNA component with dehydration.

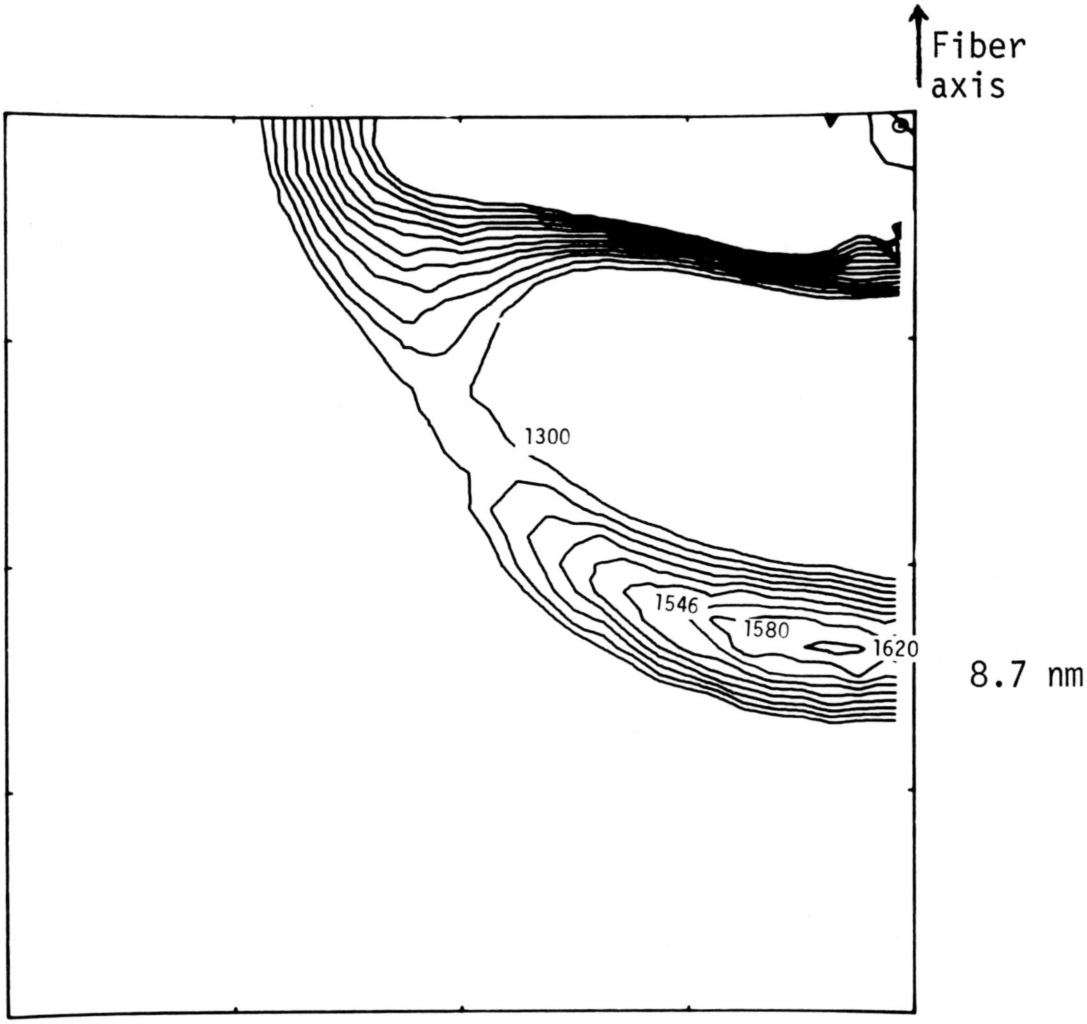

(B)

The 10–11 nm diffraction peak is well known to have semimeridional orientation. If its origin is the pitch of the supercoil of nucleosomes, as shown in Fig. 9C, then the intensity maxima would be expected to lie off the meridian (*46*). With neutron-scattering instruments it is possible to obtain large specimen-to-detector distances, and an analysis has been made of the distribution of neutron intensities within the semimeridional arc at 98% r.h. D_2O (40% w/w) for total chromatin (*55*) and at 32% and 90% r.h. D_2O for H1 depleted chromatin (*46*). For total chromatin (Fig. 15A), the contours of neutron intensities within the 10.2-nm semimeridional arc show off-meridian maxima; for H1 depleted chromatin (Fig. 15B) at 32% r.h., the contours of neutron intensities at 8.7 nm folded into one quadrant also how an off-meridian maxima, and the

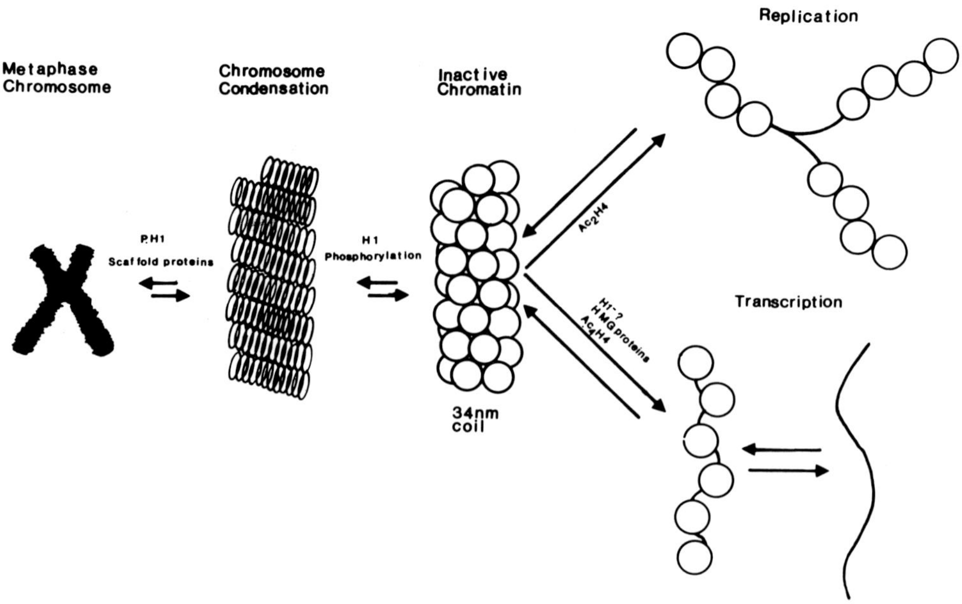

Fig. 16. Scheme showing chromatin structure-function changes and some of the factors associated with these events. (Modified from Chahal et al., *Nature* (1980) **287,** 76–79, with permission.)

same result is found at 98% r.h. (*46*). For total chromatin, the off-meridian maxima strongly support a model of a supercoil of nucleosomes with the parameters given in Fig. 9C. That the H1-depleted chromatin also gives off-meridian intensity maxima at both high- and low-states of hydration, show that the 34-nm supercoil or solenoid can be stabilized at high concentrations in the absence of histone H1. H1 probably stabilizes the 34-nm supercoil and is required to generate the 30-nm filament at low concentrations in the 10 → 30-nm filament transition. H1 binds in the mode shown in Fig. 7, with the apolar globular region sealing off two turns of DNA and is further involved in stabilizing the 34-nm supercoil and generating higher-order structures through interactions of the basic, flexible N- and C-terminal regions and modulation of these interactions by H1 hyperphosphorylation in late G2-phase and metaphase of the cell cycle (*3*).

SUMMARY

Neutron scatter and diffraction techniques have made substantial contributions to our understanding of (i) the structure of the nucleosome, (ii) the structure of the 10-nm filament, (iii) the "10-nm → 30-nm" filament transition, and (iv) the structure of the "34-nm" supercoil or solenoid of nucleosomes. Neutron techniques are unique in their properties, which allows for the separation of the spatial arrangements of histones and DNA in nucleosomes and chromatin. They have equally powerful applications in structural studies of any complex two-component biological system.

A major success for the application of neutron techniques was the first clear proof that DNA was located on the outside of the histone octamer in the core particle. A full analysis of the neutron-scatter data gave the parameters of Table 3 and the low-resolution structure of the core particle *in solution* shown in Fig. 6. Initial low-resolution X-ray diffraction studies of core particle crystals gave a model with a lower DNA pitch of 2.7 nm (*24*). Higher-resolution X-ray diffraction studies now give a structure with a DNA pitch of 3.0 nm and a hole of 0.8 nm along the axis of the DNA supercoil. The neutron-scatter solution structure and the X-ray crystal structure of the core particle are thus in full agreement within the resolution of the neutron-scatter techniques. The model for the chromatosome (Fig. 8) is largely based on the structural parameters of the DNA supercoil in the core particle, nuclease digestion results showing protection of a 168-bp DNA length by histone H1 and H1 peptide, and the conformational properties of H1. The path of the DNA outside the chromatosome is not known, and this information is crucial for our understanding of higher chromatin structure. The interactions of the flexible basic and N- and C-terminal regions of H1 within chromatin and how these interactions are modulated by H1 phosphorylation are not known. The N- and C-terminal regions of H1 represent a new type of protein behavior, i.e., extensive protein domains that are designed not to fold up into secondary and tertiary protein structures. This behavior is increasingly observed in DNA and chromatin binding proteins, and in the case of the high-mobility group proteins HMG 14 and 17, the entire polypeptide chain is a flexible random coil over a wide range of solution, ionic, and pH conditions (*56*). It follows that the native conformations are probably imposed on these flexible domains and molecules by their binding sites in chromatin.

Above the level of the chromatosome it is unlikely that crystals of oligonucleosomes and pieces of chromatin will be grown for diffraction studies. Neutron techniques will be the most powerful for studies of the hydrated forms of chromatin structures, conformation transitions induced by histone modifications, and the interactions of nonhistone chromosomal proteins, such as HMG proteins 14 and 17. For the extended form of chromatin at low-ionic strengths, the structural parameters obtained from the neutron-scatter data are given in Table 4. The model that accords with this data is given in Fig. 9A. Other models are acceptable, providing they give calculated structural parameters in agreement with those observed. In the salt-induced transition from the 10-nm filament to the 30-nm filament, a transition occurs at about 20 nm NaCl to a loosely coiled form that becomes more tightly coiled with increasing ionic strength until the 34-nm supercoiled form is reached. The structural parameters of this form are given in Table 5 and the model that most closely accords with this data is given in Figs. 9C and 12. Other models are permissible if they accord with the parameters of Table 5. The neutron-diffraction data, particularly the off-meridian maxima in the 10–11 nm semimeridional arc, strongly support the model for the 34-nm supercoil of nucleosomes. A feature of this model, for which there is as yet no experimental data, is the hole from the axis of the supercoil. This may be because of the location of histone H1 and nonhistone proteins inside of the supercoil partially filling the hole, the supercoiled form may be irregular, or the underlying structural motif of the 34-nm supercoil may be modulated by other factors. This irregularity would also explain the breadth of the 10–11-nm semimeridional arc in Fig. 14.

The pitch of the 34-nm supercoil depends on ionic conditions in solution and on its state of hydration in films and fibers. The most contracted salt-induced 34-nm supercoil in solution has the same pitch as the most hydrated form in fibers (35% w/w) and probably corresponds to the form found in vivo. With increasing concentration of chromatin in fibers, the pitch decreases and reaches 8.4 nm in the relatively dry form. This can be achieved by the faces of the nucleosome disks tilting further away from the axis of the supercoil; to give a pitch of 8.4 nm, the angle of tilt would have to be 40°. Assuming that the number of nucleosomes/turn remains unchanged, then the mass/unit length would be increased from the equivalent of 0–8 nucleosomes/nm for the hydrated form to 1.05 nucleosomes/nm for the dehydrated form of chicken-erythrocyte chromatin. This value, not the value for hydrated chromatin, should be compared with mass/unit length estimates from electron microscopy.

Reversible chemical modifications of histones and the binding of subgroups of nonhistone proteins, e.g., HMG 14 and 17, have been associated with different chromatin functions; core histone acetylation and the binding of HMG 14 and 17 with "active" chromatin; ubiquitination of histones H2A and H2B with the labeling of potentially active chromatin; and H1 and H3 phosporylation with chromosome condensation. A scheme showing proposed structure–function changes is given in Fig. 16. We now have a sound enough understanding of the structure of the nucleosome and the different orders of chromatin structure to pursue studies of the effects of these chemical modifications and HMG proteins on nucleosome and chromatin structure. Neutron scatter and diffraction techniques are by far the most powerful tools available for such studies.

ACKNOWLEDGMENTS

This work is supported by research grants from the Science and Engineering Research Council of the United Kingdom to J. P. B; from the National Institute of Health, USA (GM26901) and from the National Science Foundation USA (PCM8319215) to E. M. B. We acknowledge particularly the collaboration of our colleagues, Drs. G. W. Braddock, B. G. Carpenter, P. P. Hjelm, K. Ibel, G. G. Kneale, G. Sibbet, and Mr. S. P. Dunn, in these neutron studies of nucleosome and chromatin structure.

REFERENCES

1. Compton, J. L., Bellard, M., and Chambon, P. (1976), *Proc. Natl. Acad. Sci. USA* **73,** 4382.
2. Stein, A., and Binou, M. (1984), *J. Mol. Biol.* **178,** 341.
3. Bradbury, E. M., and Matthews, H. R. (1980), Cell Growth, in *NATO Advanced Study Institutes Series* (Nicolini, C., ed.) Plenum, NY, vol. 38, pp. 411–454.
4. Allfrey, V. G. (1976), *Chromatin and Chromosome Structure* (Li, H. J., Eckhardt, R., eds.) Academic, FL, pp. 167–191.
5. Goldknopf, I. L., Taylor, C. W., Baum, R. W., Yeoman, L. C. Olsen, M. O. J., Prestayko, A. W., and Busch, M. (1975), *J. Biol. Chem.* **250,** 7182.
6. West, M. H. P., and Bonner, W. M. (1980), *Nucleic Acids. Res.* **8,** 4671.
7. Wu. R. S., Kohn, K. W., and Bonner, W. M. (1981), *J. Biol. Chem.* **256,** 5916.
8. Langan, T. A. (1978), *Methods in Cell Biology* (Stein, G. S., and Stein, J., eds.), Academic, FL, vol. 19, pp. 127–142.
9. Gurley, L. R., Tobey, R. A., Walters, P. A., Hilderbrand, C. E., Hohman, P. G., D'Anna, J. A., Barham, S. S., and Deavan, L. L. (1978), *Cell Cycle Regulation* (Jeter, J. R., Cameron, I. L., Padilla, G. M., and Zimmerman, R. M., eds.) Academic, FL, pp. 37–60.
10. Levinger, L., and Varshavsky, A. (1982), *Cell* **28,** 375.
11. Mueller, R. D., Yasuda, H., Hatch, C. L., Bonner, W. M., and Bradbury, E. M. (1985), *J. Biol. Chem.* **260,** 5147.
12. Weisbrod, S., and Weintraub, H. (1979), *Proc. Natl. Acad. Sci. USA* **76,** 630.
13. Weisbrod, S., Groudine, M., and Weintraub, H. (1980), *Cell* **19,** 289.
14. Kneale, G. G., Baldwin, J. P., and Bradbury, E. M. (1977), *Q. Rev. Biophys.* **10,** 485.
15. Stuhrmann, H. B. (1974), *J. Appl. Cryst.* **7,** 173.
16. Stuhrmann, H. B. (1975), *Brookhaven Symposia in Biology* **27,** 3–19.
17. Ibel, K., and Stuhrmann, H. B. (1975), *J. Mol. Biol.* **93,** 255.
18. Moore, P. B., Langer, J. A., Schoenborn, B. P., and Engelman, D. M. (1977), *J. Mol. Biol.* **112,** 199.
19. Engelman, D. M. (1979), *Methods in Enzymology,* vol. 59, (Moldave, K., and Grossman, L., eds.), Academic, FL, pp. 656–669.
20. Bradbury, E. M., Baldwin, J. P., Carpenter, B. G., Hjelm, R. P., Hancock, R, and Ibel, K. (1975), *Brookhaven Symp. Biol.* **27,** 97.
21. Pardon, J. F., Worcester, D. C., Wooley, J. C., Tatchell, K., van Holde, K. E., and Richards, B. M. (1975), *Nucleic Acids Res.* **2,** 2163.
22. Hjelm, R. P., Kneale, G. G., Suau, P., Baldwin, J. P., Bradbury, E. M., and Ibel, K. (1977), *Cell* **10,** 139.
23. Suau, P., Kneale, G. G., Braddock, G. W., Baldwin, J. P., and Bradbury, E. M. (1977), *Nucleic Acids Res.* **4,** 3769.
24. Finch, J. T., Lutter, L. C., Rhodes, D., Brown, R. S., Rushton, B., Levitt, M., and Klug, A. (1977), *Nature* **269,** 29.
25. Finch, J. T., Brown, R. S., Rhodes, D., Richmond, T. J., Rushton, B., Lutter, L. C., and Klug, A. (1981), *J. Mol. Biol.* **145,** 757.
26. Bentley, C. A., Finch, J. T., and Lewit-Bentley, A. (1981), *J. Mol. Biol.* **145,** 771.

27. Richmond, T. J., Klug, A., Finch, J. T., and Lutter, L. C. (1982), *Proc. 2nd SUNYA Conversations,* (Sarma, R. H., ed.), Academic, FL, vol. 11, pp. 109–123.
28. Guinier, A., and Fournet, G. (1955), *Small Angle Scattering of X-rays,* Wiley and Sons, New York.
29. Sibbet, G., Carpenter, B. G., Ibel, K., Kneale, G. G., Bradbury, E. M., and Baldwin, J. P. (1983), *Eur. J. Biochem.* **133,** 393.
30. Braddock, G. W., Baldwin, J. P., and Baldwin, E. M. (1981), *Biopolymers* **20,** 327.
31. Cary, P. D., Moss, T., and Bradbury, E. M. (1979), *Eur. J. Biochem.* **89,** 475.
32. Simpson, R. T. (1978), *Biochemistry* **17,** 5524.
33. Thoma, F., Koller, Th., and Klug, A. (1979), *J. Cell. Biol.* **83,** 403.
34. Chapman, G. E., Hartman, P. G., and Bradbury, E. M. (1976), *Eur. J. Biochem.* **61,** 69.
35. Hartman, P. G., Chapman, G. E., Moss, T., and Bradbury, E. M. (1977), *Eur. J. Biochem.* **77,** 45.
36. Chapman, G. E., Aviles, F. X., Crane-Robinson, C., and Bradbury, E. M. (1978), *Eur. J. Biochem.* **90,** 287.
37. Cary, P. D., Hines, M. L., Bradbury, E. M., Smith, B. J., and Johns, E. W. (1981), *Eur. J. Biochem.* **120,** 371.
38. Crane-Robinson, C., Bohm, L., Puigdomenech, P., Cary, P. D., Hartman, P. G., and Bradbury, E. M., (1980), *FEBS DNA-Recombination Interaction and Repair,* Pergamon, Oxford and NY, pp. 293–300.
39. Allan, J., Hartman, P. G., Crane-Robinson, C., and Aviles, F. X. (1980), *Nature* **288,** 675.
40. Laemmli, U. K., Cheng, S. M., Adolph, K. W., Paulson, J. R., Brown, J. A., and Baumbach, W. R. (1978), *Cold Spring Harbor Symp. Q. Biol.* **42,** 351.
41. Igo-Kemenes, T., Greil, W., and Zachau, H. G. (1977), *Nucleic Acids Res.* **4,** 3387.
42. Igo-Kemenes, T., and Zachau, H. G. (1978), *Cold Spring Harbor Symp. Q. Biol.* **42,** 109.
43. Ris, H., and Kubai, D. F. (1970), *Annu. Rev. Genet.* **4,** 263.
44. Woodcock, C. L. F. (1973), *J. Cell Biol.* **59,** 737a.
45. Olins, A. L., and Olins, D. E. (1974), *Science* **183,** 330.
46. Carpenter, B. G., Baldwin, J. P., Bradbury, E. M., and Ibel, K. (1976), *Nucleic Acids Res.* **3,** 1739.
47. Finch, J. T., and Klug, A. (1976), *Proc. Natl. Acad. Sci. USA* **73,** 1897.
48. Suau, P., Bradbury, E. M., and Baldwin, J. P. (1979), *Eur. J. Biochem.* **97,** 593.
49. Baldwin, J. P., Dunn, S. P., Rattle, H. W. E., Staynov, D. Z., Boulter, E. M., and Carpenter, B. G. (1984), Personal communication.
50. McGhee, J. D., Nicol, J. M. Felsenfeld, G., and Rau, D. C. (1983), *Cell* **33,** 831.
51. Mitra, S., Sen, D., and Crothers, D. M. (1983), *Nature* **308,** 247.
52. Pardon, J. F., and Wilkins, M. H. F. (1972), *J. Mol. Biol.* **68,** 115.
53. Baldwin, J. P., Boseley, P. G., Bradbury, E. M., and Ibel, K. (1975), *Nature* **253,** 245.
54. Baldwin, J. P., Braddock, G. W., Carpenter, B. G., Kneale, G. G., Simpson, G. K., Suau, P., Hjelm, R. P., and Bradbury, E. M. (1978), *J. Appl. Cryst.* **11,** 483.
55. Baldwin, J. P., Carpenter, B. G., Crespi, H., Hancock, R., Stephens, R. M., Simpson, J. K., Bradbury, E. M., and Ibel, K. (1978), *J. Appl. Cryst.* **11,** 484.
56. Bradbury, E. M. (1982), *Non-Histone Chromosmal Proteins:HMG Proteins,* (Johns, E. W., ed.) Academic, NY, pp. 89–110.

Nuclear Structure

From the Pores to the High-Order Gene Structure

C. Nicolini

Chair of Biophysics, School of Medicine, University of Genova, Viale Benetto, Genova, Italy and Temple University, Philadelphia, PA

INTRODUCTION

Early in this symposium, distinguished scientists summarized the experimental evidence for the orderly organization of the genetic material within mammalian cells, from the primary (base sequences, as discussed by Baltimore, Watson, and Temin, in reference to gene structure and expression) to the secondary (the various forms of the DNA double helix discussed by Rich) and tertiary (the wrapping of DNA around octamer histones to form the nucleosome, as discussed by Bradbury) structures. Large uncertainty—because of the inherent difficulties of an *in situ* characterization—still exists as to the exact three-dimensional geometry of the higher-order DNA foldings (from the supernucleosomes on); suggestions and evidences for the quaternary and higher-order structures are hereby summarized, along with other nuclear structures, with a possible clue to the higher-order gene structures.

QUATERNARY STRUCTURE

The critical role of total DNA length (*see* Table 1 for the various methods of preparation) in maintaining the native quaternary chromatin structure has recently emerged. Chromatin obtained by mild micrococcal nuclease digestion (apparently to avoid distortion induced by other methods of preparation, but causes a distribution of DNA with molecular weights of 2.5×10^7, as a maximum) yields, at low-ionic strength, fibers of 100 Å, often loosely coiled, with approximately 0.1 nucleosomes/10 Å; at 20 m*M* NaCl or 0.3 m*M* MgCl they take a more condensed form, i.e., 300–400 Å in diameter, with approximately 0.2 nucleosomes/10 Å (*1*).

Upon further increase in ionic strength, the fiber appears to increase its folding, yielding, as monitored by neutron scattering studies, A DNA packing ratio of 0.6 nucleosomes/10 Å and a radius of gyration of 95 Å.

When the H1 histone is removed, addition of MgCl fails to produce any condensation. This and the ionic-strength dependence are exactly predicted by polyelectrolyte theory for the folding and bending angle of a single nucleofilament. Shearing of this chromatin preparation, even at high-ionic strength, causes the disappearance of the quaternary "solenoidal" 300 Å fiber into the 100 Å nucleofilament (*2*), as confirmed by X-ray diffraction data in which the 110 and 55 Å peaks, reflecting the organization of the nucleosomes, are replaced by 38 and 27 Å peaks, reflecting the structure of the nucleosome itself.

At approximately (or even before) this time (*3,4*), these observations on nuclease-prepared chromatin were reported, chromatin isolated by gentle cold-water swelling of isolated nuclei, followed by sucrose sedimentation and gentle suspension—without any shearing, in contrast to most of the then dominant preparations requiring homogeneous solutions with lowlight scattering rather than preservation of DNA integrity—was shown to display, even at low-ionic strength, geometry-dependent circular dichroic (CD) spectra and signal outside the absorptive band that disappears with shearing of chromatin. Similarly, a dramatic increase in ethidium bromide binding sites and template activity, accompanied by a decrease in superhelix to helix and in helix–coil melting temperatures (Fig. 1), were reported as the consequence of shearing

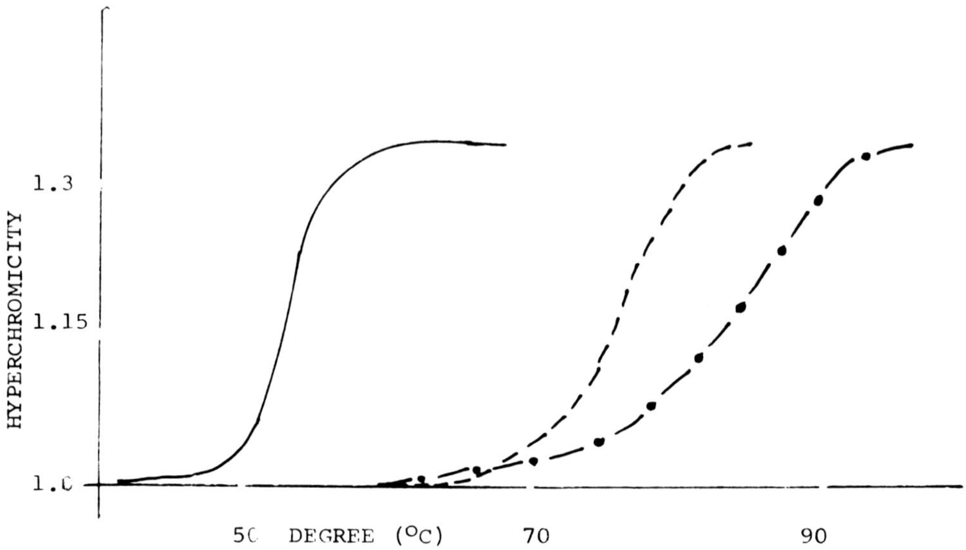

Fig. 1. Thermal denaturation profiles by absorption changes at 260 nm relative to absorption at room temperature (hyperchromicity) (top) and by molar ellipticity at 276 nm (right) of DNA from calf thymus (——) and of chromatin isolated from rat liver cells. The native chromatin was either unsheared (---) or sheared (—·—) in 1 m*M* Tris-HCl, pH 8.

on the same chromatin, pointing to structural alterations. Indeed, although the sheared chromatin significantly scatters light and contains the same amount of protein and RNA/DNA, no differential scattering of circular polarized light is present. Compatible with parallel observations on T2 bacteriophage, disrupted vs intact, and with the comfort of recent theoretical computations from first principles, this suggests that an ordered asymmetry of the scattering centers (nucleosomes), as caused by a regular or irregular coiling of the nucleofilament, is still present in native unsheared chromatin, even at low-ionic strength, as long as the average length of the DNA fragments is larger than 5×10^7 daltons (Table 1). Unsheared chromatin at high-ionic strength and isolated nuclei in which the *in situ* NaCl equivalent normality appears at about 100 m*M*, display further increase in differential light scattering, but the phenomena is already present at low-ionic strength (1 m*M*), suggesting that some kind of nucleosome organization is preserved in this chromatin preparation, contrary to nuclease methods yielding smaller molecular weight fragments (Table 1). Similar conclusions were achieved by studying the higher reactivity of certain chromatin regions, with respect to ethidium bromide (EB) intercalation, when unsheared chromatin at low-ionic strength was utilized: These sites of affinity, quite higher than free DNA, disappear after shearing, as assayed by independent fluorimetric and spectropolarimetric titration. These DNA regions appear to be maintained under physical constraints favorable to intercalation, similar to

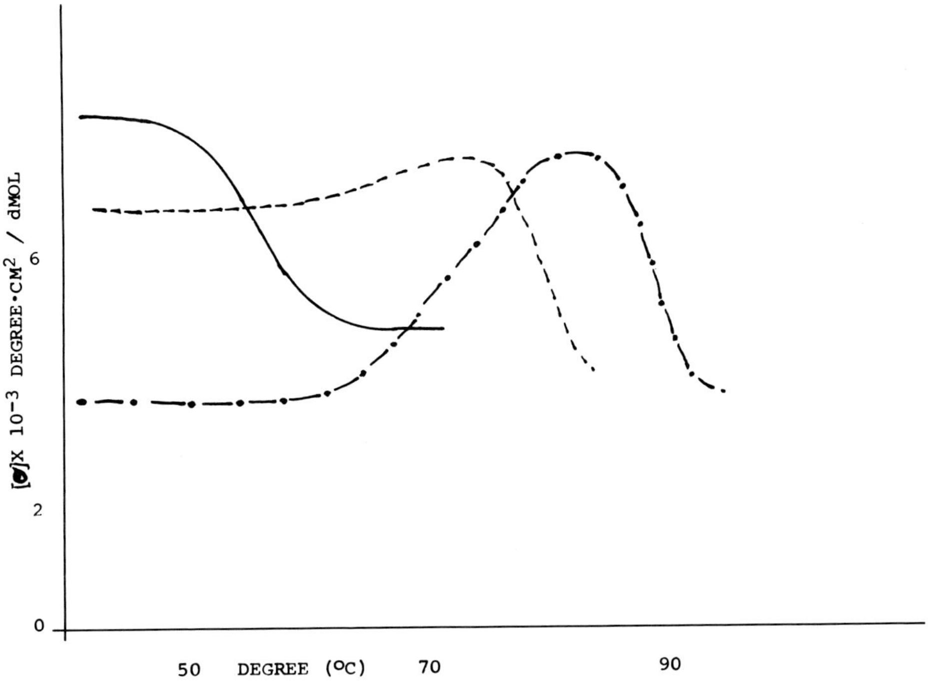

Fig. 1. *(continued)*

TABLE 1
Physical Properties of Chromatin DNA

Type of preparation	Histone DNA	Nonhistone DNA	Molecular weight, $\times 10^6$	Number of nucleosomes
Whole cell[a] (*in situ*)	—	—	3000	24,000
Intact nuclei[b]	—	—	100–3,000	800–24,000
Unsheared native chromatin[c]	0.8–1.2	0.2–5.2	10–100	80–800
Isolated native chromatin[d]	0.8–1.2	0.2–5.2	5–50	40–400
Isolated sheared chromatin[e]	0.8–1.2	0.2–5.2	4.4	30

[a]Measures were performed in alkaline sucrose gradients, calibrated with freshly prepared T4 and T7 bacteriophages and circles of SV40.

[b]Estimated from electron micrographs of chromatin DNA fibers.

[c]Contrary to teleological beliefs, better results are obtained by lysis of native nuclei in cold water, with the sedimented viscous chromatin pellet being resuspended without any shearing.

[d]Typically, 2–5 x 10^8 nuclei/mL are digested with 100–600 U/mL of micrococcal nuclease for a time ranging between 5 to 40 s. Obviously, the final length depends on the digestion time and nuclease concentration, but under no circumstances is the upper limit larger than that indicated here (in daltons).

[e]Shearing is typically accomplished by sonication at 50 W for 20 s or more or in a motor driven homogenizer for 15 s or more. This yields a quite homogeneous solution that lacks differential scattering of circularly polarized light and displays a reduced scattering of the unpolarized light. The maximum molecular weight refers to shearing following either method of chromatin isolation, by nuclease (d) or lysis (c).

that observed at the beginning of the binding process of EB to covalently twisted DNA. Our early observations with unsheared chromatin from various mammalian cells show, indeed, that, although for low concentrations the EB affinity is larger for chromatin, there is a critical amount of dye for which chromatin has equal affinity for DNA; in the absence of the dye, the superhelix density, σ_0 [number of superhelices/10 base pairs (bp)], of the chromatin can then be obtained from the relationship $\sigma = \sigma_0 + 1.45\,\nu$, assuming that the intercalation of an EB molecule unwinds the DNA complex by an angle of 26°; at the critical value of bound dye/DNA (ν_0), the superhelix disappears totally and $\sigma_0 = -1.45\,\nu_0$.

The progressive disappearance of the superhelix is compatible with an increased CD signal at 271 nm up to a maximum toward relaxed protein-free DNA, with a subsequent decrease whenever further intercalation behind ν_0 takes place: Exactly this has been found in the same chromatin, even after subtraction of the hidden EB band, derived from a conservative Cotton effect. Whenever EB intercalates into a closed DNA molecule, such as circular DNA, positive superhelical turns are generated, with an increase in the rate of sedimentation; if the closed molecule (i.e., chromatin-DNA with fixed ends) possesses negative superhelical turns, the intercalation first causes a loss of these turns, relaxing the molecule and then introduces positive turns, causing the molecule to recoil.

This, compatible with the EB studies on unsheared chromatin, is exactly what is observed in *Drosophila* and mammalian genomes, isolated as compact masses 10^{11} daltons long by means of gentle lysis-containing detergent. Under these conditions, the DNA association with histone octamers and certain nonhistone proteins (*see* later in this paper) is preserved; only DNAse I nicks, whenever they appear more frequent than every 6×10^7 daltons (or 400 nucleosomes), seem capable of fully relaxing the genome to a form that sediments more slowly, without any negative superhelical turns. Since nucleosomes are preserved, even in the presence of DNAse I, it would then appear that 6×10^7 daltons is the minimum length of DNA that can be considered a closed structure with highly-negative superhelical turns.

In summary, all of the above data suggest that:

(a) The conformation and properties of pieces of chromatin, dissected by mild nuclease digestion, are not representative of the native quaternary DNA structure that is preserved with its negative superhelical turns only when average the DNA length is 5–6×10^7 daltons or more.

(b) Topological constraints on the chromatin fiber must be present at regular intervals (about 5–6×10^7 daltons) to prevent the DNA strand from rotating; open ended strands will never yield negative superhelical turns. The presence of topological DNA constraints is also suggested by the fact that soluble chromatin isolated by the mildest nuclease or restriction enzyme digestion has a maximum length of 75 kilobase pairs (kbp) or 4.7×10^7 daltons.

(c) Even for nucleofilaments smaller than 2.5×10^6 daltons, such as that obtained following nuclease digestion, regular or irregular folding can take place in the presence of the H1 histone and in increasing ionic strength, as a result of spontaneous DNA bending by proper neutralization of the negatively charged phosphates.

The critical role of histones and ions in "modifying" the lower- and higher-order structures and controlling gene expression (*see* later in this paper) remain confirmed, but their roles in maintaining the basic skeleton of the native quaternary and quinternary structures are highly doubtful. Histone-depleted chromosomes, visualized by electron microscopy in the presence of 2.0*M* NaCl (removing all but 8% of the chromosomal proteins, virtually all of the nonhistones, as has been long known), appear to consist of single nucleoprotein fibers and reveal the presence of a skeleton, i.e., a scaffold of a dense network of fibers, surrounded by a "halo" of loops of DNA anchored at both ends at the scaffold: These results seem to be reproducible both in metaphase and interphase chromosomes, leading to the nuclear matrix concept.

Keeping all of the above evidence in mind, how does DNA go from the tertiary-nucleosome structure, constituting the basic repeating unit of chromatin, to the higher-order structure? How are nucleosomes arranged under native conditions to yield the frequently observed 300-Å quaternary fiber? What is the exact arrangement of the two-order superhelical chromatin structure suggested earlier by differential light scattering and EB binding studies, and how can it be compatible with all of the above observations?

The answers may come from recent electron-microscopic observations obtained on a phospholipid monolayer (*5*), which suggest that the 300 Å fiber from calf-thymus nuclei is formed not by a coiled single nucleofilament, but by up to four nucleofilaments properly interlinked at periodically fixed sites, uninfluenced by DNAse or very high-ionic strength. This original observation, made by a group of investigators at a CNR center in Genova (*5*), which originally left me quite skeptical because of the large numbers of artifacts present in any electron micrograph of chromatin fibers (in which most material remains as a compact mass), has been confirmed by successive work on liver nuclei (*6*) and correlated with a wide range of newly developed or upgraded biophysically nondestructive probes (*7,8*).

The multifibrillar nature of chromatin does not necessarily imply a departure from the canonic concept of a single giant DNA molecule making the genome (*9*) because it could well be the same fiber cycling back and forth a few times.

The unfolding of a giant DNA molecule (we have measured fibers up to 60 μm in length) appears confined to narrow regions, mostly of 4600 Å (but frequently quite larger), although even at low-ionic strength at each of the two extremes, fixed chromatin regions of variable lengths (2000–9000 Å) remain highly packed in the native 300-Å tetrafilament form. This explains the presence of differential light scattering (DLS) and negative superhelical turns in the unsheared chromatin similarly isolated, even at 1 m*M* Tris-HCl. The complexity of such a quaternary structures (Fig. 2) explains also why "reconstituted chromatin" from free DNA, nonhistone, and histone proteins fails to reproduce the structural and functional properties of native chromatin, despite occasional claims to the contrary.

Figure 3 shows the heat capacity (dq/dT) vs temperature, in the range of 300–400 K, for isolated unsheared chromatin in 600 m*M* NaCl, pH 8. Three transition (I, II, III), states are indeed clearly and reproducibly identifiable, at this and other (1, 10, 30, 50, and 100 m*M* NaCl) ionic strengths, whose temperature values change only slightly with increasing salt concentrations. However, as summarized in Fig. 3 and ref. (*8*), enthalpy for isolated chromatin dramatically changes with salt concentration: The enthalpy changes decrease between 1 and 10 m*M*; re-

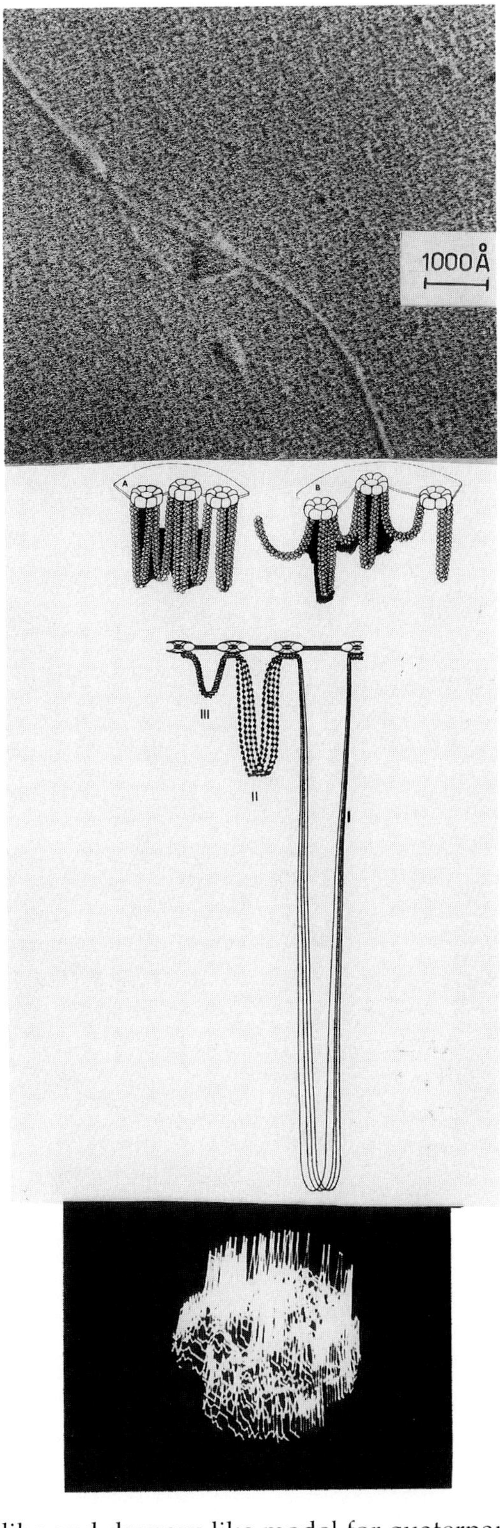

Fig. 2. Rope-like and drapery-like model for quaternary and quinternary chromatin-DNA organization in the intact nuclei during interphase, as described in the text. Examples of the transition states I, II, and III of chromatin superpacking, defined in the text, are also given in the bottom portion of this figure.

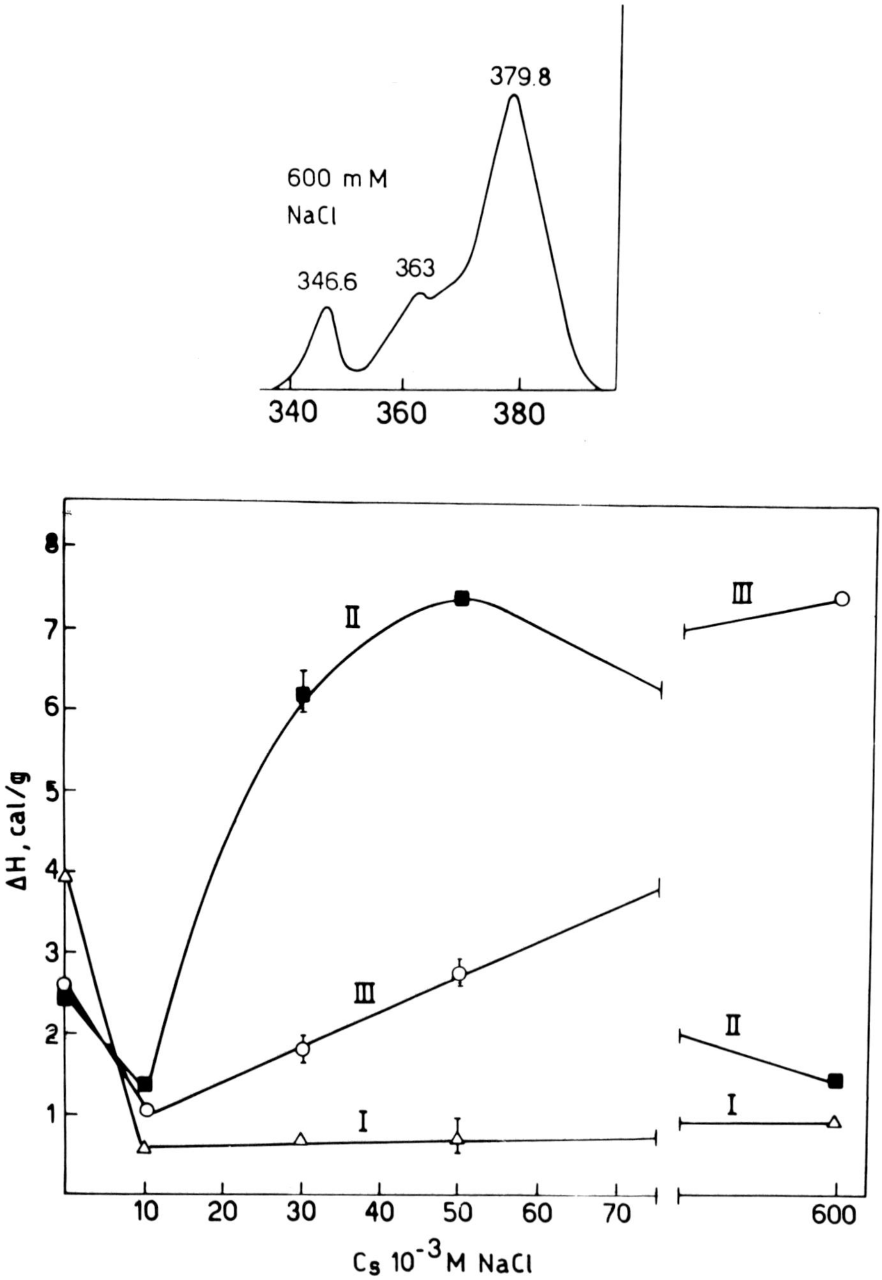

Fig. 3. (Top) Heat capacity vs temperature of calf thymus chromatin in 1 m*M* Tris-HCl, pH 8, at 600 m*M* NaCl and at varying or various ionic strength. (Bottom) The ionic strength dependence of heat capacity for I (~347 K), II (~363 K), and III (~380 K) peaks present in isolated chromatic suspended in 1, 10, 30, 50, and 600 m*M* NaCl.

main constant for the transition state I; steadily increase for the transition state III; sharply increase up to 50 m*M*; and then sharply decrease at 600 m*M* for the transition state II.

The effect of a 600 m*M* NaCl wash, known to remove the H1 Histone from chromatin and to disrupt the quaternary structure, combined with the critical dependence on ionic strength, as is also expected from polyelectrolyte theory, points to a possible assignment for transition state II to the quaternary rope-like structure, which is present even at 1 m*M* Tris-HCl. For the third (III) transition state, its melting temperature, combined with the salt dependence of the enthalpy (*8*) and the effect of carcinogens (*see* later in this paper)—which alters chromatin structure toward a more disperse uncoiled state (*7*), apparently causing a rearrangement of part of the genome from the rope-like structure (II transition) to a lower-order structure (III transition)—point to a possible assignment of the third (III) transition state to the nucleosomal DNA, i.e., the helix–coil is stabilized in the nucleofilament. This assignment is compatible with the CD and optical dichroic (OD) thermal profiles, which show that above 365 K, helix–coil dominates the denaturation process.

The first (I) transition state in isolated chromatin cannot be related to the linker-DNA, as previously suggested, not only because of its independence of salt concentration (which should stabilize this type of structure), but also because of the pronounced increase of the enthalpy of transition in isolated nuclei in which the DNA present in the linker conformation, if any, should decrease. Tentatively, this transition and the other (0), consistently observed at a lower temperature (332 K) only in native nuclei, could then be related to the quinternary, drapery-like (*see* below) chromatin structure and to the melting of macromolecules (i.e., membrane lipoproteins and nonhistone chromosomal proteins) strongly bound to native chromatin (I).

A detailed analysis, carried out by continuous measurement of viscosity, performed at 30 min intervals over a period of more than 48 h [Fig. 3A and ref. (*7*)], yields for the first time a direct experimental evaluation of the three discrete unfoldings, likely corresponding to the three step-like changes in higher-order superstructures. Similar higher-order DNA foldings and topological constraints are apparent in histone-depleted interphase nuclei and "nucleoid" cells from which much protein has been released by identical lysis in 2*M* NaCl, detergent, and chelating agents.

QUINTERNARY STRUCTURE

Interesting enough, the folding and unfolding occurring in the giant chromatin-DNA molecule (we have measured up to several micron-long fibers) never propagate along the entire length, but remain confined to narrow regions of about 4600 Å or larger, limited by fixed points identifiable as irregular pieces of "nuclear membrane" periodically attached along the chromatin fiber (Fig. 2). As previously shown, these highly irregular pieces may be recomposed by successive attempts to yield a continuum, suggesting that they likely came from the same cell component (i.e., nuclear envelope) to which chromatin fibers were linked and that were broken during the swelling of the nucleus. A suggestion could actually be made that the variable regions, which consist of constant, highly twisted tetrafilament forms and are not affected by high salt concentration, are maintained by the same strongly bound nonhistone chromosomal proteins still present in the chromosome scaffold at 2.0*M* NaCl. Puzzlingly enough, nontranscribed-DNA sequences alternating with transcribed sequences are rich in A–T bases; A–T polymers were shown some time ago, to strongly bind the "resistant" nonhistone chromosomal proteins. The presence of alternating regions of highly packed insoluble chromatin, unattackable by DNAse, in the native fiber may explain the finding that soluble chromatin obtained by mild nuclease digestion has a maximum length of about 75 kbp.

The freeze-etching data reported on the liver nuclei (*6*) corroborate the drapery-like model for the regular folding of the 300 Å fiber within the nucleus, clearly showing the attachment sites for these fibers in correspondence with nuclear pores, as originally suggested (*9*) by the annular ring of high optical density around the border of Feulgen-stained nuclei that is apparent with high-resolution image analysis (Fig. 2). This perhaps explains the presence of many nonhistone proteins bound to the chromatin and nerve growth factor (NGF) hormones bound to the nuclear membranes.

The above electron-microscopic (EM) data, obtained either by phospholipid monolayer or freeze-etching techniques, combined with viscoelastometry and microcalorimetry, suggest that each interphase chromosome is organized into one long DNA double-strand, folded in a few successive higher-order structures, from a linear nucleofilament (tertiary) to a rope-like tetrafilament (quaternary) consisting of alternating regions of differential stability and maintained at regular intervals as drapery (quinternary) near the nuclear envelope at the pores. Changes in ionic strength, metabolic activation, mechanical shearing, mild DNAse dissection, or selective H1 removal induces quantal transition among the various states, i.e., from nucleofilament to tetrafilament or vice versa. A regular three-dimensional architecture of the interphase chromatin (extended to regions around the nucleoli), qualitatively suggested in earlier studies, is substantially confirmed by an independent study conducted on gently isolated nuclei stripped of their envelope and analyzed by polarized light microscopy and three-dimensional stereo electron-microscopic techniques (*10*).

NUCLEAR PORES MEMBRANE AND CHROMOSOME SCAFFOLD

The double membrane surrounding the nucleus contains, as its most conspicious feature, the nuclear pore complexes, which are assumed to be the sites of molecular and ionic exchange between the nucleus and the cytoplasm. Curiously, these channels in the nuclear envelope are referred to as pores, even though they are usually filled with a dense plug (*11*). Biochemical characterization suggests that the nuclear pore complex is composed of only a few major polypeptides and some RNA, but no direct identification has yet been possible within the boundaries of this structure, and all speculations about its function are based on scanty, circumstantial, and at times, contradictory, data obtained by limited ultrastructural and autoradiographic methods. Although the general features of the complexes have long been known to consist of a cylindrical assembly spanning inner and outer nuclear membranes, arranged with octagonal symmetry around a central axis, detailed new information has been recently extracted from electron micrographs of an intact oocyte nuclear envelope by modern image-processing methods at the resolution of 90 Å.

According to these high-resolution methods, the pore structure appears to be a three-layered sandwich-like structure, consisting of a disk between two thin rings, coplanar with one half of the double membrane. The disk, consisting of a central plug and eight broad spokes, is suspended between the two rings by the extension of the spokes, at the periphery of which (about 1000 Å diameter) the two membranes fuse. These findings, which also provide evidence for the attachment of eight ribosomal particles to the cytoplasmic face of this structure, raise further numerous questions about the structure and function of the nuclear pores. It is not easy to see how ribosomal particles (about 200 Å diameter) assembled in the nucleus can pass through a pore filled with a plug and how the assembly of such large structures (spokes, rings, plug) can occur spontaneously from their small protein subnits without the guiding activity of other cellular components. Pore complexes are no longer considered as randomly distributed, but as being present at specific sites. Furthermore, rapid changes in pore pattern and number have been found to occur on nuclear membranes from various cell types in different functional states: A biphasic increase in pore frequency, associated with sharp transitions from a clamped to a uniform pore distribution, has been found in synchronized HeLa cells, and in PHA-stimulated human lymphocytes in the very early part of G1 and prior to DNA replication. Curiously, this biphasic change correlates with abrupt changes in the higher-order chromatin structure, from a clumped to a uniform distribution, occurring at the same time in the same cells.

Suggestive evidence for the attachment of chromatin fibers to the nuclear envelope at the pores has recently been provided by freeze-etching of rat liver nuclei (*8,12*) and fluorescence microscopy (*10*), but conclusive direct proof of the presence or absence of thin chromatin threads near or in the pore complexes is difficult with any present electron microscopic technique. Only recently a computer-based system for recording and analyzing light microscopic images has been developed that, when combined with classical cytogenetic identification of a large number of loci along each of the chromosomes, allows the spatial layout of the genome to be studied precisely, with respect to the nuclear envelope. By analyzing intact, unfixed, *Drosophila* salivary glands with this latest technique (*10*), each of the chromosome arms appears to fold up in a characteristic way along its length, to contact the inner nuclear surface via specific sites, and to reside in topological isolation from all of the other arms.

In a mammalian cell, the nuclear pores are absent in metaphase, are rapidly formed in early telephase in which membrane pieces are seen attached to the chromosomes, and are present in varying numbers and distribution along the entire interphase (G1–S–G2). Whenever pore clusters appear, is there then any correlation between the number of these clusters and the number of chromosomes (even under conditions in which chromatin and pore distributions noticeably change)? Do a similar number of clusters occur in the interphase-chromatin distribution near the envelope? Both of these questions have been answered positively (*12*).

A detailed analysis of the pore distribution on the nuclear membrane of the rat liver nuclei (Fig. 4) shows that about 10 clusters are apparent in the nuclear hemisphere, yielding a total number of 20 for the overall nuclear envelope, close to the total number of interphase chromosomes present in the same liver cells; rat diploid nuclei have been reported to contain 19–21 pairs of homologous chromosomes. Interesting enough, about 22 islands are present in similar Feulgen-stained and three-dimensionally reconstructed cells.

It would appear that the clusters of nuclear pores, corresponding to chromosome-sized clustering of chromatin fibers (up to 60 μm long, or about 10 daltons) and identified on the nuclear envelope in a number very close to the actual number of chromosomes in rat liver cells, give the possible linkage between the highly ordered "drapery" structure of interphase chromatin-DNA and the typical metaphase spread; upon proper condition, the former can indeed collapse in the latter.

Histone-depleted metaphase chromosomes display a central "scaffold" of darkly stained fibers, containing strongly bound nonhistones (the 10% still present after 2*M* NaCl), and surrounded by a "halo" of free loops of DNA anchored at both ends at the scaffold. It is this scaffold, still present in interphase chromosomes, that is apparently responsible for maintaining chromosome structure during the cell cycle, warranting con-

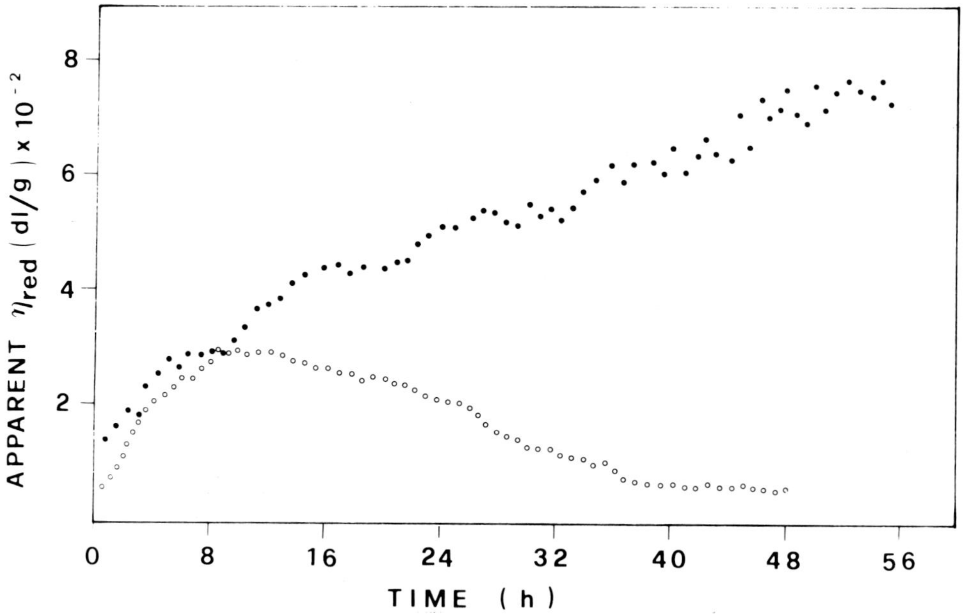

Fig. 4. Apparent value of reduced viscosity, as a function of incubation time in lysing neutral (●) and alkaline (○) solutions at high ionic strength, of rat liver nuclei.

tinuity between metaphase and interphase, even in the absence of a nuclear envelope. The drapery arrangement of native chromatin fibers during interphase in two distinct alternating regions of folded versus unfolded tetrafilaments would indeed yield loops of DNA in histone-depleted metaphase chromosomes.

The existence of a long-range order of the DNA in interphase was recently confirmed in sedimentation studies (*13*), revealing two levels of DNA folding in histone-depleted HeLa interphase nuclei. Both types of structure are capable of retaining the spherical shape of nuclei, with different degrees of folding and an expended DNA halo, and are destroyed by proteolytic agents, but not by RNAse. The proteins associated with the more folded structure, stabilized by Ca^{2+} and Cu^{2+}, are three nuclear lamina proteins of 60–70 kdaltons and numerous high molecular-weight species.

Although these latter proteins are selectively lost after DNA unfolding during the transition to type II structure, induced by histone extraction in the presence of 2-mercaptoethanol or of metal chelators, the three-lamina nonhistone proteins remain present, pointing to a critical role of the nuclear envelope, in at least providing the major DNA binding proteins of the chromosome scaffolds. This is compatible with the fragments of nuclear envelope that are observed periodically along the native chromatin fiber.

Fig. 5. Freeze-etching electron microscopy of a rat-liver nuclear membrane, showing the pores and their distribution.

NUCLEAR MATRIX AND CONTROL OF NUCLEAR VOLUME

When isolated rat-liver nuclei are sequentially extracted in the presence of low magnesium, 2*M* NaCl, detergent, and nucleases, a histone-, phospholipid-, and nucleic acid-depleted structure remains. This structure, that, according to some authors, was believed to be composed of only a peripheral lamina, has been shown to also extend within the nuclear volume, yielding a fibrilar network of proteins called the nuclear matrix (*14*).

A parallel line of thought has suggested that the nuclear protein matrix is the determinant of the structural and functional modification of nuclei and chromatin; chromatin fibers emerging as loops from the skeleton would be constrained and modulated by the network of the nuclear protein matrix.

We therefore designed a very simple experiment to test which component, chromatin or protein matrix, controls the shrinking and swelling of the overall nuclear volume. If changes in chromatin structure are just a consequence of changes in nuclear volume, being controlled by modifications in the nuclear matrix (whose protein aggregation and network contraction is also dependent on ion concentration), selective removal of DNA by DNAse I enzyme should not affect the shape and volume of liver nuclei and should still decrease their volume with increasing Ca^{2+} concentration, therefore displaying an abrupt transition quite similar to the undigested control liver nuclei. This did not occur, and digested nuclei displayed the same constant volume with varying Ca^{2+} concentration, whereas the digested control nuclei reduced their volume by several folds when exposed to the identical procedure, temperature, and buffer.

Single (DMNA) and double (AAF) DNA strand brakes, induced by large doses of chemical carcinogens (4 h after their administration), caused an immediate increase in nuclear volume and a substantial decrease in the degree of nuclei shrinkage associated with the increased ion concentration.

It then appears that the chromosome-sized DNA structure was altered at the quaternary level, probably in the direction of a decreased pitch in the rope-like fiber (Fig. 2), by the ion neutralization of phosphate charges. This controls the shrinkage of overall nuclear volume through the mediation of drapery-like chromatin fiber attachment to the nuclear membrane at the pores. Removal (DNAse) or simple disruption (chemical carcinogens) of this long chromatin-DNA fiber would cause the elimination or the reduction of the existing geometric coupling between chromatin structure and nuclear volume.

The nuclear matrix could then be a consequence of chromatin intranuclear distribution, where the location of nonhistones forming the protein network was originally determined by the extracted chromatin-DNA fibers extending from the nuclear envelope and the nucleoli. It has not escaped my attention that isolation of the nuclear matrix is critically dependent on the sequence of isolation steps and that others have found that, under similar exposure to nuclease and Triton-X, only a peripheral lamina remains, with a shape and size similar to the original nucleus. This fraction contains 18% nucleic acid, 1% lipids, and 82% nonhistones, which latter reside at the nuclear pores within the lamina and whose locations become diffuse during mitosis. These are proteins possibly involved in the attachment of chromatin fibers to the pores and in maintaining the highly packed form of the tetrafilament.

LEVELS OF DNA ORGANIZATION *IN SITU*

In recent times, analytical image analysis of Feulgen-stained cells has been profitably utilized to characterize nuclear DNA *in situ*. The most recent findings have been summarized at length in recent reviews (*9,15*), to which the reader is referred for more details. Our intention here is to present only the implications and the latest work related to chromatin structure emerging from the acquisition of optical-density values for each of the several thousand individual pixels that comprise a single nuclear image (*16*).

From such high-resolution image analysis of nuclear DNA, both at the light- and electron-microscopic level, before and after nuclease digestion, chromatin-DNA appears to exist in three resolvable states (Fig. 2) of higher-order structures, from a very dispersed one (0.03–0.07 OD), in which "active" genes attacked early by DNAse I appear located, to two others, not preferentially digested by DNAse, of a progressively larger degree of supercoiling, 0.14–0.20 (disperse "euchromatic" DNA or mononucleofilaments) and 0.30–0.51 OD (condensed "heterochromatic" DNA or tetrafilament rope), respectively.

These conclusions are encouraged by recent findings on similar levels of condensation (average optical density) of individual interphase chromosomes reconstructed from 1-μm thick serial sections of Feulgen-stained melanoma nuclei and, more significantly, on the apparent levels of condensation of individual chromatin fibrils at the electron-microscopic level.

Strikingly, the differences in OD level (2.8:1) is exactly what is expected from the difference in the DNA-packing ratio between a rope-like tetrafilament and four adjacent nucleofilaments, both maintained at the extremity by fixed attachment sites near the nuclear envelope.

Interestingly enough, any other quaternary folding so far proposed would yield an increase in packing ratio of six (solenoid) or higher. If the nucleofilaments were free to move within the nucleus and did not have periodically fixed attachments every few thousand Å to maintain them close to each other (following the structural transition III-II), the increase in packing ratio upon rope formation would have been 11.6.

HIGH-ORDER GENE STRUCTURE

Based on the wide range of experimental findings so far presented, I am now in a position to postulate a three-dimensional model of DNA structure, from the nuclear level down to the gene. This model not only accounts for many of the independent observations on the physical and chemical properties of chromatin DNA, but also accounts for the most recent discoveries on eukaryotic gene structure, mapping, and coding, suggesting a mechanism by which gene expression and cell function are controlled.

The regular three-dimensional organization of chromatin DNA within the nucleus, near the border, or around the nucleoli, leads to the identification within the 300-Å chromosome-sized fiber of a fundamental repeating unit in the drapery-like segment, usually about 4600 Å long, and capable of undergoing transitions between three discrete states of DNA packing ratios (1, 7, and 19). This unit appears regularly interspersed between regions—2000–9000 Å long—highly coiled and attached to the nuclear envelope near the pores (Fig. 2). It must be emphasized that occasionally the alternating regions of native fiber unfolding in separate nucleofilaments appear to be of a larger length, up to 25,000–30,000 Å in rat liver nuclei, suggesting that the repeating unit can also be of a size different from the 4600 Å fundamental segment. Whatever the exact size of the repeating unit (more studies are called for to clarify this point), the above findings imply an upper limit for the length of soluble chromatin obtainable by the mildest nuclease digestion of nuclei and a significant decrease in chromatin solubilization with the use of restriction enzymes, which cut the DNA with quite less frequency, with respect to a nonspecific nuclease. Both predictions are indeed satisfied; namely, a maximum length of about 5×10^7 daltons for the nuclease-soluble segment and 40% limit digestion for endo-*E.coli* enzymes have been reported in rat liver nuclei, values quite compatible with the observed rope-like arrangement of the native 300 Å chromatin fiber and with the size of its alternating regions of different degrees of foldings and compactness. Although the basic unit is critically and locally dependent on its supercoil from H1, ions, and other chromosomal proteins, most of the highly coiled regions—consisting of the same multifilament rope—are not, yielding a skeleton that cannot be attacked by nuclease, high salt concentration, and even detergent, is associated with the pores at the nuclear envelope, and is possibly maintained by lipoproteins, Ca^{2+} or Cu^{2+}, and by a fraction of nonhistone proteins (as the three nuclear lamina proteins still present at a 2.0*M* NaCl) characterized by strong binding with A–T rich regions of the genome.

The availability of purified histone mRNA and specific hybridization probes has recently confirmed this prediction in sea urchins, giving indirect support to the postulated role of nonhistones in maintaining the highly coiled regions of the fiber. It is intriguing that the varying size of these coiled regions (8–32 kbp) corresponds to the size of the nontranscribable-DNA sequences of human genome, i.e., 20–38 kbp for ribosomal genes and is intercalated by corresponding transcribable sequences of an approximately constant length, 14 kbp, that in turn compares well to the 17 kbp size of the basic "modulating" rope. The tissue dependence is present in all genes, i.e., for histone genes the size of the repeating units is 7 kbp for sea urchins and 15 kbp for chickens. Electron micrographs of properly stained ribosomal DNA in the process of transcription (*17*), in which the core of DNA is unwound from the nucleoli at low-ionic strength, reveal thin axial fibers of duplex DNA, covered along their length with a christmas-tree-like matrix of about 100 thin fibrils (RNA precursors associated with proteins) of progressively increasing length, terminating on a spherical granule (presumably RNA polymerase). Each matrix represents a transcription unit engaged in the synthesis of a 40S precursor and is separated by an alternating inactive stretch of axial core, with a DNA packing ratio of 1.66 bp/Å, as compared to the 0.24 bp/Å of the transcription unit.

All of the above prompt the postulation of the identification of the repeating unit (each one of the four nucleofilaments protruding from the nucleoli and the nuclear envelope, hereby called "fibrosome") with the coding gene. The length of such units, about 17 kbp, is, however, an immediate cause of concern: If the gene was a contiguous region of DNA, colinear with its protein product (until recently a dogma of genetics), i.e., one nucleotide sequence codes for one RNA and one protein without interruption, and if "genes" (few are known to be as small as several hundred base pairs) could be synthesized only as single entities separated in space (along the genome) and time, and never as a family being simultaneously and contiguously transcribed, then the equation fibrosome = gene would prove to have little foundation. Following the complete unfolding of the fibrosome that takes place during metabolic activation, only a few transcriptional units could indeed be synthesized:

(1) One single mRNA of 17 kbp, an exact copy of the coding DNA sequence.
(2) One mRNA of smaller size, but lacking some of the fibrosome sequence because of the presence of nontranscribable intervening bases.
(3) Several mRNAs of different size, but transcribed simultaneously and contiguously to yield a total of 17 kbp.
(4) Several mRNA coded by the same fibrosome sequence, but each one coding for different proteins of differing size, depending on the initiation and termination sites.

Strikingly, the above predictions are compatible with the most recent "yet unexplained" findings of genetic engineering. Traditional genetics, basing the identification of a gene on a genetic complementation test or on the nature of the gene product, whether protein or RNA, has now proven to be misleading. For instance, nucleic acid sequences present in more than one polypeptide have been identified: A full-length protein is translated from a complete gene, but a shorter protein, utilizing only part of this sequence, is also translated, using initiation sites or termination sites that lie within the same "gene." The "fibrosome model" also implies that much of the DNA, lying in the "coiled" alternating regions, does not code for protein and that its function is purely structural. This, again, appears to resolve the paradoxical relationship between DNA complexity and gene number, known as the C paradox: Much more genome is available than that required to code for all proteins that it specifies, and changes in DNA size within any species is up to 100-fold, whereas the number of protein products remains similar. An early explanation was that the number of nonrepetitive sequences decreases with increasing genome size, although the highly repetitive component, as satellite DNA (short sequences repeated several times), increases, leading to a multiplication in the number of copies of every gene.

However, this proves to be false since nonrepetitive sequences are present in small, but similar numbers in both low- and high-C species, and they are the major source of RNA; they vary widely between species and are the most likely candidates for correlation with gene number. All of the above is then compatible with the view that most DNA sequences do serve the sole function of maintaining the quaternary and quinternary chromatin-DNA structures, being located mostly in the nontranscribable spacers (highly repetitive in A–T) and partly within the transcribable "fibrosome" itself, where intervening sequences are nontranscribable and serve as initiator–terminator sites for specific proteins. Their function appears specifically related to maintaining the integrity of the nuclear skeleton and to creating fixed sites for the attachment of nucleofilaments that can then locally yield DNA with highly negative superhelical turns: Increasing the number of negative superhelical turns has been shown to lead, *per se*, to an higher rate of transcription in protein-free DNA and to the expression of the transformed phenotype in chromatin.

Finally, considering that the average length of each human gene, when equated to individual "fibrosomes" (45% transcribable plus 55% nontranscribable), would be about 40 kbp, the total number of genes for a typical mammalian cell (with 3×10^9 bp genome) would be approximately 75,000. The total number of mRNA present in mammalian cells is 15,000–20,000, as determined by restriction enzyme or hybridization techniques; comparison of the complexity of nuclear RNA and cytoplasmatic RNA led, however, to the discovery that only about 20% of the RNA transcript reaches the cytoplasm (partly because of processing from introns), making the total number of different RNA species 75,000–100,000. This, however, is only an approximate estimate of the number of functioning genes; to determine the actual total number, we would have to assay each cell type during its developmental stages and determine the extent of overlap between the RNA transcript population: For instance, several RNAs could be coded for by the same unit of gene expression, depending on the initiation and termination sites. For a more meaningful comparison we have to rely on indirect estimates based on spontaneous or induced mutation rates that yield, strikingly, a number of about 100,000 genes similar to that predicted from the fibrosome model.

The number of recessive lethals generated is the product of the deterious mutation rate and the number of susceptible genes. Since the rate of mutation in mammalian cells is about 0.00001–0.000001/locus/yr, and assuming a total number of genes equal to 2,000,000 (as derived from the mammalian DNA molecular weight and an average 1500 bp/gene, apparent from average protein size), the number of mutated loci would be 2–20/genome, which is considered unacceptable if all mutations are deterious to coding sequences. The expected number of recessive lethals existing in humans is then likely to be 0.1–0.5/genome, yielding a total number of 100,000 susceptible loci or genes, meaning that, as predicted by the "fibrosome" model, a large part of the genome does not code for proteins, and the gene size is one order of magnitude larger than that expected from the average gene product size.

CONCLUSIONS

In conclusion, it appears that several recent, unexplained observations made by genetic engineers can be accounted for to a degree unsuspected and unforeseeable from the "fibrosome model," based on numerous (some very recent) biophysical characterizations of native chromatin. Despite enormous advances, the basic questions "What is a gene?" and "How is it expressed?" often seem incapable of resolution. This leads to an increasingly sophisticated jargon and an increasing spread of empirical utilization (witnessed by the boom of genetic engineering), behind which the basic structural and functional questions remain unanswered. The fibrosome = gene equation—in which each of the 75,000 fibrosomes has its own degree of supercoil locally maintained and has the capability to undergo locally abrupt structural transitions leading to gene expression (*18*)—does offer a coherent and not circumstantial answer, both in structural and functional terms, which calls for further probative experiments.

At the same time, the overall structure of the nucleus, from the pores down to its molecular constituents, appears progressively clarified by the increasing resolution permitted by the modern, highly sophisticated and noninvasive biophysical techniques. The new structural and functional insights already acquired at the single cell level—as summarized earlier in this chapter—may indeed represent only the tip of the iceberg produced by the new frontiers at the crossing between physical and life sciences.

ACKNOWLEDGEMENTS

This work was supported by a grant from the CNR Finalized Project "Oncology" (contract 84.00702.44). This lecture presented at San Remo for the Nobel celebration has been partly published elsewhere.

REFERENCES

1. Bradbury, E. M., and Matthews, H. (1982), *Cell Growth,* in *NATO-Life Sciences Series* (Nicolini, C., ed.), Plenum, NY, 411–454.
2. Finch, J., and Klug, A. (1976), *Proc. Natl. Acad. Sci. USA* **73,** 1897.
3. Nicolini, C., Kendall, F., and Baserga, R. (1976), *Science* **192,** 796.
4. Nicolini, C., and Kendall, F. (1977), *Physiol. Chem. Phys.***9,** 265.
5. Cavazza, B., Ricci, B., Trefiletti, V., and Pratone, E. (1983), *J. Cell Sci.* **62,** 89.
6. Nicolini, C., Beltrame, F., Cavazza, B., Trefiletti, V., and Pratrone, E. (1983), *J. Cell Sci.* **62,,** 103.
7. Nicolini, C., Carlo, P., Martelli, A., Finollo, R. and Brambilla, G. (1982), *J. Mol. Biol.* **161,** 155.
8. Nicolini, C., Trefiletti, V., Cavazza, B., Pratone, E., and Carlo, P. (1983). *Science.* **219,** 176.
9. Kendall, F., Beltrame, F., Belmont, A., and Nicolini, C. (1980), *Cell Biophys.* **4,** 19.
10. Agard, A., and Sedat, J. (1983), *Nature* **302,** 676.
11. Unwin, P., and Milligan, R., (1984), *J. Cell Biol.* **93,** 63.
12. Nicolini, C., Maraldi, N., Capitani, S., and Vernazza, R. (1984), *J. Cell Sci.* **72,** 75.

13. Lebrowski, A., and Laemmli, U. (1982), *J. Mol. Biol.* **156,** 309.
14. Pardoll, D., Volgenstein, B., and Coffey, D. (1980), *Cell* **19,** 527.
15. Nicolini, C. (1980), *J. Cell Sci.* **12,** 475.
16. Belmont, A., Kendall, F., and Nicolini, C. (1984), *J. Cell Sci.* **65,** 123.
17. Miller, O. L., and Beatty, B. R. (1969), *Science* **164,** 955.
18. Nicolini, C. (1986), *Biophysics and Cancer,* Plenum, New York, NY.

Section III BRAIN AND VISION

Blobs and Color Vision

DAVID H. HUBEL

Department of Neurobiology, Harvard Medical School, Boston, MA 02115

Two main topics will thread their way through this paper: The structure of the monkey striate cortex, including its inputs and outputs, and the physiological basis of color vision. When Margaret Livingstone and I began this work four years ago we did not set out to study color, but the structures in the cortex that we have been looking at turn out to be strongly involved in that sense. The work to be described here was done entirely in macaque and squirrel monkeys.

First, let me remind you of the structure and connections of the monkey striate cortex. In the classical visual pathway in mammals, information is transmitted from the eyes to the lateral geniculate body and from there to the primary visual cortex. Each geniculate layer receives its input from one eye only, and a cell in the lateral geniculate is consequently supplied by a single eye, the left or right, depending on the layer in which that cell is situated (Fig. 1). The magnocellular geniculate layers project to layer 4Cα and the parvocellular to 4Cβ, with small contributions to 4A and 6 (Fig. 2). The output of the cortex is from pyramidal cells of the upper (2 and 3) and the lowermost (5 and 6) layers; layer 4 cells do not project out of the cortex. Layers 2 and 3 send their axons mainly to other cortical regions, and it is these pathways that are presumably involved in perception. Layers 5 and 6 project mainly to subcortical areas, especially to the superior colliculus and, in a recurrent path, back to the lateral geniculate.

When responses were studied at different levels in the path from retina to cortex, it was found that up to and including layer 4C cells responded best to spots of light, and usually had concentric, center-surround receptive fields, whereas most or all cells in the upper (2 and 3) and lower (5 and 6) layers responded best to specifically oriented slits, bars, or edges of light (*1*). Taken literally, this would mean that the entire output of the striate cortex consisted of orientation-specific cells, and, in particular, that the entire projection of the upper layers to other cortical regions was in the form of orientation-specific cells. As we shall see, these generalizations turn out to require some revision.

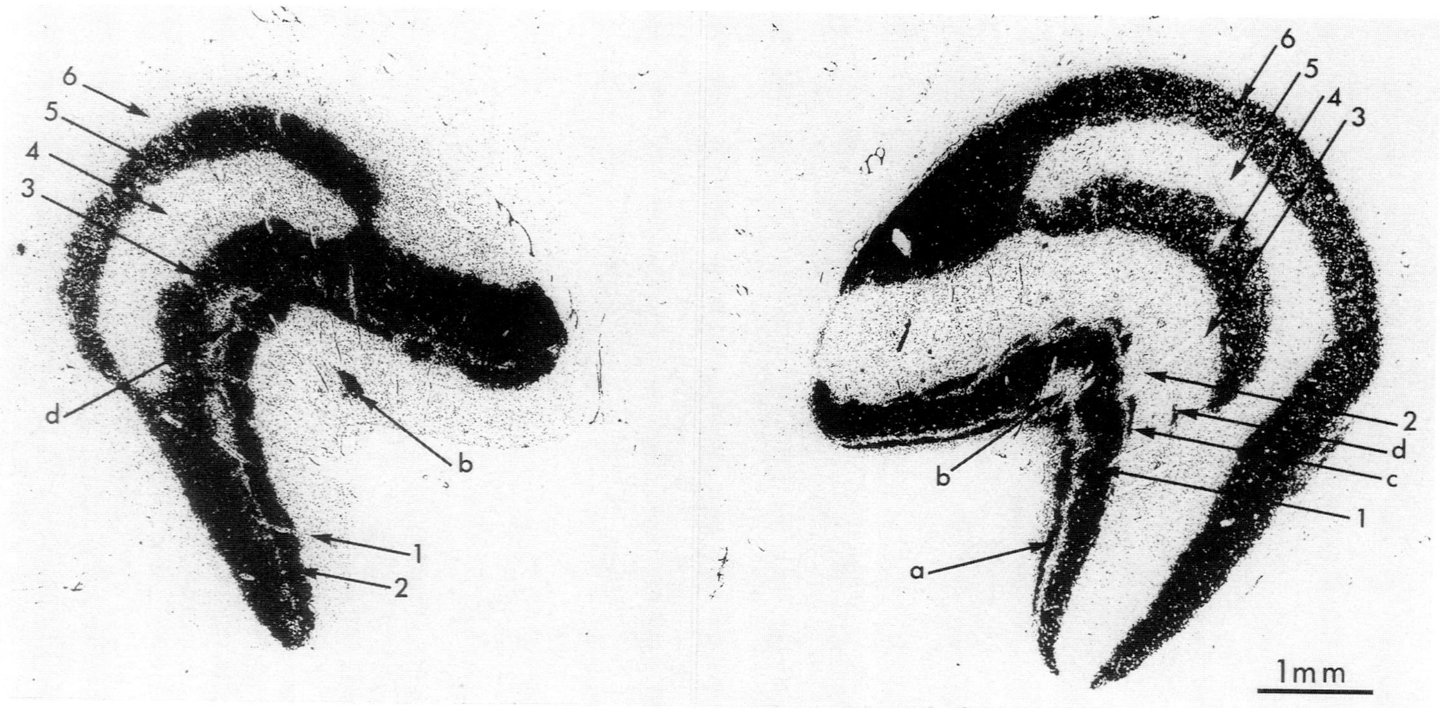

Fig. 1. Autoradiographs of left and right lateral geniculate bodies of the 1-wk-old monkey whose left eye was injected at birth with tritiated fucose. Light field: Labeled regions are black. The classical six layers are 1–6, numbered from ventral dorsally, and represent four additional regions of input: Areas a and c are supplied from the contralateral eye, b from ipsilateral, and d is apparently supplied by both eyes, possibly as alternating band-like monocular regions (from ref. 3).

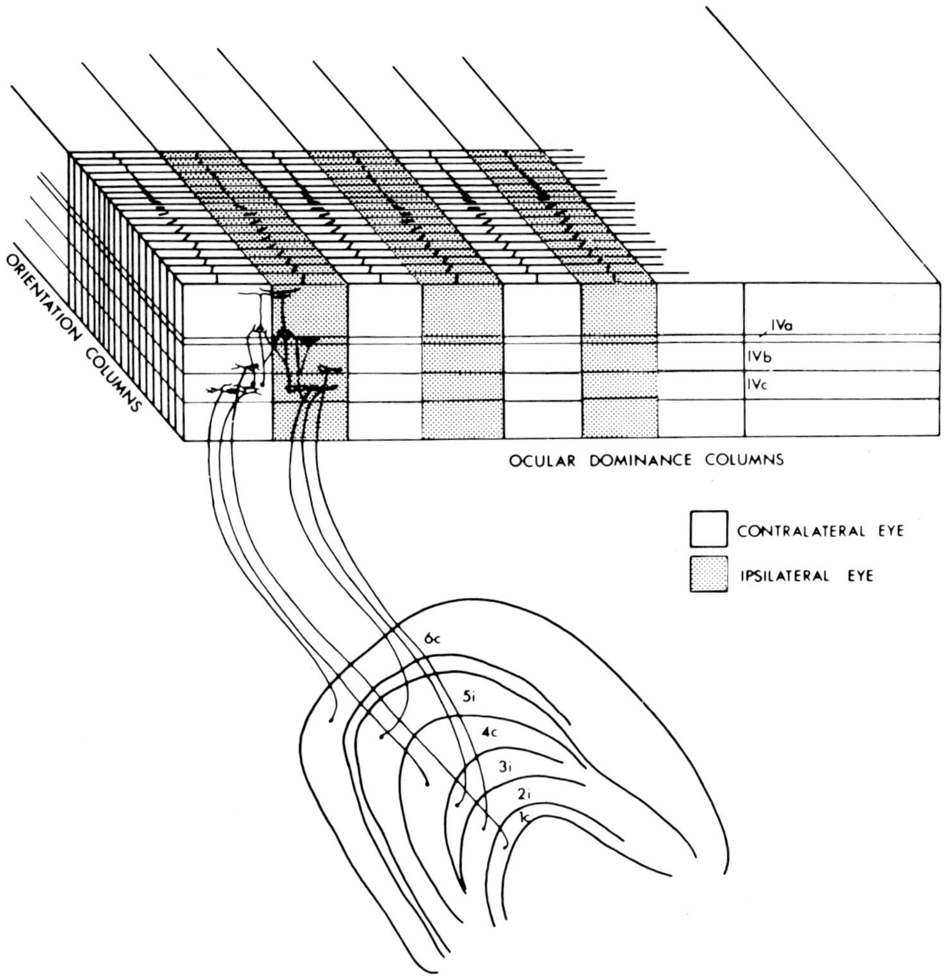

Fig. 2. Diagram of geniculate input to striate cortex [From Hubel, D. H., and Wiesel, T. N. (1972), *J. Comp. Neurol.* **146,** 441–450, Fig. 18].

For years one has known that cortical cells are functionally segregated in two different ways. A fiber from a single layer of the lateral geniculate body, and hence from one eye, breaks up as it approaches layer 4 of the cortex into patches of terminals with terminal-free gaps between them (2). All the fibers belonging to that eye supply these same patches, and fibers from the other eye supply the gaps. This is the basis of the partial eye segregation in the form of eye-dominance columns (3). Such a segregation is absent or poorly developed in the squirrel monkey (4). The second system concerns receptive-field orientation. As one threads an electrode along the cortex parallel to the surface, the optimum stimulus orientation changes in small steps in a very regular way—about 10°/50 microns (5). If, instead, one goes downward through the cortex, orientation stays constant. It follows from this orderliness that cells of similar orientation specificity are grouped in narrow vertical slabs arranged perpendicular to the the cortical surface (*5,6*).

These two systems are shown in what we now term the Ice-Cube Model, Mark I (*3*) (Fig. 3). Classical anatomical staining methods, like the Golgi, or stains for Nissl substance or myelin, give no hint of these subdivisions. A Nissl stain of a tangential section through layers 2 and 3 shows no subdivision of any kind. The ocular dominance columns have been demonstrated anatomically by a number of more modern methods, for example, by making geniculate lesions in single layers and using the Nauta technique for staining degenerating fiber terminals or by putting radioactive proline in one eye and letting it be transported by way of the geniculate to the cortex, where it can then be seen in autoradiographs in layer 4. So far the only way of seeing orientation columns has been through the use of the 2-deoxyglucose method (*7*). Figure 3 is necessarily a caricature: The two sets of columns are not known to be orthogonal—in fact, they probably do not have any constant relationship—and the orientation columns are almost certainly not straight for any great distance.

Around 1977 there appeared reports of another kind of periodicity. Weber et al. (*8*) and, independently, Hendrickson et al. (*9*) and Hubel and Wiesel (*4*) all observed faint puffs of label in layer 3 of squirrel monkey cortex after injection of proline into the eye or geniculate; one wondered (wrongly) if these might in some way be concerned with ocular dominance. In 1978 Margaret Wong-Riley stained sections of cortex for the mitochondrial enzyme cytochrome oxidase and observed similar labeled aggregations in layers 2 and 3 and also, faintly, in 5 and 6 (personal communication). She commented on the similarity of these structures to the proline puffs. It was not until 1980 that the cytochrome oxidase technique was used to stain the upper layers of striate cortex in tangential sections (*10,11*). Astonishingly, the densities were then seen to form a quasiregular polka-dot pattern of patches, which we have designated, by the legitimate English term, "blobs" (*12*). These are shown in Fig. 4, a flat mounted section through layer 3 of macaque striate cortex, stained for cytochrome oxidase. The blobs are roughly 200 microns in diameter and are spaced at 0.5 mm intervals. (Area 18 also shows a series of coarser periodicities that we shall say more about later.) In macaques, they line up in parallel rows that intersect the 17–18 border at 90°, suggesting some relationship to ocular dominance columns, and it has in fact been established by several independent methods that the blobs are aligned along the centers of ocular dominance columns (*13,14*). No such rows appear in the squirrel monkey, in which ocular dominance columns are lacking (Fig. 5a). In both species the blobs extend through layers 2 and 3, are not seen in 4, but reappear faintly in 5 and 6, where they lie in register with the upper layer blobs. In Fig. 5b, a deeper section through the same block of cortex, one can make out the blobs in layers 5 and 6, lying directly below those in 2 and 3. Layers 4A and 4C stain uniformly and densely. Figure 5, a and b, also shows a set of regular, much coarser, p eriodicities in area 18; these have been of great interest to us in recent months.

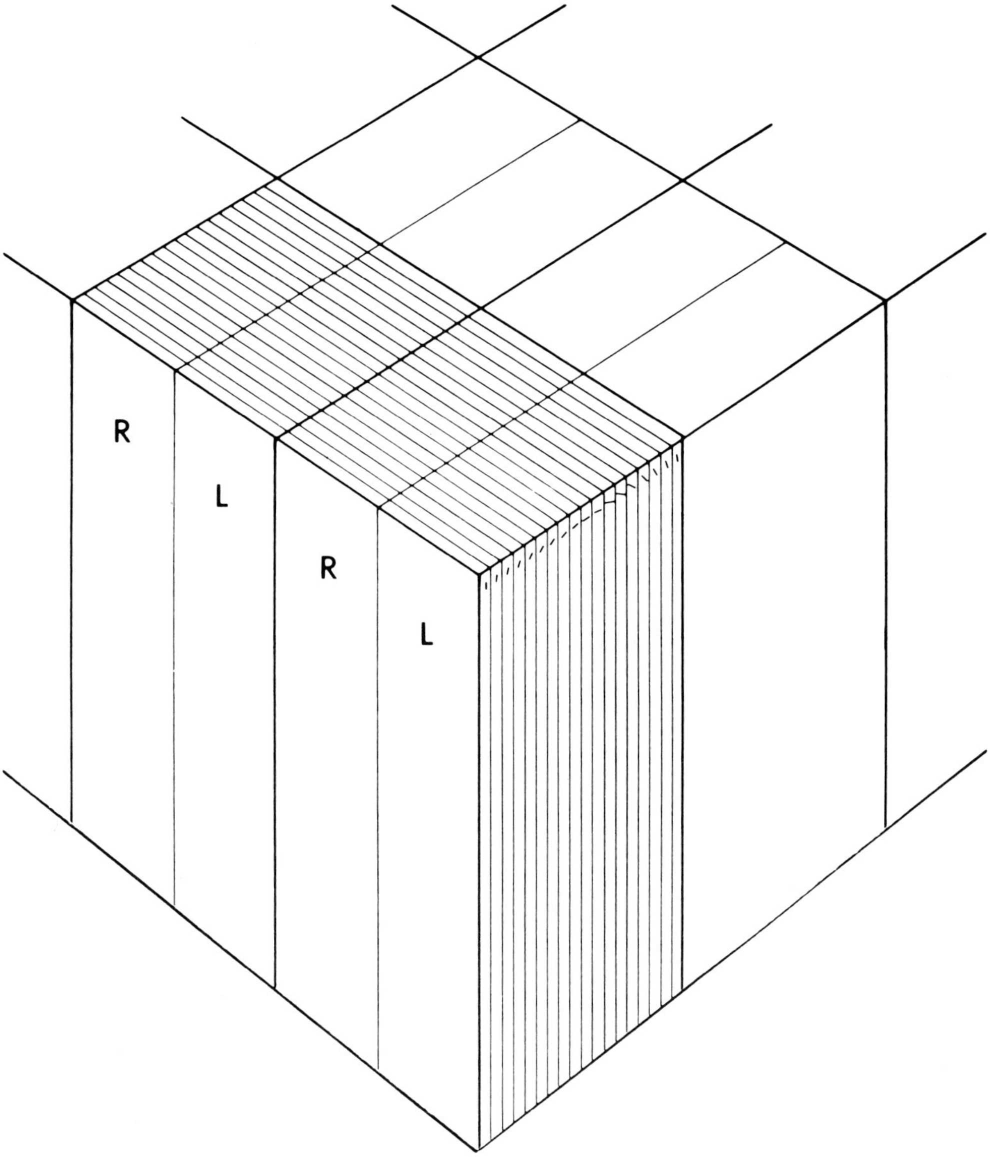

Fig. 3. Model of the striate cortex, to show roughly the dimensions of the ocular dominance slabs (L, R) in relation to the orientation slabs and the cortical thickness. Thinner lines separate individual columns, thicker lines demarcate hypercolumns, two pairs of ocular dominance columns, and two sets of orientation columns. The placing of these hypercolumn boundaries is of course arbitrary; one could as well begin at the horizontal or any of the obliques. The decision to show the two sets of columns as intersecting at right angles is also arbitrary, since there is at present no evidence about the relationship between the two sets. Finally, for convenience, the slabs are shown as plane surfaces, but, whereas the dominance columns are indeed more or less flat, the orientation columns are not known to be so, and may, when viewed from above, have the form of swirls. (From ref. 3).

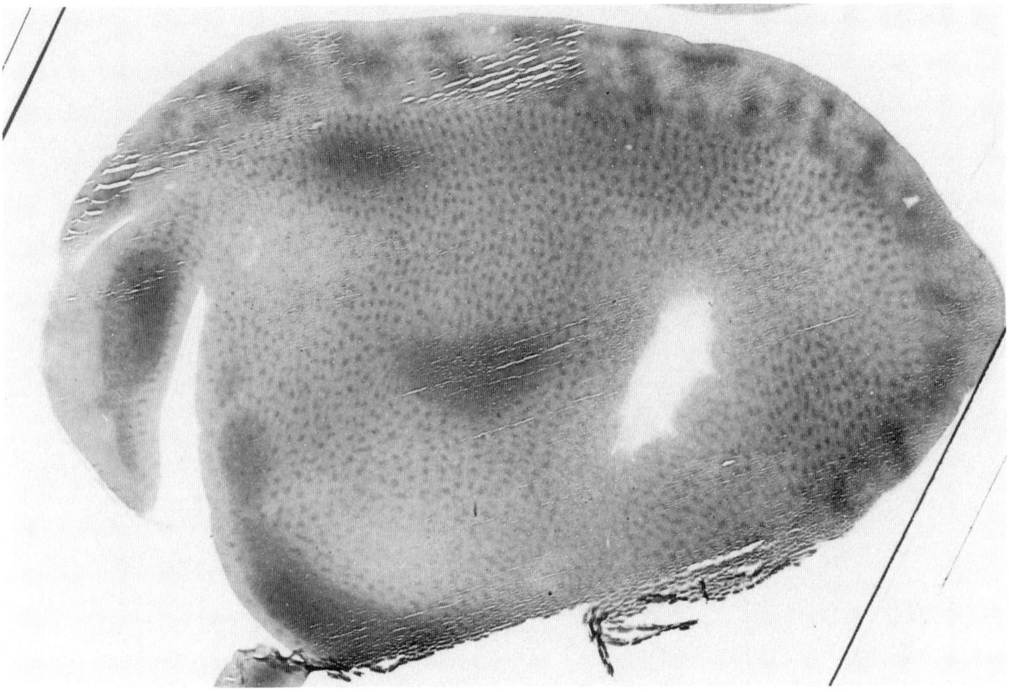

Fig. 4. This macaque visual cortex was stained for cytochrome oxidase after it had been sliced tangentially. The section passes through layers 2 and 3.

We naturally wondered whether cells in the blobs might have physiological properties different from the cells between blobs, and two years ago we set out to correlate physiology with histology by recording from upper-layer cells in tangential penetrations. We very soon observed that cells between blobs were sharply tuned for orientation, whereas cells inside blobs showed a complete lack of orientation selectivity (*15*). Cells on the fringes of blobs showed a slight, but clear, orientation preference. We were at first skeptical, but have penetrated through and recorded from some 78 mm of cortex, traversed 72 blobs, and have seen nonoriented cells in all but nine of these (*16*).

Figure 6, for example, shows a section of cortex through layer 3, stained for cytochrome oxidase, and indicates the positions of two parallel, closely spaced electrode tracks. Polar histograms of response vs slit movement direction are shown for cells recorded at various points along these tracks. Polar histograms are simply plots in polar coordinates of average response vs direction of movement of a slit oriented at 90° to this direction. An elongated graph represents a cell sharply tuned for orientation, whereas a circle represents a cell with no orientation specificity. Between blobs the cells were well-oriented; in blobs, all of the cells we recorded showed no hint of orientation preference, responding actively as a line swept across the receptive field, whatever its orientation. We thus

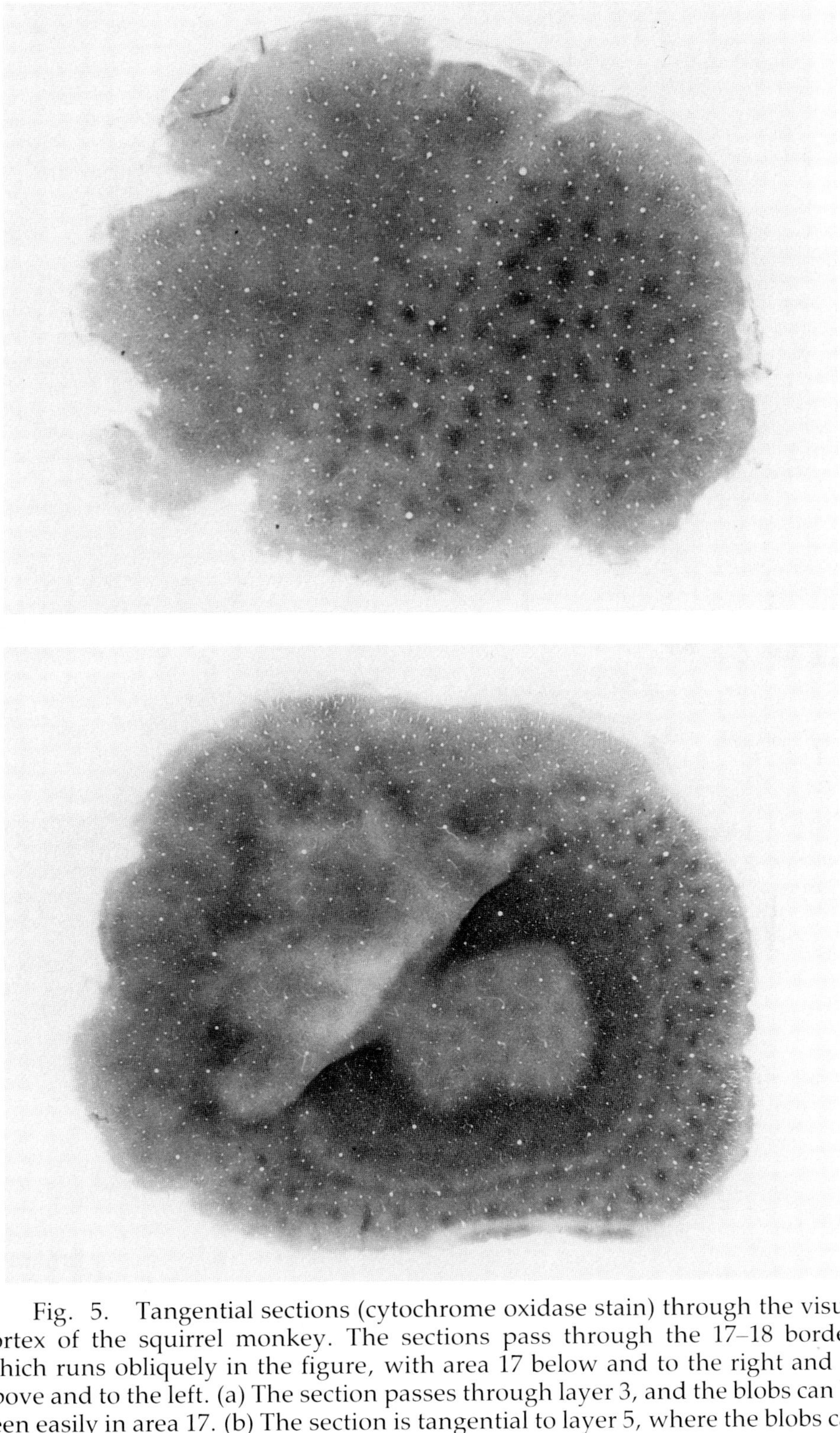

Fig. 5. Tangential sections (cytochrome oxidase stain) through the visual cortex of the squirrel monkey. The sections pass through the 17–18 border, which runs obliquely in the figure, with area 17 below and to the right and 18 above and to the left. (a) The section passes through layer 3, and the blobs can be seen easily in area 17. (b) The section is tangential to layer 5, where the blobs can again be seen, though faintly; these lie in register with the upper-layer blobs (From ref. *1*).

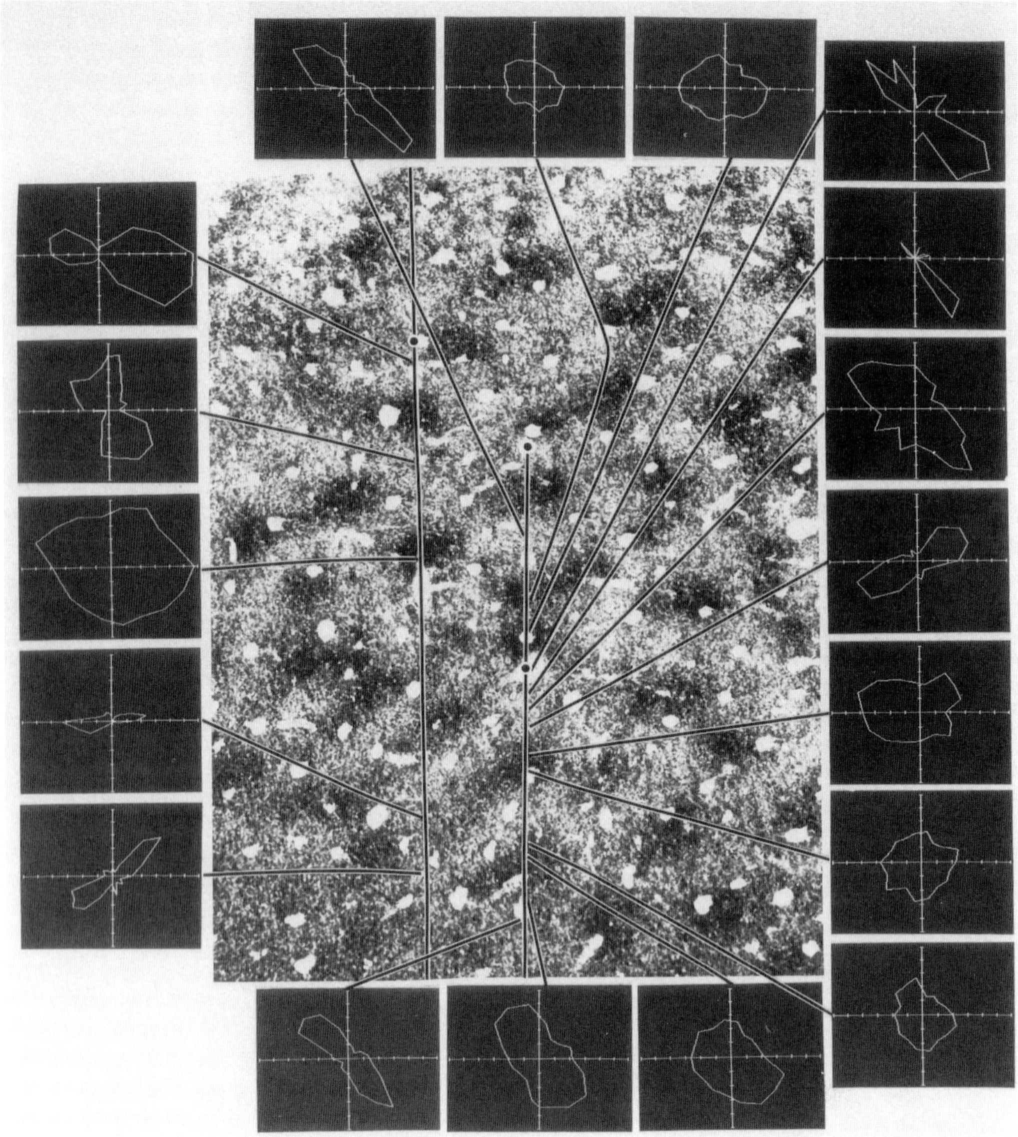

Fig. 6. Tangential section of macaque striate cortex showing two electrode tracks and polar-coordinate histograms for some of the cells studied in the penetration.

have little doubt that all of the cells in any blob are insensitive to line orientation.

We have already said that, as one moves across the cortex, the optimal orientation changes in regular steps. We wondered whether the iso-orientation lines (e.g., the close-spaced lines in the Ice-Cube Model of Fig. 3) are thrust aside by the blobs, or are unperturbed by them. Figure 7 helps answer this. Here, as the electrode traversed some 5 mm of cortex, we plotted orientation shifts (y-axis) through five entire cycles of 180°.

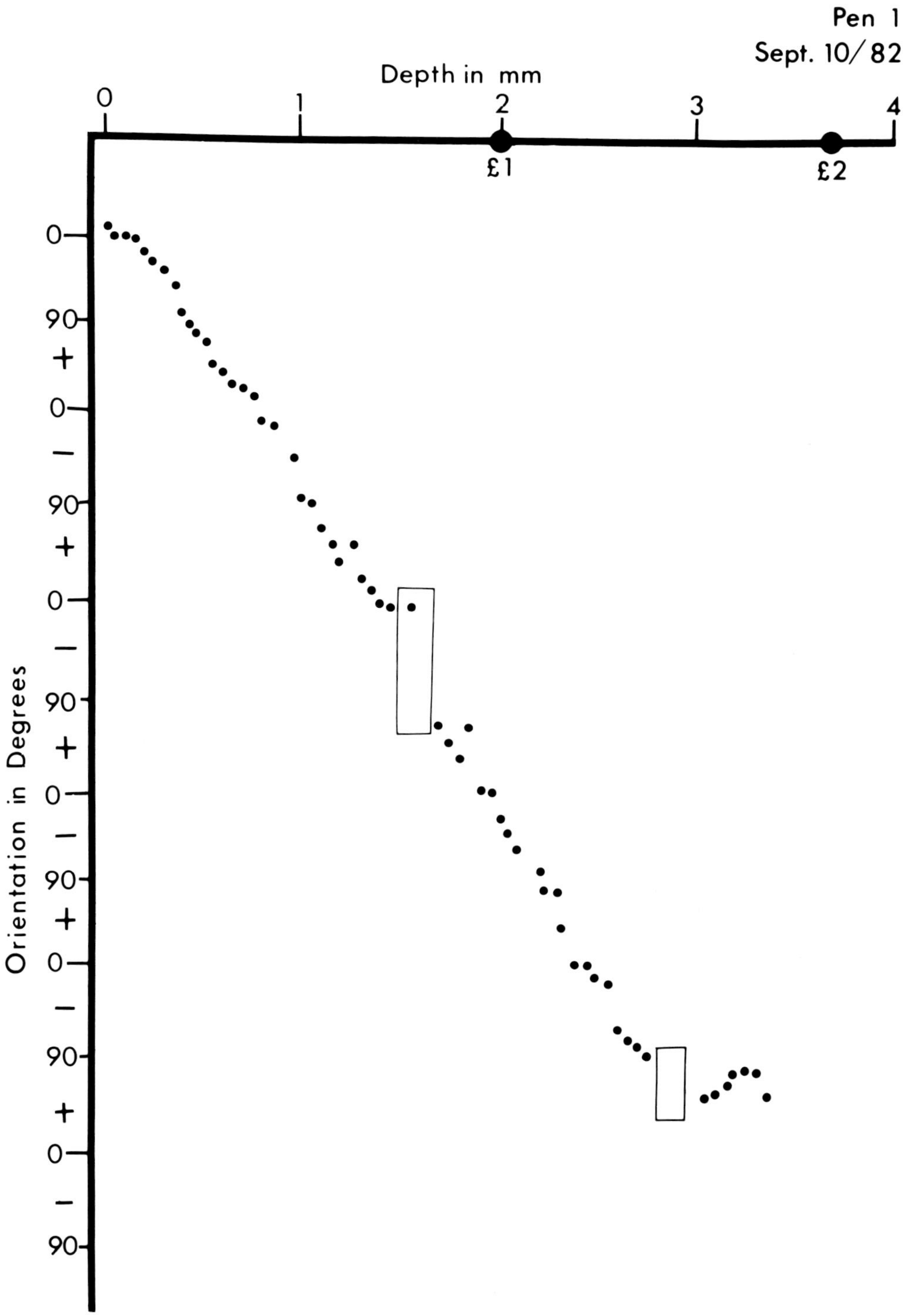

Fig. 7. Plot of preferred orientation against distance along electrode track for a single 5-mm penetration through the upper layers of macaque striate cortex. The rectangles indicate blobs, within which the cells showed no preferred orientation. Note that the linearity of the plot is unperturbed by the presence of the first blob.

The rectangle indicates a place where a blob, was encountered, and of course no orientations could be plotted for about 200 μm, but on leaving the blob, the sequence resumed as though nothing had happened. The orientation columns thus seem to be unperturbed by the presence of the blobs.

These results can be summarized by showing our modified Mark II Ice-Cube Model (Fig. 8). The blobs spanning the layers, interrupted only by layer 4, are thrust into the columnar systems as though by a twist drill. So, ironically, one ends up, for historical reasons, calling the ocular dominance and orientation groupings "columns," though they are really slabs, whereas the blobs really are columns, in the Greek sense.

A lack of orientation in blob cells could have one of two interpretations: The cells might receive convergent input from neighboring cells of all orientations, and hence be higher-order cells, though lacking orientation tuning; or they could be simpler, perhaps receiving direct input from layer 4C or from the lateral geniculate body. When we looked more closely at the receptive fields of these cells, it was immediately clear that they are simpler, and that most have concentric center-surround fields. They have much in common with geniculate cells and layer 4 cells, although they do show some striking differences.

Over the past year, the main thrust of our research has been in further exploring the physiological properties of blob cells and their connections. Both sets of results are fascinating, but at the moment it is too early to describe them, except in the most tentative fashion. Unlike the rest of the cells in the upper layers, blob cells do seem to receive input from the lateral geniculate body directly rather than by way of layer 4C—several anatomical experiments point strongly in this direction, although none so far provide definitive proof (*16–18*).

Blob cells and nonblob cells seem to have different outputs, the blob cells projecting selectively to a set of much coarser blobs in area 18, and the nonblob cells to the nonblob parts of 18 (personal observation). This only serves to emphasize the extreme specificity with which the nervous system is wired, with 17 projecting to 18, not only in topographic correspondence, but also with correspondence of the finer-grained blob–nonblob subdivisions.

Perhaps most exciting to us, at present, is the increasingly strong impression we are gaining that a high proportion (but perhaps not all) of the blob cells are specifically concerned with color. Color-specific cells have been studied for some years at more peripheral levels in the visual system—in the retina and especially in the geniculate—and the behavior of these cells has helped explain the physiological basis for the complex and often counterintuitive laws of color mixing, that blue light added to yellow produces white, for example. The cortical blob cells seem to take the process a step further; they begin to explain the physiology behind the psychophysics explored by Edwin Land (*19,20*) and Jameson and Hurvich (*21*) over the past twenty years, especially the fact that colors in

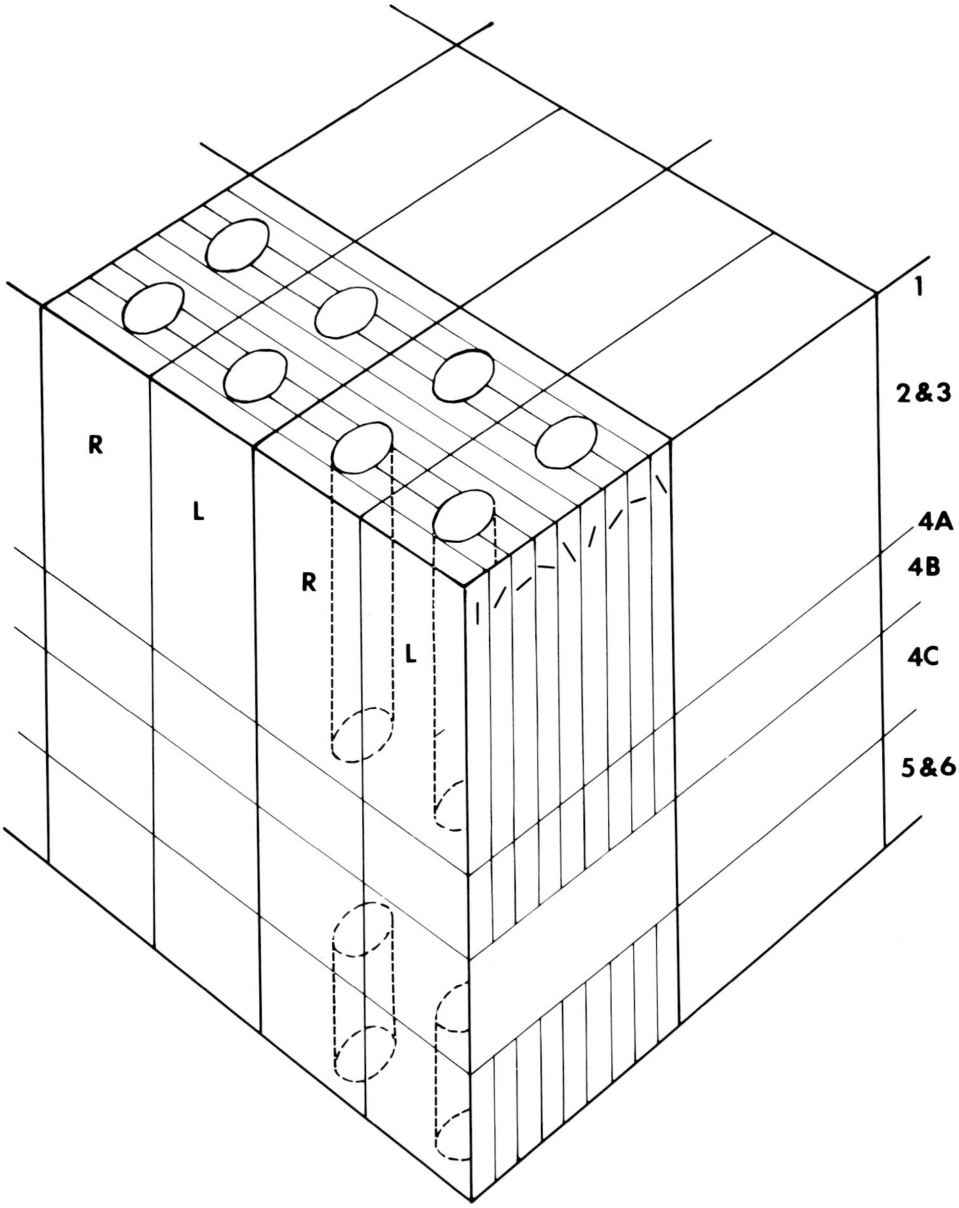

Fig. 8. Our revised model of the striate cortex.

a scene are so incredibly constant despite marked changes in the spectral content of the light source (*19–21*). This is something our visual system does that a photographic film cannot do: An outdoor film used inside with tungsten light gives the all-too-familiar pinks or oranges instead of whites. A knowledge of color photography may indeed explain why Land recognized what color vision experts had either missed or not emphasized for many centuries. The blob cells behave in a way, in response to color stimuli, that suggests that they may form the building blocks for this system.

As is typical in much research, we thus find ourselves plunged into a field that we originally had no definite plans to study. We began with an interest in blobs, and that led us to color because color is what the blobs seem to be largely concerned with. Perhaps the most exciting thing about studying the brain is the feeling of unexpectedness, the inability to predict exactly what one may be doing in even a few months time. This makes it hard to apply for research money—granting agencies and one's fellow scientists expect to be presented with concise, long-range plans—but it does make for variety and a feeling of exhilaration.

REFERENCES

1. Hubel, D. H. (1982), (Nobel Lecture) *Nature* **299,** 515.
2. Ferster, D., and LeVay, S. (1978), *J. Comp. Neurol.* **182,** 923.
3. Hubel, D. H., and Wiesel, T. N. (1977), (Ferrier Lecture) *Proc. R. Soc. Lond. B.* **278,** 377.
4. Hubel, D. H., and Wiesel, T. N. (1978), *Soc. Neurosci. Abst.* **4,** 632.
5. Hubel, D. H., and Wiesel, T. N. (1974), *J. Comp. Neurol.* **158,** 295.
6. Hubel, D. H., and Wiesel, T. N. (1968), *J. Physiol.* **195,** 215.
7. Hubel, D. H., Wiesel, T. N., and Stryker, M. P. (1977), *Nature* **269,** 328.
8. Weber, J. T., Kaas, J. H., Huerta, M. F., and Harting, J. K. (1977), *Soc. Neurosci. Abst.* **3,** 580.
9. Hendrickson, A. E., Wilson, J. R., and Ogren, M. P. (1978), *J. Comp. Neurol.* **182,** 123.
10. Horton, J. C., and Hubel, D. H. (1980), *Soc. Neurosci. Abst.* **6,** 315.
11. Humphrey, A. L., and Hendrickson, A. E. (1980), *Soc. Neurosci. Abst.* **6,** 315.
12. *The Oxford English Dictionary (1978),* (Murray, J. A. H., Bradley, H., Craigie, W. A., and Onions, C. T., eds.), Oxford University, Oxford, vol. 1, p. 926.
13. Horton, J. C., and Hubel, D. H. (1981), *Nature* **292,** 762.
14. Hendrickson, A. E., Hunt, S. P., and Wu, J. -Y. (1981) *Nature* **292,** 605.
15. Hubel, D. H., and Livingstone, M. S. (1982), *Soc. Neurosci. Abst.* **8,** 706.
16. Livingstone, M. S., and Hubel, D. H. (1982), *Proc. Natl. Acad. Sci. USA* **79,**6098.
17. Fitzpatrick, D., Itoh, K., and Diamond, I. T. (1983), *J. Neurosci.* **3,** 673–702.
18. Weber, J. T., Huerta, M. F., Kaas, J. H., and Harting, J. K. (1983), *J. Comp.Neurol.* **213,** 135.
19. Land, E. H. (1959), Parts I and II *Proc. Natl. Acad. Sci. USA* **45,** 115 and 636.
20. Land, E. H. (1964) *Am. Scientist* **52,** 247.
21. Jameson, D., and Hurvich, L. M. (1961), *J. Opt. Soc. Am.* **51,** 46.

Neuron Learning to Brain Organization

LEON N COOPER

Department of Physics and Center for Neural Science, Brown University, Providence, Rhode Island 02912

INTRODUCTION

The sequence of articles having been changed so that Professor Hubel's article could appear first allows experiment, appropriately, to precede theory. Whether theory is at all possible in this area (not to say fruitful) is indeed one of the questions we must consider.

What you have read in Professor Hubel's discussion gives an indication of the enormous complexity of the network of neurons. And the work of Professors Hubel and Wiesel over the last generation has been in the forefront in illustrating some of the complexity of cortical structure as well as some of the marvelous details of the architecture and organization.

The problem we are concerned with here is how the complex network of neurons is organized so that it can do some of the functions it seems to perform? For example, where and how might memory be stored?

The brain, of course, is composed of biological elements, but the problem of organization is not entirely a biological problem. And it is just here that the opportunity for theory exists. But useful theory must be constructed with close attention to experiment.

Let me begin by attempting a very quick description of the elements of this problem. In Fig. 1, we see a view of the human brain. It is an incredibly complex piece of machinery involving many individual elements—the most relevant of which are known as neurons or nerve cells. It is believed that information processing, memory storage, logical thinking, and so on occurs among the neurons. Neocortex (new cortex)—generally thought to be the thinking part of the brain—is on the surface: This sheet of neurons if spread out is rather large—perhaps several square meters. To fit it into a reasonably sized skull, it had to be folded;

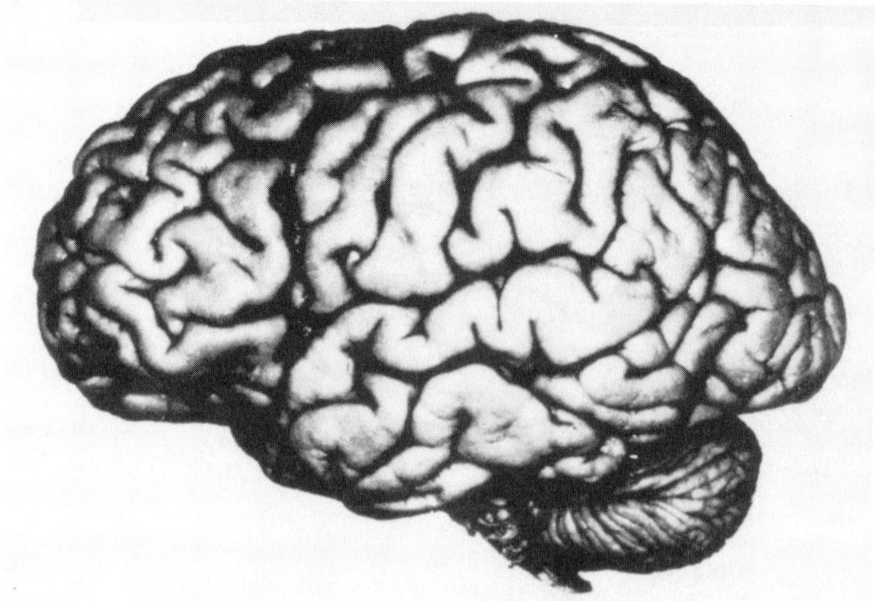

Fig. 1. Side view of the human brain (From DeArmend, Fusco, and Dewey *Structure of the Brain*).

typical folds on the surface of the cortex are seen in Fig. 1. In Fig. 2, a portion of the neural network (in visual cortex) is shown. We see suggested here some of the complexity of the cellular circuitry.

The new part of the brain, cortex, the special biological gift of higher mammals, evolved very rapidly in only a few million years. In contrast, other parts, such as the brainstem, which we share with reptiles, is mostly hard-wired and performs a great variety of control functions that took hundreds of millions of years to develop. This is a very suggestive fact.

The neurons are a marvellous piece of machinery. Like most cells, they share basic structures that keep themselves alive, but have become extremely specialized. Their primary function is to transmit (and probably also to store) information. The fundamental device utilized by these cells is an excitable membrane. The cell is capable of altering normal ion concentrations in its interior. The proportions of ions such as Na^+, Ca^{2+}, K^+ and so on, in squid blood are almost the same as those found in sea water, which, by the way, suggests strongly where the blood comes from (Table 1).

Inside a neuron there is an excess of K^+ and too little Na^+. This is because of a metabolic pump that slowly pumps Na^+ out and K^+ in (Fig. 3). The pump can be thought of, for all practical purposes, as slowly charging a battery. A channel is left open for K^+ ions. Because of the concentration difference, the K^+ ions try to get outside the membrane. But since they carry a positive charge, they build up an electric potential difference across the membrane as they move along the concentration gra-

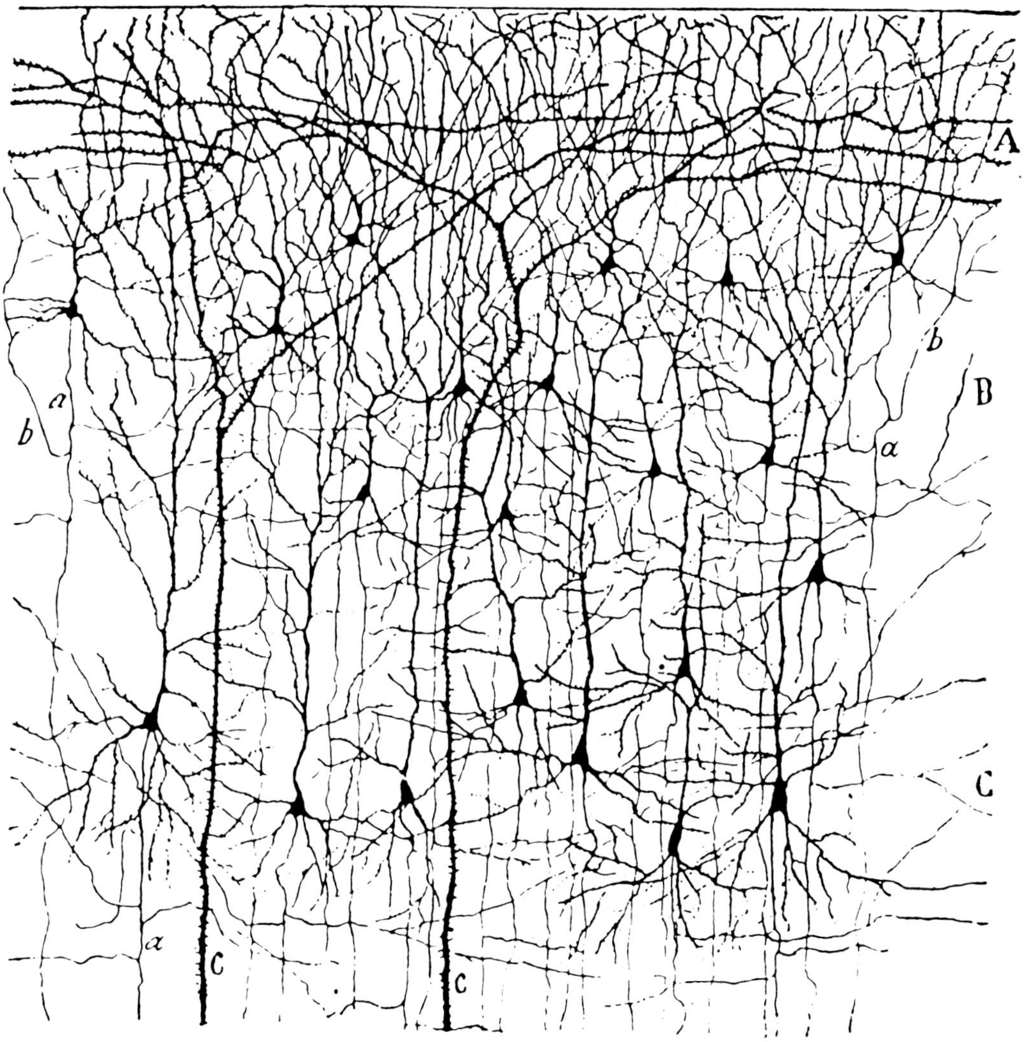

Fig. 2. A portion of the neural network in visual cortex (From R. Cajal, *Histologie du Systeme Nerveux*).

dient. This potential difference balances the concentration difference when there is a potential difference of about 70 mV across the membrane.

Electrical or chemical disturbances on some region of the membrane open the Na^+ channel; there is a rush of ions locally in that region; the negative resting potential, of about -70 mV, may jump to a positive potential (about $+55$ mV); this is known as the action potential. Eventually, equilibrium is reestablished in the original region. But the potential disturbance has spread a bit; this opens the Na^+ channel a bit down the line; then one has another action potential. So the action potential moves down the membrane. Depending on how much depolarization, in a way

TABLE 1
Concentrations of Ions Inside and Outside Freshly Isolated Axons of Squid

	Concentration, m*M*		
Ion	Axoplasm	Blood	Seawater
Potassium	400	20	10
Sodium	50	440	460
Chloride	40 – 150	560	540
Calcium	0.3×10^{-3}[a]	10	10

[a]The precise value of ionized intracellular calcium is not known. Data from Hodgkin (1964) and Baker, Hodgkin and Ridgway (1971) (From Kuffler and Nicholls *Neuron to Brain*).

that I do not have time to go into, one can convert a signal of a given intensity into a given number of action potentials per second. Thus, the information is frequency modulated and can be transmitted with essentially no loss over large distances. It is quite remarkable, considering the materials available, that information can be transmitted with such accuracy.

The transmission of information from one neuron to another is perhaps even more remarkable. An axon of one neuron will terminate, in general, near the dendrite of another, in a structure known as a synaptic junction (Fig. 4). Action potentials arriving from the axon initiate release of transmitter substances that diffuse across the synaptic cleft. Upon arrival at the postsynaptic membrane, these transmitters produce changes in membrane conductivity, initiating a flow of ions that alters the dentrite potential. The dendrite potentials propagate to the cell body, where they are integrated and determine the firing rate of the postsynaptic cell. Thus, the information flow continues.

It is now commonly thought that the synaptic junction may be a means by which to store information (memory, for example), as well as to transmit it from neuron to neuron. Large networks of neurons connected to other neurons via modifiable synaptic junctions are what are used to try to construct entities that are capable of holding memory and performing mental acts.

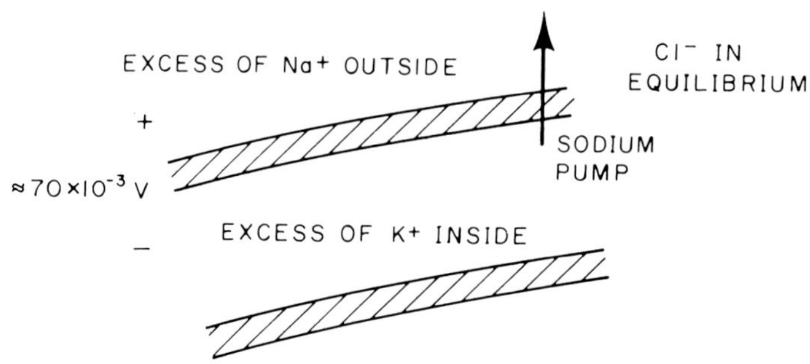

Fig. 3. The neuron membrane.

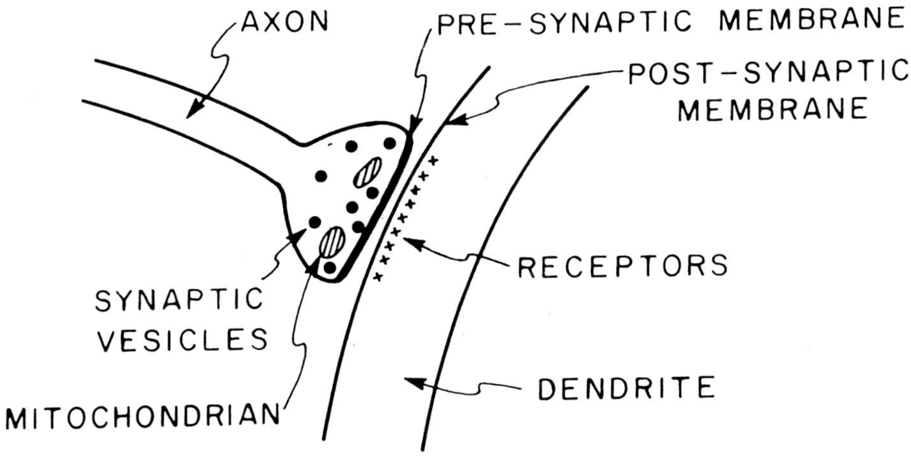

Fig. 4. A synaptic junction.

Although the central nervous system contains something of the order of 10–100 billion neurons, it is somewhat depressing to learn that these cells are so specialized that they do not reproduce. Thus, when the embryo is given its store of neurons, these are the only ones we will ever have. But cells die. We lose an estimated hundred thousand neurons each day, so that, at the end of a long lifetime, we have lost a very substantial number of neurons. Any theory that purports to explain the functioning of the central nervous system must explain how it continues to function in spite of the loss of individual neurons.

This is one of the facts that makes the sometime employed analogies between the brain and current computers really very bad. Modern computers perform large numbers of sequential operations very rapidly and very accurately. It is a technological miracle that so many operations can be performed with so few mistakes. The central nervous system works slowly, with cycle times that cannot be shorter than a few milliseconds. It is not very accurate: Neurons may not fire if they are tired, depressed or just too lazy—in addition, they die.

It may be that current computers can do some of the things that humans can do. However, most of the things a computer can do, humans do not do very well. For example, we are very poor at arithmetic and sequential logical operations; This, current computers do very well. We are very good at recognizing things and getting a sense of what is going on. For this, computers are very bad. So, if computers can mimic human beings by doing some of the things we do, it is very likely that the same thing is being done in a rather different way with rather different hardware.

Although this is a very complex problem, in a certain sense there is much that is known. A set of coupled nonlinear differential equations, including time delays, can be written down that in effect summarizes everything known about the transmission of electrical signals along excitable membranes and from axon to dendrite in a large coupled and reentrant network of neurons. Such systems can be stationary or can evolve in time by various learning algorithms. Obviously, such a set of equations with no simplification is extremely complex and beyond the capacity of current analysis or numerical techniques. The essential point is to make the appropriate approximations and to clearly illuminate the paths connecting assumptions and consequences. Various approaches, such as those of Amari (*1*), Wilson and Cowan (*2*), Edelman (*3*), Edelman and Reeke (*4*), Grossberg (*5*), have been applied to this problem.

In our work we emphasize the transfer of information between neuron sets to neuron sets; we propose that there is much parallel processing in the central nervous system, and that in contrast with machine memory, which is at present local (an event stored in a specific place) and addressable by locality (requiring some equivalent of indices and files), our memory is distributed and addressable by content or by association. In addition, there need be no clear separation between memory and "logic," which is a result of association and an outcome of the nature of the memory itself.

Such distributed memories are more like a hologram than a photograph. An individual synaptic junction holds superimposed information concerning many events. In order to obtain a single event, one has to gather information from many junctions. In a system like this, loss of individual neurons decreases the signal-to-noise ratio, but does not lose individual items of information. Therefore, if it is overbuilt in the first place, one can retain a complete memory with an acceptable signal-to-noise ratio, even with the loss of neurons.

These ideas are described in more detail in the following discussion. Although these initial attempts are clearly oversimplifications, our hope is to capture some of the more important qualitative features of a very complex phenomenon in a piece of structure that is clear enough so that we can say what follows from what, and in an explicit enough way so that we can make contact with experiment as soon as possible.

1. THEORETICAL BACKGROUND

Distributed Memory

That most intriguing aspect of memory—its persistence in spite of continual loss of individual neurons over the lifetime of the individual—led us early to the concept of distributed, rather than local, memory storage. Distributed storage possesses in a very natural way the property of relative invulnerability to the loss of individual storage units. We have been analyzing a class of neural models for the acquisition and storage of distributed memories that displays on a primitive level features such as recognition, association, and generalization, and that suggests some of the mental behavior associated with animal memory and learning (*6,7*).

The mechanisms we employ seem to be biologically plausible and are not inconsistent with known neurophysiology. In addition, the networks that result seem to be a reasonable outcome of evolutionary development under the pressure of survival. Some of our ideas are related to or are generalizations of earlier concepts, such as perceptrons or similar models (*8–10*). In addition, holographic or nonlocal memories have been explored previously (*11,12*).

Although the concept of distributed mappings and memory storage is less familiar than those of local storage, distributed mappings and their properties have been discussed (*6,13–16*) and probably have already been observed. An example is superior colliculus. In spite of the fact that the retinal afferents that project to the colliculus form a very precise, fine-grained map, cells that are just a few mm below the very precise cells respond to stimuli over a wide area of visual space. Thus, we have the apparently paradoxical situation—that seems to be true of other parts of the brain as well—that great precision of response is generated by systems composed of cells that progressively show less and less selectivity as the motor output of the system is approached (*17*).

We believe that much of the learning and resulting organization of the central nervous system occurs because of modification of the efficacy or strength of at least some of the synaptic junctions between neurons, thus altering the relation between presynaptic and postsynaptic potentials. It is known that small but coherent modifications of large numbers of synaptic junctions can result in distributed memories. Whether and how such synaptic modification occurs, what precise forms it takes, and what the physiological and/or anatomical bases of this modification are, rank among the most interesting questions in this area.

There is direct experimental evidence that at least some modification of synaptic strength occurs in invertebrates (*18*), and there are various indications that synaptic modification is a rather general phenomenon (*19*). In recent years, many conjectures have been made concerning the kind of modification that might occur at synaptic junctions. Kandel and coworkers (*18*) have shown that modification of the synapse between a sensory and motor neuron of the marine mollusk Aplysia is the basis of habituation and sensitization. The synaptic modification they have observed can be dependent only on presynaptic information. In our work, we have had to assume that synaptic modification is a function of more general variables; local, quasilocal, and global. The presence of quasilocal variables leads to forms of synaptic modification (denoted as Hebbian) that depend on information not immediately available at the synaptic site (e.g., cell-firing rates).

These hypotheses have been developed in some detail (*20–22*) and applied to experimental results that have been obtained in visual cortex by many workers over the last generation, as well as to higher-level network properties. As will be explained more fully, we have been able to obtain agreement with classical visual cortex experimental results and in addition, have suggested new experiments. These theoretical results have been obtained with a minimum of anatomical details and have been primarily concerned with single neurons. It is very likely, however, that interactions between cortical neurons play an important role in cortical function, as well, perhaps, as in selectivity of individual cortical cells (*23,24*). In addition, individual synapses are generally either inhibitory or excitatory and not both, as some theoretical work has assumed for simplicity. One objective of our current research is to extend our results, taking into account a more realistic anatomy so that a more detailed comparison between theory and experiment can be made.

The theoretical ideas mentioned above have also led to several suggestions for new experiments in visual cortex. In these experiments, we attempt to verify predictions concerning the connection between specificity and ocular dominance of cortical cells under various rearing conditions. In addition, we investigate connections between ocular dominance and the variety of visual input allowed to the open eye. We expect the results to give us further information about the detailed mechanism of synaptic modification among cortical cells, as well as to enable us to determine various system parameters.

For a distributed memory, it is the simultaneous or near simultaneous activities of many different neurons (the result of external or internal stimuli) that are of interest. Thus, a large spatially distributed pattern of neuron discharges, each of which might not be very far from spontaneous activity, could contain important, if hard to detect, information. Let us consider the behavior of an idealized neural network (that might be regarded as model component of a nervous system) to illustrate some of the important features of distributed mappings.

Consider *N* neurons, 1,2 . . . *N*, each of which has some spontaneous firing rate r_{j0}. (This need not be the same for all of the neurons nor need it be constant in time.) We can then define an *N*-tuple whose components are the difference between the actual firing rate r_j of the *j*th neuron and the spontaneous firing rate r_{j0}:

$$f_j = r_j - r_{j0} \qquad [1.1]$$

By constructing two such banks of neurons connected to one another (or even by the use of a single bank that feeds signals back to itself), we arrive at a simplified model, as illustrated in Fig. 5.

The actual synaptic connections between one neuron and another are generally complex and redundant; we have idealized the network by replacing this multiplicity of synapses between axons and dendrites with a single ideal junction, which summarizes logically the effect of all of the synaptic contacts between the incoming axon branches from neuron *j* in the *F* bank and the dendrites of the outgoing neuron *i* in the *G* bank (Fig. 6). Each of the *N* incoming neurons in *F* is connected to each of the *N* outgoing neurons in *G* by a single ideal junction.

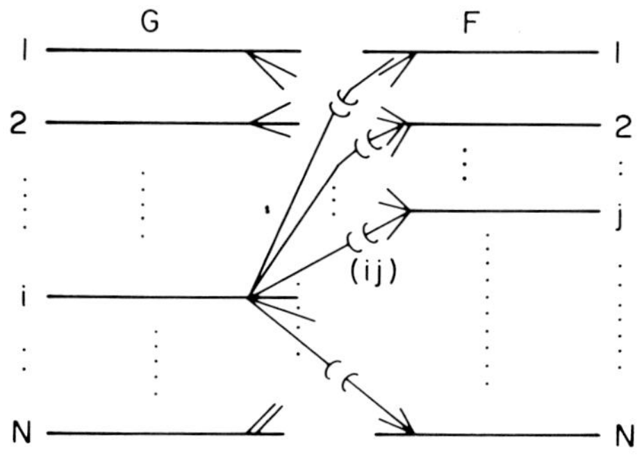

Fig. 5. An ideal distributed mapping. Each of the N input neurons in F is connected to each of the N output neurons in G by a single ideal junction. (Only the connections to i are drawn.)

Although the firing rate of a neuron depends, in a complex and nonlinear fashion, on the presynaptic potentials, there is usually a reasonably well-defined linear region. Some very interesting network properties are already evident in this linear region. We therefore focus our attention, for the moment, on the region above threshold and below saturation, for which the firing rate of neuron i in G, g_i, is mapped from the firing rates of all of the neurons, f_j, in F by:

$$g_i = \sum_{i=1}^{N} A_{ij} f_j \qquad [1.2]$$

In doing this, we are regarding as important average firing rates and time averages of the instantaneous signals in a neuron (or perhaps a small population of neurons). We are further using the known integrative properties of neurons.

We may then regard $[A_{ij}]$ (the synaptic strengths of the N^2 ideal junctions) as a matrix or a mapping that takes us from a vector in the F space to one in the G space. This maps the neural activities $f = (f_1, f_2 \ldots f_N)$ in the F space into the neural activities $g = (g_1 \ldots g_N)$ in the G space and can be written in the compact form:

$$g = Af \qquad [1.3]$$

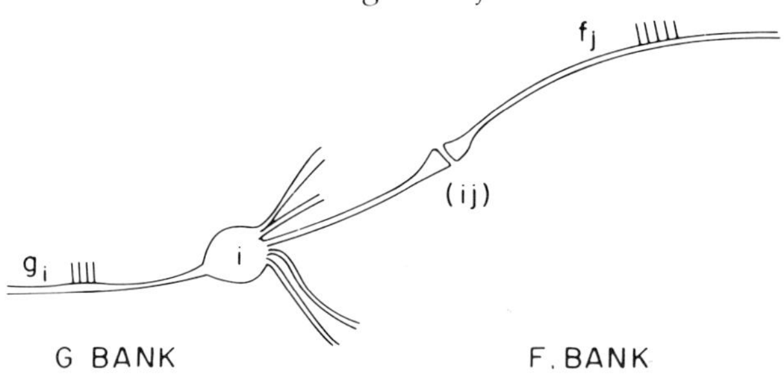

Fig. 6. An ideal synaptic junction.

We propose that it is in modifiable mappings of the type A that human memory is stored. Presently, machine memory is local (an event stored in a specific place) and addressable by locality (requiring some equivalent of indices and files). In contrast, human memory is more likely to be distributed and addressable by content or by association. In addition, for such a memory, there need be no clear separation between memory and "logic."

It is convenient to write the mapping A in the basis of vectors the system has experienced:

$$A = \sum_{\mu\nu} c_{\mu\nu} g^{\mu} \times f^{\nu} \qquad [1.4]$$

Here g^{μ} and f^{ν} are output and input patterns of neural activity, whereas the $C_{\mu\nu}$ are coefficients reflecting the degree of connection between various inputs and outputs. The symbol $\times$ represents the "outer" product between the input and output vectors. Although [1.4] is a well known mathematical form, its meaning as a mapping among neurons deserves some discussion. The ijth element of A gives the strength of the ideal junction between the incoming neuron j in the F bank and the outgoing neuron i in the G bank (Fig. 6.).

Thus, if only f_j is nonzero, G_i, the firing rate of the ith output neuron is:

$$g_i = A_{ij} f_j \qquad [1.5]$$

Since

$$A_{ij} = \sum_{\mu\nu} c_{\mu\nu} g_i^{\mu} f_j^{\nu} \qquad [1.6]$$

the ijth junction strength is composed of a sum of the entire experience of the system, as reflected in firing rates of the neurons connected to this junction. Each experience or association ($\mu\nu$), however, is stored over the entire array of $N \times N$ junctions. This is the essential meaning of a distributed memory: Each event is stored over a large portion of the system, and at any particular local point many events are superimposed.

We show below that the nonlocal mapping A can serve in a highly precise fashion as a memory that is content addressable and in which "logic" is a result of association and an outcome of the nature of the memory itself.

Long- and Short-Term Memory

The N^2 junctions, A_{ij}, contain the content of the distributed memory. It could be that a particular junction strength, A_{ij}, is composed of several different components with different lifetimes, e.g.:

$$A = A_{ij}^{(1)} + A_{ij}^{(2)} + \ldots + A_{ij}^{(s)} \qquad [1.7]$$

where the individual $A_{ij}^{(l)}$ might be thought of as corresponding to different physiological or anatomical effects (e.g., changes in the number of presynaptic vesicles, and/or postsynaptic receptors; changes in Ca^{2+} levels and/or availability; and anatomical changes such as might occur in growth or shrinkage of spines). We then have the possibility that the actual memory content (even in the absence of additional learning) will vary with time. For a two-component system, we might have:

$$A_{ij}^{(t)} = A_{ij}^{(long)}(t) = A_{ij}^{(short)}(t) \qquad [1.8]$$

where $A_{ij}^{(t)}$ represents the memory at some time, t, and $A_{ij}^{(long)}$ and $A_{ij}^{(short)}$ have long and short lifetimes. Thus, in time, $A_{ij}^{(short)}$ will decay, leaving $A_{ij}^{(t)} = A_{ij}^{(long)}$. Whether what is in the short-term memory component is transferred to the long-term component might be determined by some global signal—depending on the interest of the information contained in the short-term component. The existence of such global signals, as well as possible anatomical or physiological correlates of short- or long-term memory, are the subject of some of our current research.

From this point of view, the site of long- and short-term memories can be essentially identical. At any given time there is a single memory. The distinction between long- and short-term memories is contained in the lifetime of the different components of A_{ij}.

Recognition and Recollection

The fundamental problems posed by a distributed memory are the address and accuracy of recall of the stored patterns. Consider first the "diagonal" portion of A:

$$(A)_{diagonal} \equiv \mathcal{R} \equiv \sum_{\nu} c_{\nu\nu} g^{\nu} \times f^{\nu} \qquad [1.9]$$

An arbitrary event, e, in the external world mapped by the sensory apparatus into the pattern of neural activity, f, will generate the response in G:

$$g = Af$$

(The pattern, f, might also be the result of some other internal pattern of neural activity.) If we equate recognition with the strength of this response, say the inner product (g,g), then the mapping A will distinguish between those events it contains, the f^{ν}, $\nu = 1,2 \ldots K$ and other events separated from these.

The work "separated" in the above context requires definition. Suppose the vectors, f^{ν}, are thought to be independent of each other, and to satisfy the requirements that, on the average:

$$\sum_{i=1}^{N} f_i^{\nu} = 0 \qquad \sum_{i=1}^{N} (f_i^{\nu})^2 = 1 \qquad [1.10]$$

Any two such vectors have components that are random with respect to one another so that a new vector, f, presented in the F bank, as above, gives a noise-like response in the G bank, since, on the average, (f^{ν},f), is small. The presentation of a vector seen previously, f^λ, however, gives the response in the G bank:

$$\mathcal{R}f^\lambda = c_{\lambda\lambda}g^\lambda + \text{noise} \qquad [1.11]$$

It can be shown that if the number of imprinted events, K, is small compared to the dimensionality, N, the signal-to-noise ratios are reasonable.

If we define separated events as those that map into orthogonal vectors, then, clearly, a recognition matrix composed of K orthogonal vectors $f^1, f^2, \ldots f^K$:

$$\mathcal{R} = \sum_{\nu=1}^{K} c_{\nu\nu}g^\nu \times f^\nu \qquad [1.12]$$

will distinguish between those vectors contained and all vectors separated from (perpendicular to) these. Furthermore, the response of the system to a vector previously recorded is unique and completely accurate:

$$\mathcal{R}f^\lambda = c_{\lambda\lambda}g^\lambda \qquad [1.13]$$

In this special situation, the distributed memory is as precise as a localized memory.

In addition, this type of memory has the interesting property of recalling an entire associated vector g^λ even if only part of f^λ is presented. Let:

$$f^\lambda = f_1^\lambda + f_2^\lambda \qquad [1.14]$$

if only part of f^λ, say f_1^λ, is presented, we obtain:

$$\mathcal{R}f_1^\lambda = c_{\lambda\lambda}(f_1^\lambda, f_1^\lambda)g^\lambda + \text{noise} \qquad [1.15]$$

the result is the entire response to the full f^λ, with a reduced coefficient plus noise.

Association

If we now take the point of view that presentation of the event e^ν, which generates the vector f^ν, is recollected if:

$$\mathcal{R}f^\nu = c_{\nu\nu}g^\nu + \text{noise} \qquad [1.16]$$

then the off-diagonal terms:

$$\mathcal{A} \equiv \sum_{\mu \neq \nu} c_{\mu\nu}g^\mu \times f^\nu \qquad [1.17]$$

may be interpreted as containing associations between events initially separated from one another.

For such terms, the presentation of event e^ν will generate not only g^ν (which is equivalent to the recollection of e^ν), but also, and perhaps more weakly, g^μ, which should result with the presentation of e^μ. Thus, for

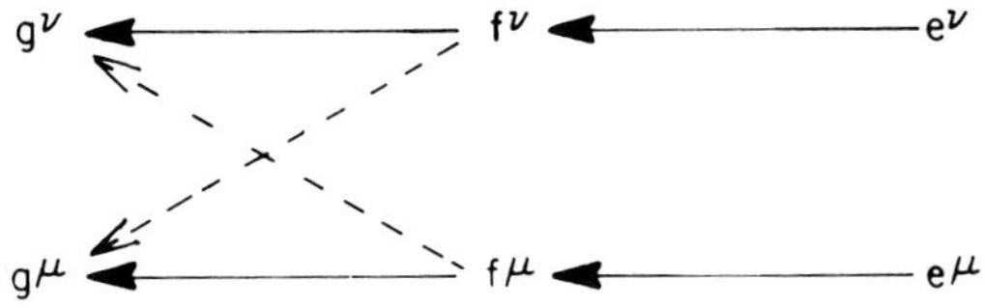

Fig. 7. An ideal association.

example, if g^μ will initiate some response, originally a response to e^μ, the presentation of e^ν when $c_{\mu\nu} \neq 0$ will also initiate this response (*see* Fig. 7).

We can thus divide the association matrix A into two parts:

$$A = \sum_{\mu\nu} c_{\mu\nu} g^\mu \times f^\nu = \mathcal{R} + \mathcal{A} \quad [1.18]$$

where

$$\mathcal{R} = (A)_{\text{diagonal}} \equiv \sum_{\nu} c_{\nu\nu} g^\nu \times f^\nu \quad [1.19]$$

$$\mathcal{A} = (A)_{\text{off-diagonal}} \equiv \sum_{\mu \neq \nu} c_{\mu\nu} g^\mu \times f^\nu \quad [1.20]$$

The $c_{\mu\nu}$ are then the direct recollection and association coefficients. Some of the consequences of the properties discussed in the last two sections are the subject of some of our current and continuing research and are further discussed in the following subsections: Network Modification, Learning; Passive Modification; and Feature Abstraction.

Network Modification, Learning

The properties described above require coherence among many synaptic junctions. We therefore ask: According to what rule and by what means do neurons modify themselves to form a matrix of junctions with the properties of memory? A major effect of our research is to elucidate this question.

Such a modification rule can be cast in the form of stochastic or deterministic differential equations, dependent on variables that we classify as local, quasilocal, and global:

$$\mathring{A}_{ij} = \phi(g_i, f_j, A_{ij}, t, \ldots) \quad [1.21]$$

Different such rules lead to various types of memories. In the following sections, several rules for plasticity will be examined. The recollection–association memory [1.18] described above is obtained from the following simple, bilinear modification rule:

$$\delta A_{ij} \sim g_1 f_j \quad [1.22]$$

This δA_{ij} is proportional to the product of the differences between the actual and the spontaneous firing rates in the pre- and postsynaptic neurons, i and j. [This is one realization of Hebb's form of synaptic modification (*25*)]. The addition of such changes to A for all associations $g^{\mu} \times f^{\nu}$ results finally in a mapping with the properties discussed in the previous sections.

Synaptic modification, dependent on inputs alone, of the type already directly observed in Aplysia (*26,27*) is sufficient to construct a simple memory—one that distinguishes what has been seen from what has not, but does not easily separate one input from another. To construct a mapping of the form above, however, requires synaptic modification dependent on information that exists at different places on the neuron membrane, what we call two- (or higher) point modification.

In order for this to take place, information must be communicated from, for example, the axon hillock to the synaptic junction, to be modified. This implies the existence of a means of internal communication of information within a neuron—in the above example, in a direction opposite to the flow of electrical signals (*6*). The junction ij, for example, must have information of the firing rate, f_j (which is locally available), as well as the firing rate g_i, which is somewhat removed (Fig. 8).

One possibility could be that the integrated electrical signals from the dendrites produce a chemical or electrical response in the cell body that controls the spiking rate of the axon and at the same time communicates (by backward spiking, for example) to the dendrite ends the information of the integrated slow potential. Another possibility is that dendritic shafts act as somewhat independent units so that the local integrated dendritic potentials interacting with the potentials incoming to the individual spines combine to produce changes in spine shape and resistivity. Such changes might be observable in anatomical studies and are the subject of one of our current research projects.

One might guess that once the physiological mechanism for such communication was available, different types of two- (or higher) point modification evolved in various ways. It is tempting to conjecture that a liberating evolutionary step was just the development of this means of

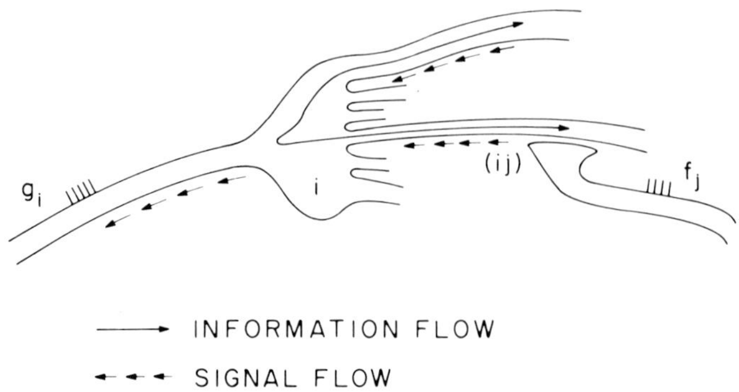

Fig. 8. Two point modification.

internal communication that, coupled with the ability of synapses to modify, created the possibility for a new organization principle.

There is a variety of means by which the coefficient A_{ij} might be modified, given that the necessary information is available at the *ij*th junction. Among these might be the growth of additional or a change in the electrical properties of dendrite spines, addition of new synaptic junctions, activation of synaptic junctions previously inactive, changes in membrane resistivity, and/or changes in the amount of transmitter or receptor in a synapse. Although some structural changes have been observed, there is little evidence yet to choose among the possibilities mentioned above. This is the subject of much current research.

Passive Modification

To make the modification:

$$\delta A \sim g^{\mu} \times f^{\nu} \qquad [1.23]$$

by any of the mechanisms suggested above, the system must have the signal distribution f^{ν} in its F bank and g^{μ} in its G bank. It is easy to obtain f^{ν} since this is mapped from either an external event or is some internal pattern. But to get g^{μ} in the G bank is more difficult since this, in effect, is what the system is trying to learn.

In what we denote as active learning, the system is presented with some f^{λ}, searches for a response, and is given some indication of when it is coming closer. When by some procedure or another it finds the "right" response, say g^{ω}, it is "rewarded" and responds to the reward by printing into A the information:

$$\delta A_{ij} = \eta g_i^{\omega} f_j^{\lambda} \qquad [1.24]$$

(The information is available at the time of the reward since at that time the system is mapping f^{λ}, responding g^{ω}, and thus has just the desired spiking frequencies in the F and G banks of neurons.) Active learning probably describes a type of learning in which a system response to an input is matched against an expected or desired response and judged correct or incorrect.

However, there is a type of learning that does not seem, from visible external indications, to require this type of a search procedure. It is the type of learning in which, as far as can be seen, an animal is placed in an environment and seems to learn to recognize and to recollect in a far more passive manner.

To arrive at an algorithm that produces what we call passive learning, we utilize a distinction between forming an internal representation of events in the external world as opposed to producing a response to these events that is matched against what is expected or desired in the external world.

The simple but important idea is that *the internal electrical activity that in one mind signals the presence of an external event is not necessarily (or likely to be) the same electrical activity that signals the presence of that same event for another mind.* There is nothing that requires that the same external event be mapped into the same neural patterns by different animals. The event e^{ν}, which for one animal is mapped into the signal distributions f^{ν} and g^{ν}, in another animal is mapped into f'^{ν} and g'^{ν}. What is required for eventual agreement between animals in their description of the external world is not that the electrical signals mapped be identical, but, rather, that the relation of the signals to each other and to events in the external world be the same (Fig. 9).

If we now allow the output of a cell to be determined by the input to that cell and the already existing synaptic junction strengths, as well as by possible noise-like fluctuations (making no prior requirement on what the output should be), we arrive at a mathematical formulation of what we call passive modification (6).

$$\delta A_{ij} \sim g_i f_j \sim \sum_{k=1}^{N} A_{ik} f_k f_j \qquad [1.25]$$

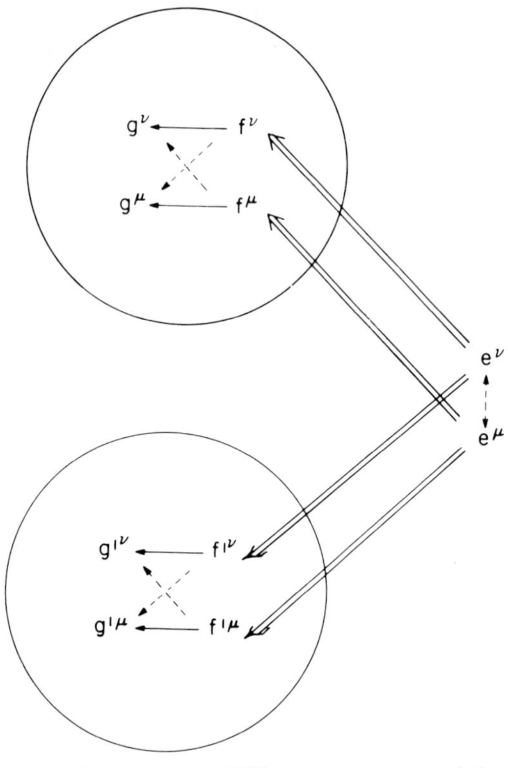

Fig. 9. Representations in two different systems of the same external fabric of events. The two representations are not identical, but they each stand in a one-to-one relation to the external fabric and to each other.

It has been shown in the above reference that with a simple form of passive modification a system generates its own response to incoming patterns in such a way as to construct distributed mappings that can function as memories capable of recognition and association. To a limited extent, these mappings can be regarded as internal representations of what has arrived from the outside world. It has further been shown (*20*) that a form of passive modification can result in the formation of feature detectors or threshold response units that learn to respond to repeated patterns even in the absence of any initial bias. Such units can serve to perform some nonlinear separations.

More detailed discussions of the consequences of these modification procedures and the properties of some of the mappings that result are contained in the references cited above. The application of these ideas to visual cortical cells is discussed in the section entitled Application to Visual Cortex: Comparison of Theory with Experiment.

Feature Abstraction

Some networks of neurons must have the ability to extract meaningful information from a broad range of input environments. In the case of sensory input to cortex, for example, the system's range is internally constrained by the response characteristics of the sensory neurons and externally by the nature of the stimulus environment. This stimulus environment depends a great deal upon the nature of the creature's surroundings. Precise statements regarding aspects of environmental structure relevant to mathematical models are given in the next section.

Consider the recognition–association memory [1.18] described above. In actual experience, the events to which the system is exposed are not in general highly separated nor are they independent in a statistical sense. There is no reason, therefore, to expect that all vectors, f^ν, printed into A according to the modification rule [1.25] would be orthogonal or even very far from one another. Rather, it seems likely that often large numbers of these vectors would lie close to one another. Under these circumstances, a distributed memory might be "confused" in the sense that it will respond to new events as if they were old, if the new event is close to an old one. It will "recognize" and "associate" events never, in fact, seen or associated before.

The memory will tend to categorize stimuli on the basis of the past history of the system. For example, suppose a number of vectors in the memory are of the form:

$$f^\nu = f^0 + n^\nu \qquad [1.26]$$

where n^ν varies randomly; f^0 will eventually be recognized more strongly than any particular f^ν actually presented. This, of course, is reminiscent of psychological properties called "generalization" or "abstraction." From such a point of view, generalization grows from the loss of detail of individual instances, a trade-off that seems characteristic of distributed systems.

We have here an explicit realization of feature abstraction. This generalizing quality might be described as the result of a built-in directive for inductive logic. The associative memory by its nature takes the following step:

$$f^0 + n^1,\ f^0 + n^2,\ \ldots\ f^0 + n^k,\ \ldots \rightarrow \mathrm{f} \qquad [1.27]$$

which one perhaps attempts to describe in language as passing from particulars: cat^1, cat^2, cat^3 . . . to general cat.

How fast this step is taken depends on the parameters of the system. By altering these parameters, it is possible to construct mappings that vary from those that retain all particulars to which they are exposed, to those that lose the particulars and retain only common elements—the central vector of any class.

In addition to "errors" of recognition, the associative memory also makes "errors" of association. If, for example, all (or many) of the vectors of the class $\{f^\alpha\}$, defined as a class of vectors not very separated from one another, associate some particular g^β so that the mapping contains terms of the form:

$$\sum_{\nu=1}^{K} c_{\beta\nu} g^\beta \times f^\nu \qquad f^\nu \in \{f^\alpha\} \qquad [1.28]$$

with $c_{\beta\nu} \neq 0$ over much of $\nu = 1,2 \ldots K$, then the new event, e^{K+1}, which maps into f^{K+1}, also in the class $\{f^\alpha\}$, will not only be recognized, the inner product $(\mathcal{R}f^{K+1}, \mathcal{R}f^{K+1})$ being large, but will also associate g^β, $\mathcal{A}f^{K+1} = cg^\beta + \ldots$ as strongly as any of the vectors $f^1 \ldots f^K$ explicitly contained in [1.28].

If errors of recognition lead to the process described in language as going from particulars to the general, errors of association might be described as going from particulars to a universal: cat^1 meows, cat^2 meows, → all cats meow.

Whatever efficacy this inductive process has will depend on the order of the world in which the animal system finds itself. If the world is properly ordered, an animal system that "jumps to conclusions" in the sense above may be better able to adapt and react to the hazards of its environment and thus survive.

By a sequence of mappings of the form above (or by feeding the output of A back to itself) one obtains a fabric of events and connections that is rich as well as suggestive. One easily sees the possibility of a flow of electrical activity influenced both by internal distributed mappings and external input. This flow is governed not only by direct association coefficients $c_{\mu\nu}$ (which can be explicitly learned), but also by indirect associations resulting from the overlapping of the mapped events. One can imagine situations arising in which direct access to an event or a class of events has been lost while the existence of this event or class of events in *A* influences the flow of electrical activity.

One problem in making the identifications suggested above is that such systems tend to form excessively large all-encompassing classes. But means have been devised to limit the extent of class formation. In fact, such mappings can be made to separate classes as well as to unite them (*16,21,22*).

Another problem is a direct consequence of the assumption of the linearity of the system. Any state is generally a superposition of various vectors. Thus, one has to find a means by which events—or the entities into which they are mapped—are distinguished from one another.

There are various possibilities; neurons are so nonlinear that it is not at all difficult to imagine nonlinear or threshold devices that would separate one vector from another. Such separation processes compliment generalization processes in that they bring out the differences in an input environment while generalizing cells tuned to the component most common to the constituent stimuli. But the occurrence of a vector in a distributed memory is a set of signals over a large number of neurons, each of which is far from threshold. A basic problem, therefore, is how to associate the threshold of a single cell or a group of cells with such a distributed signal. One way this might come about has been shown by Nass and Copper (*20*). Another possibility is the stochastic process recently discussed by Hopfield (*28*).

In addition to the appearance of "pontifical" cells or groups of cells, there will be a certain separation of mapped signals because of the actual localization of the areas in which these signals occur. For example, optical and auditory signals are subjected to much processing before they actually meet in cortex. It is possible to imagine that identification of optical or auditory signals (as optical or auditory) occurs first from where they appear and their immediate cluster associations. Connections between an optical and an auditory event might occur, as suggested in Fig. 10. Although the systems described above are relatively primitive, they suggest various psychological properties and are used in our research to construct models of some aspects of behavior and language learning.

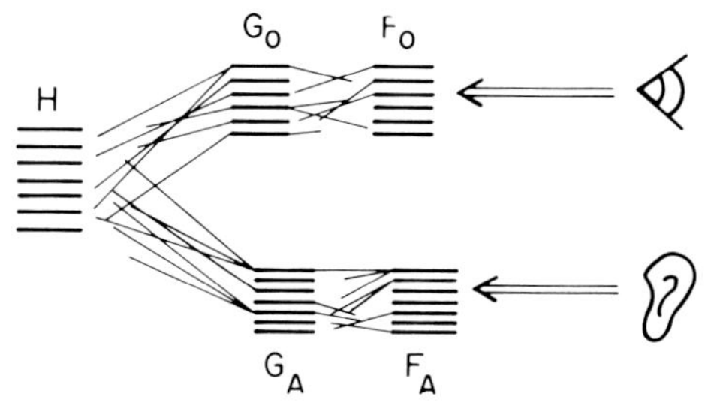

Fig. 10. A model optical–auditory system.

2. APPLICATION TO VISUAL CORTEX: COMPARISON OF THEORY WITH EXPERIMENT

Summary of Related Visual Cortex Experimental Data

The discussion above leads to a central issue: What is the principle of local organization that, acting in a large network, can produce the observed complex behavior of higher mental processes. There is no need to assume that such a mechanism—believed to involve synaptic modification—operates in exactly the same manner in all portions of the nervous system or in all animals. However, one would hope that certain fundamental similarities exist so that a detailed analysis of the properties of this mechanism in one preparation would lead to some conclusions that are generally applicable. We are interested in visual cortex because the vast amount of experimental work done in this area of the brain—particularly area 17 of cat and monkey—strongly indicates that one is observing a process of synaptic modification dependent of the information locally and globally available to the cortical cells.

Experimental work of the last generation, beginning with the pathbreaking work of Hubel and Wiesel (*29,30*), has shown that there exist cells in visual cortex (areas 17, 18, and 19) of the adult cat that respond in a precise and highly tuned fashion to external patterns, in particular bars or edges of given orientation and moving in a given direction. Much further work (*31–34*)has been done, indicating that the number and response characteristics of such cortical cells can be modified. It has been observed in particular (*35–39*) that the relative number of cortical cells that are highly specific in their response to visual patterns varies in a very striking way with the visual experience of the animal during the critical period.

Most kittens first open their eyes at the end of the first week after birth. It is not easy to assess whether orientation selective cells exist at that time in striate cortex: Few cells are visually responsive, and the response's main characteristics are generally "sluggishness" and fatigability. However, it is quite generally agreed that as soon as cortical cells are reliably visually stimulated (e.g., at 2 wk), some are orientation selective, whatever the previous visual experience of the animal (*36,37,39,40*).

Orientation selectivity develops and extends to all visual cells in area 17 if the animal is reared and behaved freely in a normal visual environment (NR): Complete "specification" and normal binocularity (about 80% of responsive cells) are reached at about 6 wk of age (*39*). However, if the animal is reared in total darkness from birth to the age of 6 wk (DR), none or few orientation selective cells are then recorded (from 0 to 15%, depending on the authors and the classification criteria); however, the distribution of ocular dominance seems unaffected (*32,35–37,39,41*). In animals whose eyelids have been sutured at birth, and which are, thus, binocularly deprived of pattern vision (BD), a somewhat higher proportion (from 12 to 50%) of the visually excitable cells are still orientation selective at 6 wk (and even beyond 24 mo of age), and the proportion of binocular cells is less than normal (*36,42–45*).

Imbert and Buisseret (*35*) have classified cortical cells that respond to visual stimuli into three groups—aspecific, immature, and specific. Fregnac and Imbert (*39*), have measured the relative propertions of these groups, depending on the visual experience of the animal. The distribution of the different cell types in three age groups is shown in Fig. 11.

Examination of these results, which were obtained from the study of 1050 cells, shows that cells having some of the highly specific response properties of adult visual cortical-neurons, especially concerning orientation selectivity, are present in the earliest stages of postnatal development, independent of visual experience (*38,39*). However, visual experience between 17 and 70 d is critical in determining the evolution of these cells. Animals reared normally showed a marked increase in the number of specific cells compared with aspecific. (The period between 17 and 28 d is usually sufficient to reach the normal adult level of specificity.) The reverse is true for animals reared in the dark. A statistical analysis of this evolution, performed by Fregnac (*46*) showed clearly the striking dependence of the ratio of sharply to broadly tuned cells, depending on the experience of the animal.

In addition, as has been shown by Imbert and Buisseret (*35*), Buisseret and Imbert (*37*), and Buisseret et al. (*47*) as little as 6 h of nor-

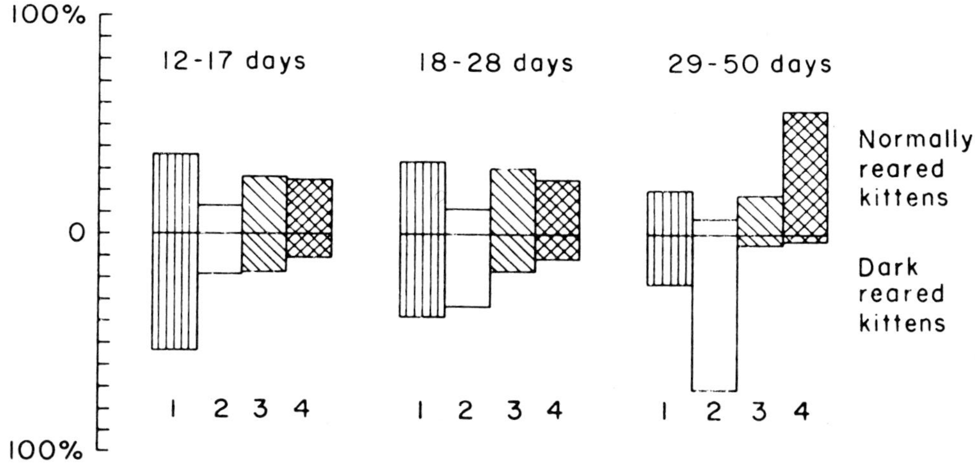

Fig. 11. Distribution of the different types of cells in three age groups in normal- (upper part) and dark-reared (lower part) kittens. The ordinate is normalized so that the heights are the percentages of cells in the various function groups. Type 1, nonactivable (vert. stripes); type 2, nonspecific (open); type 3, immature (diagonal stripes); and type 4, specific (cross-hatched). [From Fregnac and Imbert (1977, 1978.)]

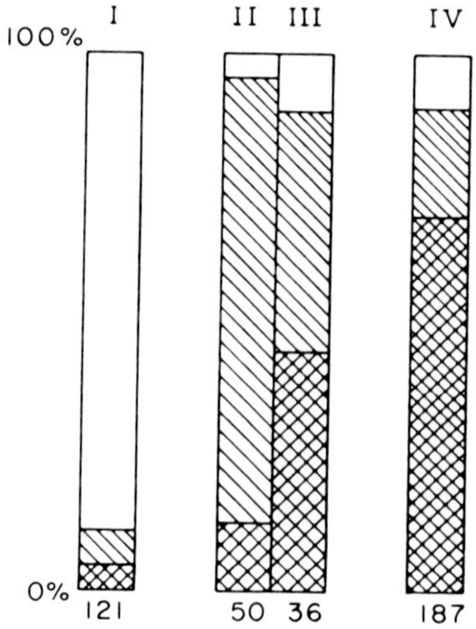

Fig. 12. Distribution in percentage of the three types of visual cortical units (area 17) recorded after 6 h of visual exposure for 6-wk-old dark-reared kittens. Columns: I, dark-reared kittens; IV, normally reared kittens; During 8 h of exposure, conditions were: in II and III, freely moving; in III, 12 h in the dark followed the 6 h of exposure. Numbers of visual cells recorded are given under each column. Specific cells (cross-hatched) are activated by oriented stimuli within a sharp angle (<60°). Immature cells (diagonal stripes) are activated by oriented stimuli within a larger angle (<150°). Nonspecific cells (open) are activated by nonoriented stimuli moving in any direction. A statistical analysis reveals no significant difference in the percentage of immature and specific units between columns III and IV. Therefore, it may be that for a 6-wk-old dark-reared kitten, a 6-h exposure to visual input followed by 12 h in the dark is sufficient to produce a distribution of cortical cells similar to that of normally reared animals. [From Buisseret et al. (1978)].

mal visual experience at about 42 d of age can alter in a striking fashion the ratio of specific or immature to aspecific cells (Fig. 12). That such a short visual experience can change the tuning ratios so markedly is clear evidence of the great plasticity of these cortical cells at the height of the critical period.

Of all visual deprivation paradigms, putting one eye in a competitive advantage over the other has probably the most striking consequences. If monocular lid-suture (MD) is performed during a "critical" period (ranging from about 3 to 12 wk), there is a rapid loss of binocularity to the profit of the open eye (*42,48*). At this stage, opening the closed eye and closing the experienced one may result in a complete reversal of ocular dominance (*49*). A disruption of binocularity that does not favor one of the eyes may be obtained, for example, by provoking an artificial strabismus (*50*) or by an alternating monocular occlusion, which gives both eyes an equal amount of visual stimulation (*51*). In what follows, we call this uncorrelated rearing (UR).

These results seem to us to provide direct evidence for the modifiability of the response of single cells in the cortex of a higher mammal according to its visual experience. Depending on whether patterned visual information is part of the animal's experience, the specificity of the response of cortical neurons varies widely. Specificity increases with normal patterned experience. Deprived of normal patterned information (dark-reared or lid-sutured at birth, for example) specificity decreases. Furthermore, even a short exposure to patterned information after 6 wk of dark-rearing can reverse the loss of specificity and produce an almost normal distribution of cells.

We do not claim, and it is not necessary, that all neurons in visual cortex be so modifiable. Nor is it necessary that modifiable neurons are especially important in producing the architecture of visual cortex. It is our hope that the general form of modifiability we require to construct distributed mappings manifests itself for at least some cells of visual cortex that are accessible to experiment. We thus make the conservative assumption that biological mechanisms, once established, will manifest themselves in more or less similar forms in different regions. If this is the case, modifiable individual neurons in visual cortex can provide evidence for such modification more generally.

Modification of Cortical Synapses: Global and Local Variables

To apply the general theoretical ideas of the previous section to visual cortex, we introduce the following notation. Consider a cortical cell, as shown in Fig. 13.

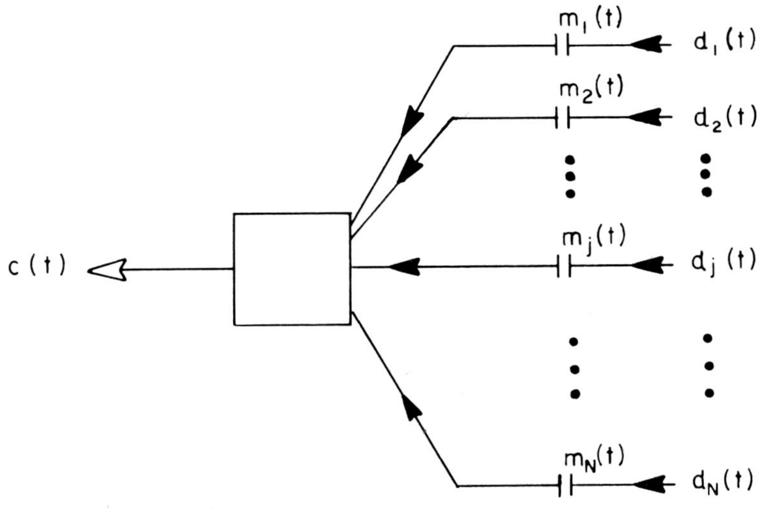

Fig. 13. A model neuron that processes the input *d(t)* according to the synaptic weights *m(t)* to yield the response *c(t)*.

Replacing Eqs. [1.1] and [1.2], we write:

$$c(t) = \sum_j m_j(t)d_j(t) \qquad [2.1]$$

where $c(t)$ is the output at time t, $m_j(t)$ is the efficacy of the jth synapse at time t, $d_j(t)$ is the jth component of the input at time t (the firing frequency of the jth presynaptic neuron), and Σ_j denotes summation over j, i.e., over all presynaptic neurons. We can then write:

$$\begin{aligned} m(t) &= [m_1(t), m_2(t), \ldots m_N(t)] \\ d(t) &= [d_1(t), d_2(t), \ldots d_N(t)] \qquad [2.2] \\ c(t) &= [m(t)] \cdot [d(t)] \end{aligned}$$

$m(t)$ and $d(t)$ are real-valued vectors of the same dimension, N, i.e., the number of ideal synapses onto the neuron, and $c(t)$ in the inner product (or "dot product") of $m(t)$ and $d(t)$. The vector of synaptic efficacies at time t, $m(t)$, is called the *state* of the neuron at time t. [Note that $c(t)$, as well as all components of $d(t)$, represent firing frequencies that are measured from the level of average spontaneous activity; thus, they might take negative as well as positive values; $m_j(t)$ is dimensionless.]

We can now formulate the question: What is the local principle of organization, by asking what is the change in time of $m_j(t)$ [the jth synapse onto the cortical cell, receiving inputs $d_j(t)$] and on what variables does this depend.

The various factors that influence synaptic modification may be divided broadly into two classes—those dependent on global, and those on local, information. Presumably, global information in the form of chemical or electrical signalling influences most (or all) modifiable junctions of a given type in a given area in the same way. Evidence for the existence of global factors that affect development has been found by Kasamatsu and Pettigrew (*52,53*), Singer, (*54,54a*), Buisseret et al. (*47*), Bear and Daniels (*55*), Bear et al. (*56*). On the other hand, local information available at each modifiable synapse can influence each junction in a different manner.

An early proposal as to how local information could affect synaptic modification was made by Hebb (*25*). His now classical principle was suggested as a possible neurophysiological basis for operant conditioning: "When an axon of cell A is near enough to excite a cell B and repeatedly or persistently takes part in firing it, some growth process or metabolic change takes place in one or both cells, such that A's efficiency, as one of the cells firing B, is increased." Thus, the increase of the synaptic strength connecting A to B is dependent upon the correlated firing of A and B. Such a correlation principle has inspired the work of many theoreticians on various topics related to learning, associative memory, pattern recognition organization of neural mappings (retinotopic projections), and development of selectivity of cortical neurons.

It is fairly clear that in order to actually use Hebb's principle, one must state conditions for synaptic decrease as specific as those for synaptic increase: If synapses are allowed only to increase, all synapses will eventually saturate; no information will be stored and no selectivity will develop (*see*, for example, refs. *57,58*). What is required is thus a complementary statement to Hebb's principle, giving conditions for synaptic decrease. Such a statement is given in what follows.

For a general form of synaptic modification, we write:

$$\mathring{m}_j = F(d_j \ldots m_j; d_k \ldots \bar{c} \ldots c; X \ldots Y \ldots Z) \qquad [2.3]$$

where the first set of variables, $d_j \ldots m_j$, are called local, the second set, $d_k \ldots c$, quasilocal, and the third set, $X \ldots Z$, global. Local variables, such as $d_j \ldots m_j$, are those directly at the synaptic site. Thus, any information would be directly available. Quasilocal variables are those such as $d_k \ldots c, \bar{c}$. These are physically connected to the synaptic site by the cell itself. However, in order that the information they contain be available, some means of internal cellular communication must be assumed. Note that we include among these such variables as $\bar{c}$ (the averaged activity of the cell over time). Global variables are called $X \ldots Y \ldots Z$.

In work done in the past few years, we have explored a form of a synaptic modification that can be written as follows: Referring to the jth synaptic junction:

$$\mathring{m}_j = \phi(c,\bar{c})d_j - \epsilon m_j \qquad [2.4]$$

Note, that as in passive modification, the output of a cell is determined by the input and the already existing synaptic strengths, as well as by noise-like fluctuations. The precise form of ϕ is not critical as long as it has certain general characteristics. Cooper et al. (*21*), showed that if the function ϕ goes through zero, then the sharpness of the tuning curve is altered by the visual experience of the animal, in agreement with what is observed. This modification might be called "Hebbian" when the output is above the modification threshold, θ_M, and "anti-Hebbian" when the output is below this threshold. The function, ϕ, is also assumed to have a dependence on global variables, not explicitly written. Cooper et al., thus assumed that the modifiability of a synaptic junction is dependent on events that occur at different parts of the same cell and on the rate at which the cell responds. They proved several theorems that show that with this form of passive modification there is an increase in the specificity of the response of a cortical cell to visual input (sharpening of its tuning curve) when that cell is exposed to stimuli that are the result of normal patterned visual experience and a loss of specificity when that cell is exposed to noise-like input, such as might be expected when an animal is dark-reared or raised with eyelids sutured. Specificity can be regained, however, with a return of input resulting from patterned vision.

In addition to this basic behavior, simulations and mathematical results on the asymptotic states of the neural network show some more subtle phenomena that depend upon values of system parameters. Of note are the rate of decay (forgetting per unit time), the strength of selective modification of synaptic junctions, the interaction of modifiable with nonmodifiable synapses, and the different statistical properties of noise factors.

The reason for the increase of selectivity is the crossover of the ϕ function from the negative to the positive region at the modification threshold, θ_M. This was recognized by Cooper et al. (*21*) to be associated with some property of the cell, possibly the average firing rate. This idea was enlarged and extended by Bienenstock et al. (*22*) and applied to a great variety of situations in visual cortex. The essential idea of Bienenstock et al. was to allow θ_m to vary nonlinearly with the average activity of the cell, $\bar{c}$. Doing this, they achieved a variety of desirable properties as well as a theoretical structure in excellent agreement with available experimental data. The crucial point in the choice of the function $\phi(c,\bar{c})$ is the determination of the threshold $\theta_M(t)$, i.e., the value of c at which $\phi(c,\bar{c})$ changes sign. A candidate for $\theta_M(t)$ is the average value of the postsynaptic firing rate, $\bar{c}(t)$. The time average is meant to be taken over a period T, preceding t much longer than the membrane time-constant τ, so that $\bar{c}(t)$ evolves on a much slower timescale than $c(t)$. This can usually be approximated by averaging over the distribution of inputs for a given state $m(t)$:

$$\bar{c}(t) = m(t) \cdot \bar{d} \qquad [2.5]$$

This results in an essential feature, the *instability of low selectivity points.* (This can be most easily seen at zero selectivity equilibrium points, where, with any perturbation, the state is driven away from this equilibrium, whatever the input.)

Therefore, if stable equilibrium points exist in the state space, they are of high selectivity. However, do such points exist at all? The answer is generally yes, provided that the state is *bounded from the origin and from infinity*. These conditions, instability of low-selectivity equilibria, as well as boundedness, are fulfilled by a single function, $\phi(c,\bar{c})$ if we define $\theta_M(t)$ to behave as a *nonlinear function* of $\bar{c}(t)$, for example, a power. The exponent should then be larger than 1. The final requirement on $\phi(c,\bar{c})$ thus reads:

$$\text{sign } \phi(c,\bar{c}) = \text{sign}\left[c - \left(\frac{\bar{c}}{c_0}\right)^p \bar{c}\right] \text{ for } c > 0 \qquad [2.6]$$

$$\phi(0,\bar{c}) = 0 \text{ for all } \bar{c}$$

where c_0 and p are two fixed positive constants. The threshold $\theta_M(\bar{c}) = (\bar{c}/c_0)^p\bar{c}$ thus serves two purposes: allowing threshold modifications when $\bar{c} \simeq c_0$, as well as driving the state from regions such that $\bar{c} \ll c_0$ or $\bar{c} \gg c_0$. The process of synaptic growth, starting near 0, to eventually end in a stable selective state, may be described as follows. Initially, $\bar{c} \ll c_0$, hence $\phi(c,\bar{c}) \gg 0$ for all inputs in the environment; the responses to all inputs grow. With this growth, $\bar{c}$ increases, thus increasing θ_M. Now, some inputs result in postsynaptic responses that exceed θ_M, whereas others—those whose direction is far away (close to orthogonal) from the favored inputs—give a response less than θ_M. The response to the former continues to grow and the response to the latter decays. This results in a form of *competition between incoming patterns* rather than competition between synapses. The response to unfavored patterns decays until it reaches 0, where it stabilizes for $\phi(0,\bar{c}) = 0$ for any $\bar{c}$. The response to favored patterns grows until the mean response $\bar{c}$ is high enough, and the state stabilizes. This occurs in spite of the fact that many complicated geometrical relationships may exist between different patterns, i.e., that they are not orthogonal since different patterns may and certainly do share common synapses.

Any function, ϕ (Fig. 14), that satisfies [2.6] will give these qualitative results. The precise form of this function (e.g., the numerical values of p and c_0) will affect the detailed behavior of the system, such as rate of convergence and height of the maximum for a selective cell, as well as a

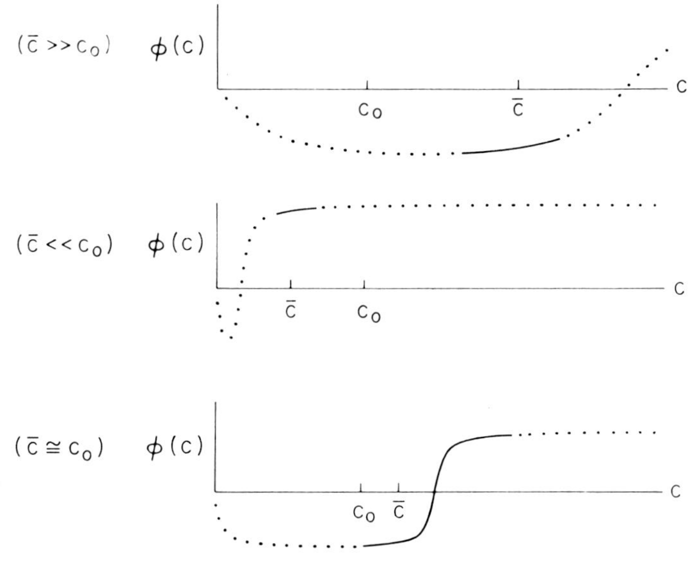

Fig. 14. A function-satisfying condition [2.6]. The three diagrams show the behavior of $\phi(c,\bar{c})$ as a function of c for three different constant values of $\bar{c}$. In each diagram, the *solid part* of the *curve* represents $\phi(c,\bar{c})$ in the vicinity of $\bar{c}$, which is the relevant part of this function.

variety of other more subtle effects. We are presently investigating the consequences of various detailed assumptions concerning the form of $\phi(c,\bar{c})$ and comparing these with existing and proposed experiments. In doing this we hope to arrive at a detailed understanding of the form of the function that controls synaptic modification.

We note also that with this form of modification, the control of θ_M by a global signal (in addition to $\bar{c}$) could produce the following results: If θ_M is set to be very large, the cell's response would diminish. This will result in a behavior that is like that described by Eric Kandel (*26*) in Aplysia habituation experiments. If θ_M is set very low, the cell will rapidly increase its response to a stimulus. This could be related to a type of sensitization in which the sensitizing signal has the effect of resetting θ_M to a very low level. For a variable, θ_M, as will be shown below, applied to visual cortex, one gets increasing and decreasing of selectivity, such as that seen in experimental results over the last generation. We thus have the possibility that a single mechanism of modification, functioning in slightly different ways, can account for a variety of experimental data in both invertebrates and vertebrates.

Some Mathematical Results

Selectivity

It is common usage to estimate the orientation selectivity of a single visual cortical neuron by measuring the half-width and half-height—or an equivalent quantity—of its orientation tuning curve. The selectivity is then measured with respect to a parameter of the stimulation, namely the orientation, that takes on values over an interval of 180°. In our work, various kinds of inputs are considered, e.g., formal inputs with a parameter taking values on a finite set of points, rather than a continuous interval. It will then be useful to have a convenient general index of selectivity, defined in all cases. We propose the following:

$$\mathrm{Sel}_d(\mathcal{N}) = 1 - \frac{\text{mean response of } \mathcal{N} \text{ with respect to } d}{\text{maximum response of } \mathcal{N} \text{ with respect to } d} \qquad [2.7]$$

With this definition, selectivity is estimated *with respect to*, or *in* an *environment for, the neuron*, that is, a random variable d that takes on values in the space of inputs to the neuron $\mathcal{N}$. The variable d represents a random input to the neuron and is characterized by its probability distribution that may be discrete or continuous. [During normal development, the input to the neuron (or neural network) is presumably distributed uniformly over all orientations. In abnormal rearing conditions (e.g., dark-reared) the input during development could be different from the input for measuring selectivity. How this should be translated in the formal space R^N will be discussed later.] This distribution defines an environment, mathematically, a random variable d. Selectivity is estimated (before, or *after* development) with respect to this same environment. Obviously, $\mathrm{Sel}_d(\mathcal{N})$always falls between 0 and 1, and the higher selectivity of $\mathcal{N}$in d, the closer $\mathrm{Sel}_d(\mathcal{N})$is to 1.

We analyze the behavior of [2.4] for $\epsilon = 0$. The behavior depends critically on the environment, that is, on the distribution of the stationary stochastic process, d. Two classes of distributions may be considered: (a) *Discrete distributions (K* possible inputs $d^1 \ldots d^K$); These are generally assumed to occur with the same probability $1/K$. The process d is then a jump process that randomly assumes new values at each time increment. The vector m is (roughly) a Markov process. (b) *Continuous distributions;* in the work of Bienenstock et al. (22), the only continuous distribution that is considered is a uniform distribution, d, over a closed one-parameter curve in the input space R^N. Although the principles underlying the convergence to selective states are intuitively fairly simple, mathematical analysis of the system is not entirely straightforward, even for the simplest d. Mathematical results, obtained only for certain discrete distributions, are of two types: (1) Equilibrium points are locally stable if and only if they are of highest available selectivity with respect to the given distribution of d; (2) given any initial value of m in the state space, the probability that $m(t)$ converges to one of the maximum selectivity fixed points as t goes to infinity is 1. Results of the second type are much stronger and require a tedious geometrical analysis. Results are stated here in a somewhat simplified form. For exact statements and proofs, the reader is referred to Bienenstock (59) or Bienenstock et al. (22). To illustrate, we study the simple case in which d takes on values on only two possible input vectors, d^1 and d^2, that occur with the same probability, and let $\epsilon = 0$ in [2.4]:

$$P[d = d^1] = P[d = d^2] = \tfrac{1}{2}$$

Whatever the real dimension, N, of the system, it reduces to two dimensions. (Any component of m outside the linear subspace spanned by d^1 and d^2 will eventually decay to 0 because of the uniform decay term.)

Analytic Results in Two Dimensions

It follows immediately from the definition that the maximum value of $\mathrm{Sel}_d(m)$ in the state space is ½. It is reached for states m that give null response when d^1 comes in (i.e., are orthogonal to d^1), but positive response for d^2, or vice versa. Minimum selectivity, namely 0, is obtained for states m, such that $m \cdot d^1 = m \cdot d^2$. Equilibrium states of both kinds indeed exist.

Lemma 1. *Let d^1 and d^2 be linearly independent and d satisfy* $P[d = d^1] = P[d = d^2] = \tfrac{1}{2}$. *Then, for any* ϕ *satisfying* [2.6], *the system* [2.4] *admits exactly four fixed points, m^0, m^1, m^2, and $m^{1,2}$, with:* $\mathrm{Sel}_d(m^0) = \mathrm{Sel}_d(m^{1,2}) = 0$, *and* $\mathrm{Sel}_d(m^1) = \mathrm{Sel}_d(m^2) = \tfrac{1}{2}$. *(Here the superscripts indicate which of the d's are not orthogonal to m [m^0 is the origin].* Thus, for instance, $m^1 \cdot d^1 > 0$, $m^1 \cdot d^2 = 0$.)

The behavior of the system depends on the geometry of the inputs, in the present case on $\cos(d^1,d^2)$. The crucial assumption that is needed here is that $\cos(d^1,d^2) > 0$. This is a reasonable assumption that is obviously satisfied if all components of the inputs are positive, as is assumed in some models (*60,61*). We may then state the following:

Theorem 1. *Assume that, in addition to the conditions of Lemma* 1, $\cos(d^1,d^2) > 0$. *Then,* m^0 *and* $m^{1,2}$ *are unstable;* m^1 *and* m^2 *are stable; and whatever its initial value, the state of the system converges almost surely (i.e., with probability* 1) *either to* m^1 *or* m^2.

Theorem 1 is the basic result in the two-dimensional setting: It characterizes evolution schemes based on *competition between patterns,* saying that the state eventually reaches maximal selectivity even when the two input vectors are very close to one another. Obviously, this requires that some of the synaptic strengths be negative since the neuron has linear integrative power. Inhibitory connections are thus necessary to obtain selectivity. Some selectivity is also realizable with no inhibitory connections—not even "intracortical" ones—if the integrative power is appropriately nonlinear. However, whatever the nonlinearity of the integrative power, Theorem 1 could not hold for evolution equations based on *competition between converging afferents.*

In Theorem 1, we have a discrete sensory environment that consists of exactly two different stimuli—a situation, although simple mathematically, not often encountered in nature. It may, however, very well correspond to a visual environment restricted to only horizontally and vertically oriented contours, present with equal probability. Theorem 1 then predicts that cortical cells will develop a selective response to one of the two orientations, with no preference for either (other than what may result from initial connectivity). Thus, on a large sample of cortical cells, one should expect as many cells tuned to the horizontal orientation as to the vertical one. So far, no assumption is made on intracortical circuitry. We discuss this later.

The proof of Theorem 1 is based on the existence of *trap regions* around each of the selective fixed points.

Theorem 2. *Under the same conditions as in Theorem* 1, *there exists around* $m^1(m^2)$ *a region,* $F^1(F^2)$, *such that once the state enters* $F^1(F^2)$, *it converges almost surely to* $m^1(m^2)$.

The meaning of Theorem 2 is the following: Once $m(t)$ has reached a certain selectivity, it cannot "switch" to another selective region. Applied to cortical cells in a patterned visual environment, this means that once they become sufficiently committed to certain orientations, they will remain committed to those orientations (provided that the visual environment does not change), becoming more selective as they stabilize to some maximal selectivity. Theorems 1 and 2 are illustrated in Fig. 15.

It is worth mentioning that when $\cos(d^1, d^2) < 0$, the situation is much more complicated: Trap regions do not necessarily exist, and peri-

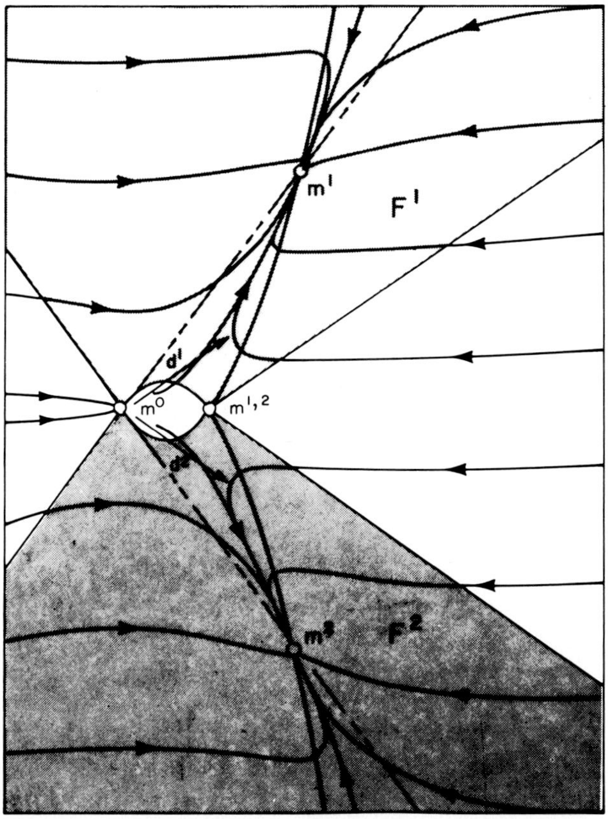

Fig. 15. The phase portrait of equation [2.4] subject to condition [2.6]. The diagram shows the trajectories of the state of the neuron starting from different initial points. The final state of the system (m^1 or m^2) is determined when the trajectory enters the corresponding "trap" (shaded) region (F^1 or F^2).

odic asymptotic behavior, i.e., limit cycles, may occur, bifurcating from the stable fixed points when $\cos(d^1, d^2)$ becomes too negative (*see* ref. *59*).

Higher Dimensions

We now turn to the case in which d takes on K values. The following is easily obtained:

Lemma 2. *Let* $d^1, d^2, \ldots d^K$ *be linearly independent and* d *satisfy* $P[d = d^1] = \cdots = P[d = d^K] = 1/K$. *Then, for any* ϕ *satisfying [2.6], [2.4] admits exactly* 2^K *fixed points with selectivities* $0, 1/K, 2/K, \ldots (K-1)/K$. *There are* K *fixed points* $m^1, \ldots m^K$ *of selectivity* $(K - 1)/K$.

Obviously, $(K - 1)/K$ is also the maximum possible selectivity, with respect to d. It means a positive response for one and only one of the inputs. The situation is now much more complicated than what it was with only two inputs: It is not obvious that whether in all cases assuming that all the cosines between inputs are positive is sufficient to yield stability of the maximum selectivity fixed points. However, we may state the following:

Theorem 3. *Assume, in addition to the conditions of Lemma 2, that $d^1, \ldots d^K$ are all mutually orthogonal or close to orthogonal. Then the K fixed points of maximum selectivity are stable, and, whatever its initial value, the state of the system converges almost surely to one of them.*

The proof of Theorem 3 also involves trap regions around the K maximally selective fixed points, and the analog of Theorem 2 is true here.

Although the general case has not yet been solved analytically, as will be seen later, computer simulations suggest that for a fairly broad range of environments, if $d^i \cdot d^j > 0$, even if $d^1, \ldots d^K$ are far from being mutually orthogonal, the K fixed points of maximum selectivity are stable.

Simulations suggest further that even if the $d^1, \ldots d^K$ are *not* linearly independent and are very far from being mutually orthogonal, the asymptotic selectivity is close to its maximum value with respect to d.

Analytic results in two dimensions and computer simulations in higher dimensions indicate that the form of synaptic modification described here leads, in general, to the evolution of maximum selectivity, with respect to the environment. We are trying to extend the linear analysis of stability performed in two dimensions to higher dimensions. Stability analysis has been attempted on systems of K dimensions for a general linearly independent environment. The same arguments that lead to statements of stability in two dimensions apply in this general case. However, the technical difficulty increases. The problem may be stated in terms of a Kth order eigenvalue equation. Local stability for an $s = 1$ fixed point will be assured if the eigenvalues of the matrix of coefficients of the K differential equations are negative. Similarly, the instability of points for which $s > 1$ would be characterized by the presence of positive eigenvalues. Since this matrix of coefficients exhibits some symmetry, there is hope that the problem could be solved analytically (for a reasonable size K, the system of equations could be solved numerically for special cases). This kind of analytic statement would confirm that the states of high selectivity observed in computer simulations are indeed stable asymptotic states.

The Monocular Problem: A Simple Circular Environment

We now apply this theory to the problem of orientation selectivity and binocular interaction in primary visual cortex. The ordinary development of these properties in mammals depends to a large extent on normal functioning of the visual system (i.e., normal visual experience) during the first few weeks or months of postnatal life. This has been demonstrated many times by various experiments, based mainly on the paradigm of rearing the animal in a restricted sensory environment. We show that the theory described above can account for both normal development and development in restricted visual environments.

Consider first a classical text environment used to construct the tuning curve of cortical neurons. This environment consists of an elongated light bar successively presented or moved in all orientations—in a random sequence—in the neuron's receptive field. Thus, all the parameters of the stimulus are constant except one, the orientation, which is uniformly distributed on a circularly symmetric closed path. We assume that the retinocortical pathways map this family of stimuli to the cortical neuron's space of inputs in such a way as to preserve the circular symmetry (as defined below). Thus, the typical theoretical environment that will be used for constructing the neuron's tuning curve is a random variable, d, uniformly distributed on a circularly symmetric, closed, one-parameter family of points in the space R^N. The parameter coding orientation in the receptive field is, in principle, continuous. However, for the purpose of numerical simulations, the distribution is made discrete. Thus, d takes on values on the points $d_1, \ldots d^K$.

The requirement of circular symmetry is expressed mathematically as follows: The matrix of inner products of the vectors $d^1, \ldots d^K$ is circular (i.e., each row is obtained from its nearest upper neighbor by shifting it one column to the right) and the rows of the matrix are unimodal. A random variable, d, uniformly distributed on such a set of points will be, hereafter, called a *circular environment*. Such a d may be roughly characterized by three parameters: N, K, and a measure of the mutual geometrical closeness of the d's, for instance the minimum value of $\cos(d^1, d^i)$ over the environment.

We are now faced with the difficult problem of specifying the stationary stochastic process that represents the time-sequence of inputs to the neuron during development. To begin, we simplify the problem by giving the stochastic process exactly the same distribution as the circular d defined above. In doing so, we assume that development of orientation selectivity is to a large extent independent of other parameters of the stimulus, e.g., contrast, shape, position in the receptive field, retinal disparity for binocular neurons, and so on. The elementary stimulus for a cortical neuron is a rectilinear contrast edge or bar. Any additional pattern present at the same time in the receptive field is regarded as random noise. [A discussion of this point is given by Cooper et al. (*21*).]

Simulations show the following behavior:

(1) The stage converges rapidly to a fixed point, or *attractor*.
(2) Various such attractors exist. For a given d and ϕ they all have the same selectivity, which is close to its maximum value in d.
(3) The aymptotic tuning curve is always unimodal. One may thus talk of the preferred orientation of an attractor.
(4) There exists an attractor in each possible orientation.
(5) If there is no initial preference, all orientations have equal probability of attracting the state. (Which one will become favored depends on the exact sequence of inputs.) This does

not hold for environments that are not perfectly circular, at least for a single neuron system, as the one studied here.

The system thus behaves exactly as expected from the results of the preceding section.

The Binocular Problem: A More Complex Environment

We now consider a binocularly driven cell. The firing rate of the neuron at time t is now given by

$$c(t) = m_l(t) \cdot d_l(t) + m_r(t) \cdot d_r(t) \qquad [2.8]$$

with evolution schemes for "left" and "right" states, m_l and m_r, straightforward generalizations of [2.4]. We have partitioned the input vector space into a left space and a right space; Hence, m goes to (m_l,m_r) and d becomes (d_l,d_r). Since d_l and d_r can be independent, the topology of the environment is potentially more complex.

Various possibilities exist for the input (d_l,d_r): One may wish to consider normal rearing (both d_l and d_r circular and presumably highly correlated) monocular deprivation, binocular deprivation, and so on. The vector (d_l,d_r) is a stationary stochastic process, whose distribution is one of the following, depending on the experimental situation one wishes to reproduce:

Normal Rearing (NR) $d_l(t) = d_r(t)$ for all t, and d_l is circular. (Noise terms that may be added to the inputs may or may not be stochastically independent.)

Uncorrelated Rearing (UR) d_l and d_r are independent identically distributed (i.i.d.): they have the same circular distribution, but no statistical relationship exists between them.

Binocular Deprivation (BD) The $2N$ components of (d_l,d_r) are i.i.d.: d_l and d_r are uncorrelated noise terms.

Monocular Deprivation (MD) d_l is circular, d_r is a noise term: $d_r = n$.

In the NR case, the inputs from the two eyes to a binocular cell are probably well correlated. We therefore assume that they are equal, which is mathematically equivalent. The BD distribution represents dark discharge.

Uncorrelated or strabismic rearing (UR) involves presenting fully two independent circular environments (a "toroidal" environment). The final state can be either monocular and specific or binocular and specific, with no correlation between the patterns perferred by the two eyes.

The results of binocular deprivation or (corrrelated) NR are just those of the monocular case. We assume that binocular stimuli presented during NR are exactly correlated so that each pattern incident to the left-eye synapses is consistently accompanied by a corresponding pattern to the right-eye synapses. Since the left and right components of each pair are identical, the cell tunes to the same pattern in each eye. Binocularly deprived input environments consist of stimulus components uniformly distributed over some range with zero mean. In this case (BD), the average response of the cell is null, and so ϕ is always nonnegative, resulting in random fluctuations of the synaptic state.

The development of a neuron receiving patterned input from only one eye (and uniform noise from the other) is somewhat surprising. The response curve goes to maximum selectivity with respect to the open eye, but, consistent with observation, the response to the closed eye does *not* fluctuate randomly. Rather, the neuron becomes nonresponsive to inputs to the deprived eye. Asymptotic convergence to this state is assured *regardless of the initial state*. The theoretical implications for the reverse suture (RS) paradigm are straightforward: A monocularly deprived neuron, having reached a monocular selective state, is driven to another monocular selective state preferring the *newly opened eye*, upon reversal of suture.

This behavior relies upon some activity, albeit purely random, to be present in the afferents from the closed eye. Such noise may be a result of diffuse light through the eyelid or spontaneous firing of LGN and/or retinal neurons. As a neuron becomes selective, with respect to the open eye, patterns that are preferred give a response near threshold, whereas the other patterns give a much lower response. In either case, ϕ is near zero. Noise accompanying a preferred pattern drives the neuron from the modification threshold, so the deprived synapses grow stronger. However, the opposite effect weakens the synapses when nonpreferred patterns are presented. A mathematical demonstration of this argument, given in Appendix C of Bienenstock et al. (22), is presented in the section entitled Comparison of Theory with Classical Experimental Results.

Comparison of Theory with Classical Experimental Results

The simulated behavior of neurons in visual cortex with binocular connectivity is illustrated in Fig. 16. The seemingly inconsistent experimental results (MD vs BD) are faithfully reproduced by computer simulation. Each of these paradigms was tested in both deterministic and stochastic simulation algorithms over several pattern sets. The model withstood considerable noisy input; indeed, successful simulation of some paradigms (RS in particular) *required* that a noise-like component accompany the "pure" inputs.

Simulations of the behavior of the system in these different environments give the following:

(1) NP; all asymptotic states are selective, binocular, and have matching preferred orientations for stimulation through each eye.
(2) BD; the motion of the state (m_l, m_r) resembles a random walk. (The small exponential decay term is necessary here in order to prevent large fluctuations.) The two tuning curves there-

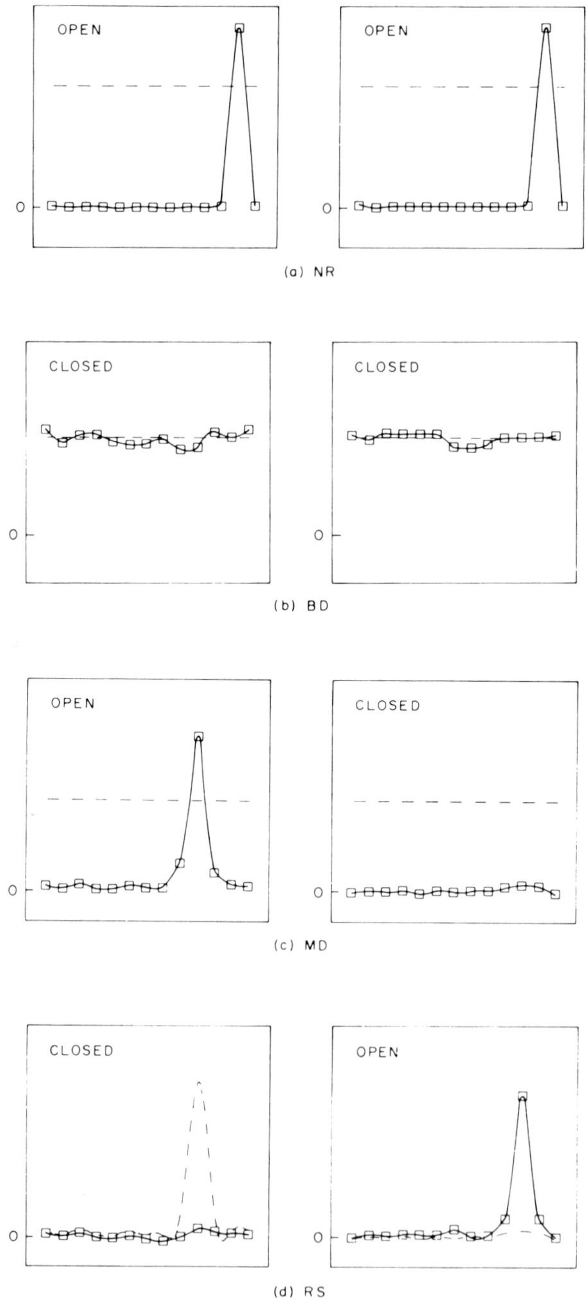

Fig. 16. Computer simulations of various rearing conditions. Initial (dashed) and final (solid) responses to the two eyes are shown separately (left/right).

fore undergo random fluctuations that are essentially determined by the second-order statistics of the input d. As can be seen from the figure, these fluctuations may sometimes result in a weak orientation preference or unbalanced ocular dominance. However, the system never stays in such states very long; its average state, on the long run, is perfectly binocular and nonoriented. Moreover, whatever the second-order statistics of d and the circular environment in which tuning curves are assessed, a regular unimodal orientation tuning curve is rarely observed, and selectivity never exceeds 0.6. We may thus conclude that orientation selectivity, as observed in the NR case (both experimental and theoretical), cannot be obtained from purely random synaptic weights. It is worth mentioning here that prolonged dark-rearing has been reported to increase response variability (*41*); a similar observation was made by Fregnac and Bienenstock (*62*).

(3) MD and RS; the only stable equilibrium points are monocular and selective. The system converges to such states whatever the initial conditions. In particular, this accounts for reverse suture experiments (*49,49a*).

(4) UP; in contrast to NR, monocular as well as binocular equilibria exist. The asymptotic state generally observed with $m_l(0) = m_r(0) = 0$ is monocular. (This should be attributed to the mismatched inputs from the two eyes, as is done by most authors.) Asymptotic states are selective, and when they are binocular, preferred orientations through each eye do not necessarily coincide. It should be mentioned here that Blakemore and Van Sluyters (*49*) report that after a period of alternating monocular occlusion, the remaining binocular cells may differ in their preferred orientations for stimulation through each eye.

Selectivity and Ocular Dominance

As an example of the kind of new and subtle effects that are contained in this theory, we consider in detail the sequence in which ocular dominance and selectivity develop in the monocularly deprived environment.

According to [2.8], the firing rate of a binocularly driven neuron at time t is given by

$$c(t) = m_l(t) \cdot d_l(t) \cdot d_r(t)$$

In a situation corresponding to monocular deprivation—patterned information to one eye (right), noise to the other (left)—we can write for the environment

$$d = (d_r, n)$$

and for the set of synaptic weights

$$m = (m_r m_l)$$

Where m_r and m_l are the synaptic weights from the right and left eyes respectively.

In this situation, m_r goes to one of its selective fixed points, as in the monocular case. The only fixed point for m_l in the noise-like environment is zero; but this is unstable in the monocular case. It is instructive to follow the behavior of m_l in this binocular case.

Let (x_r, x_l) be a small perturbation from equilibrium. The motion at point $(m_r^* + x_r, x_l)$ is given by:

$$\mathring{x}_r = \phi(m_r^* \cdot d_r + x_r \cdot d_r + x_l \cdot n,\ m_r^* \cdot \bar{d}_r + x_r \cdot \bar{d}_r)d_r \qquad [2.9r]$$

$$\mathring{x}_l = \phi(m_r^* \cdot d_r + x_r \cdot d_r + x_l \cdot n,\ m_r^* \cdot \bar{d}_r + x_r \cdot \bar{d}_r)n \qquad [2.9l]$$

where we assume that the noise has zero mean.

We analyze separately, somewhat informally, the behavior of the two equations. The stability of [2.9*r*] is immediate from the stability of the selective state, m_r^* in the circular environment d_r. To analyze [2.9*l*], we divide the range of the right eye input, d_r, into three classes: (i) d_r Is such that $m_r^* \bar{d}_r$ is either far above threshold, θ_M, and therefore, $\phi(m_r^* d_r, m_r^* d_r) > 0$, or far below threshold, θ_M (but still positive) and therefore $\phi(m_r^* d_r, m_r^* d_r) < 0$; (ii) d_r Is such that $m_r d_r$ is near threshold, θ_M, and therefore, $\phi(m_r^* d_r, m_r^* d_r) \simeq 0$; (iii) d_r Is such that $m_r^* d_r \simeq 0$ and again $\phi(m_r^* d_r, m_r^* \bar{d}_r) \simeq 0$.

For the first class of inputs, the sign of ϕ is determined by d_r alone, hence [2.9*l*] is the equation of a random walk. To investigate the behavior of [2.9*l*] in the two other cases, we neglect the term x_r and linearize ϕ around the relevant one of its two zeros. It is easy to see that case (ii) yields:

$$\mathring{x}_1 \simeq \epsilon_l(x_l \cdot n)n \qquad [2.10]$$

whereas in case (iii), one obtains:

$$\mathring{x}_1 \simeq -\epsilon_2(x_l \cdot n)n \qquad [2.11]$$

where ϵ_1 and ϵ_2 are *positive* constants, measuring, respectively, the absolute value of the slope of ϕ at the modification threshold and at zero.

Since n is a noise-like term, its distribution is presumably symmetric, with respect to x_l so that averaging [2.10] and [2.11] yields, respectively:

$$\mathring{x}_l \simeq \epsilon_1 \bar{n}_0^2 x_l \qquad [2.12]$$

$$\mathring{x}_l \simeq -\epsilon_2 \bar{n}_0^2 x_l \qquad [2.13]$$

where $\bar{n}_0^2$ is the average squared magnitude of the noise input to a single synaptic junction from the closed eye.

We thus see that input vectors from the first class move x_l randomly, and inputs from the second class drive it away from 0, whereas inputs from the third drive it toward 0. In the case in which the range of d_r is a set of K linearly independent vectors and m_r^* is of maximum selectivity, $(K - 1)/K$, case (i) does not occur at all. (The random contribution occurs only before the synaptic strengths from the open eye have settled to one of their fixed points.) Case (ii) occurs only for one input, say d_r^1, with $m_r^* \cdot d_r^1$ exactly equal to threshold, θ_M, and (iii) occurs for the other $K - 1$ vectors, which are all orthogonal to m_r^*. In the general case (d_r, any circular environment), the more selective m_r^*, with respect to d_r, the higher the proportion of inputs belonging to class (iii), the class that yields [2.13] i.e., that brings x_l back to 0.

The stability of the global system still depends on the ratio of the quantities ϵ_1 and ϵ_2 as well as on the statistics of the noise term n (e.g., its mean square norm). We may, however, formulate two general conclusions. First, under reasonable assumptions (ϵ_1 of the order of ϵ_2 and the mean square norm of n of the same order as that of d_r), $x_l = 0$ is stable on the average for a selective m_r^*. Second, the residual fluctuation of x_l around 0, essentially because of inputs d_r in classes (i) and (ii), is smaller for highly selective m_r^*'s than it is for mildly selective ones.

Thus, one should expect that in a monocularly deprived environment, nonselective neurons tend to remain binocularly driven. In addition, since it is the nonpreferred inputs from the open eye, accompanied by noise from the closed eye (case three), that drive the response to the closed eye to zero, if inputs to the open eye were restricted to preferred inputs (case ii), even a selective cell would remain less monocular.

To better confront these ideas with experiment (22), the single neuron must be placed in a network with the anatomical features of visual cortex, a network in which inhibitory and excitatory cells receive input from LGN and from each other. This has been done (63). Their conclusions are similar to those above, with explicit further statements concerning the independent effects of excitatory and inhibitory neurons on selectivity and ocular dominance. For example, shutting off inhibitory cells lessens selectivity and alters ocular dominance, giving "masked synapse" effects.

Quantitative tests of progressions, such as those shown in Fig. 17, are in progress in our laboratory. We hope that such experiments can provide detailed comparisons with theory and provide us with a sensitive tool for determining synaptic modification among various classes of neurons—a possible entry to the process by which the nervous system organizes itself.

ACKNOWLEDGMENT

The work on which this article is based was supported in part by the US Office of Naval Research, under contract #N000-14-81-K-0136.

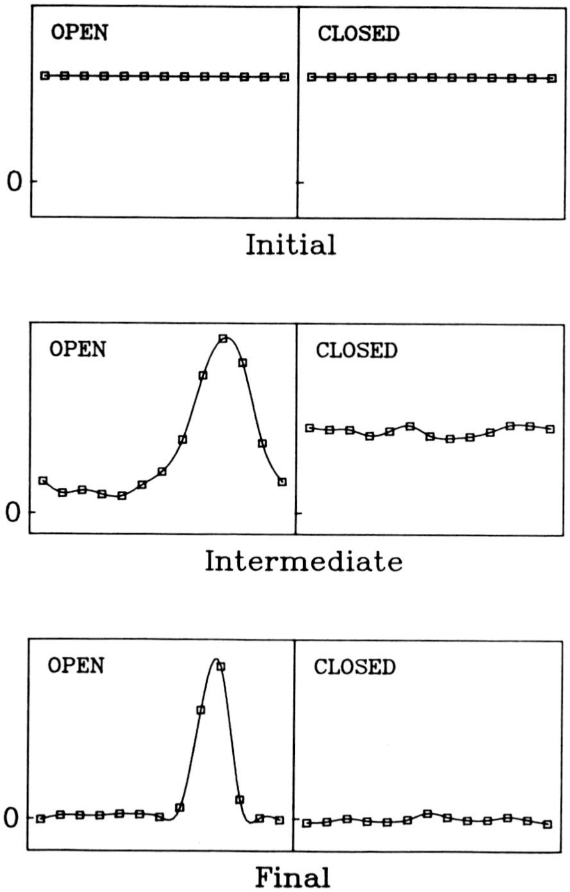

Fig. 17. Progression of development of selectivity and ocular dominance. Note that selectivity develops for the open eye *before* the response to the closed eye is driven to zero.

REFERENCES

1. Amari, S. I. (1974) *Adv. Biophys.* **6,** 75.
2. Wilson, H. R., and Cowan, J. D. (1973), *Kybernetik* **13,** 55.
3. Edelman, G. M. (1981), Group Selection as the Basis for Higher Brain Function, in *The Organization of the Cerebral Cortex* (Schmitt, F. O., Worden, F. G., Adelman, G., and Dennis, S. G., eds.) MIT, Cambridge, MA, pp, 535–563.
4. Edelman, G. M., and Reeke, Jr., G. N. (1982), *PNAS Biol.* **79,** 2091.
5. Grossberg, S. (1982), *Studies of Mind and Brain,* Reidel Dordrecht, Holland, Boston, USA, London, England.
6. Cooper, L. N (1973), A Possible Organization of Animal Memory and Learning, in *Proc. Nobel Symposium on Collective Properties of Physical Systems* (Lindquist, B., and Lindquist, S., eds.), Academic, New York, NY, pp. 252–264.

7. Anderson, J. A., and Cooper, L. N (1978), *Biological Organization of Memory,* Encyclopaedia Universalis France S. A. (Pluriscience), pp. 168–175.
8. Block, H. D., *Rev. Mod. Phys.* (1962), **34,** 123.
9. Block, H. D., Knight, Jr., B., and Rosenblatt, F. (1962), *Rev. Mod. Phys.* **34,** 135.
10. Minsky, M., and Papert, S. (1969), *An Introduction to Computational Geometry,* MIT, Cambridge, MA.
11. Longuet-Higgins, H. C. (1968), *Nature* **217,** 104.
12. Longuet-Higgins, H. C. (1968), *Proc. R. Soc. Lond. (Biol.)* **171,** 327.
13. Anderson, J. A. (1970), *Math. Biosci.* **8,** 137.
14. Anderson, J. A. (1972), *Math. Biosci.* **14,** 197.
15. Kohonen, T. (1972), Correlation Matrix Memories, IEEE Trans. on Computers, C. **21,** 353.
16. Kohonen, T. (1977), Associative Memory: A System Theoretic Approach, Springer, Berlin.
17. McIlwain, J. T. (1976), Large Receptive Fields and Spatial Transformations in the Visual System, in *International Review of Physiology, Neurophysiology* II., vol. 10, (Porter, R., ed.), University Park, Baltimore, MD.
18. Kandel. E. R. (1976), *Cellular Basis of Behavior: An Introduction to Behavioral Neurobiology,* Freeman, San Francisco, CA.
19. Levy, W. B., and Steward, O. (1979), *Brain Res.* **175,** 233.
20. Nass, M. M., and Cooper, L. N (1975), *Biol. Cybern.* **19,** 1.
21. Cooper, L. N , Liberman, F., and Oja, E. (1979), *Biol. Cyber.* **33,** 9.
22. Bienenstock, E. L., Cooper, L. N , and Munro, P. W. (1982), *J. Neurosci.* **2,** 32.
23. Creutzfeldt, O. D., Kuhnt, U., and Benevento, L. A. (1974), *Exp. Brain Res.* **21,** 251.
24. Sillito, A. M. (1975), *J. Physiol.* **250,** 304.
25. Hebb, D. O. (1949), *The Organization of Behavior,* Wiley, New York, NY, P. 62..
26. Kandel, E. R., and Tanc, L. (1965), *J. Physiol.* **181,** 28.
27. Castellucci, V. F., and Kandel, E. R. (1974), *Proc. Natl. Acad. Sci.* **71,** 5004.
28. Hopfield, J. J. (1982), *PNAS* **79,** 2554.
29. Hubel, D. H., and Wiesel, T. N. (1959), *J. Physiol.* (Lond.) **148,** 574.
30. Hubel, D. H., and Wiesel, T. N. (1962), *J. Physiol.* (Lond.) **160,** 106.
31. Blakemore, C., and Cooper, G. F. (1970), *Nature* **228,** 477.
32. Blakemore, C., and Mitchell, D. E. (1973), *Nature* **241,** 467.
33. Hirsch, H. V. B., and Spinelli, D. N. (1971), *Exp. Brain Res.* **12,** 509.
34. Pettigrew, J. D., and Ferman, R. D. (1973), *Science* **182,** 599.
35. Imbert, M., and Buisseret, Y. (1975), *Exp. Brain Res.* **22,** 2.
36. Blakemore, C., and Van Sluyters, R. C. (1975), *J. Physiol.* **248,** 663.
37. Buisseret, P., and Imbert, M. (1976), *J. Physiol.* **255,** 511.
38. Fregnac, Y., and Imbert, M. (1977), *J. Physiol.* (Paris) **6,** 73.
39. Fregnac, Y., and Imbert, M. (1978), *J. Physiol.* (Lond.) **278,** 27.
40. Hubel, D. H., and Wiesel, T. N. (1963), *J. Neurophysiol.* **26,** 994.
41. Leventhal, A. G., and Hirsch, H. V. B. (1980), *J. Neurophysiol.* **43,** 1111.
42. Wiesel, T. N., and Hubel, D. H. (1965), *J. Neurophysiol.* **28,** 1029.
43. Kratz, K. E., and Spear, P. D. (1976), *J. Comp. Neurol.* **170,** 141.
44. Leventhal, A. G., and Hirsch, H. V. B. (1977), *Proc. Natl. Acad. Sci. USA* **74,** 1272.
45. Watkins, D. W., Wilson, J. R., and Sherman, S. M. (1978), *J. Neurophysiol.* **41,** 322.

46. Fregnac, Y. (1978), Cinetique de Development du Cortex Visual Primarie Chez le Chat. Effets de la Privation Visuelle Binoculaire et Modele de Maturation de la Selective a l'Orientation, Doctoral Thesis, University Rene Descartes, France.
47. Buisseret, P., Gary-Bobo, E., and Imbert, M. (1978), *Nature* **272,** 816.
48. Wiesel, T. N., and Hubel, D. H. (1963), *J. Neurophysiol.* **26,** 1003.
49. Blakemore, C., and Van Sluyters, R. C. (1974), *J. Physiol.* (Lond.) **237,** 195.
49a. Movshon, J. A. (1976), *J. Physiol.* (Lond.) **261,** 125.
50. Hubel, D. H., and Wiesel, T. N. (1965), *J. Neurophysiol.* **28,** 1041.
51. Blakemore, C. (1976), *J. Physiol.* **261,** 423.
52. Kasamatsu, T., and Pettigrew, J. D. (1976), *Science* **194,** 206.
53. Kasamatsu, T., and Pettigrew, J. D. (1979), *J. Comp. Neurol.* **185,** 139.
54. Singer, W. (1979), Central-Core Control of Visual-Cortex Functions, in *The Neurosciences; Fourth Study Program,* (Schmit, F., and Worden, R., eds.), MIT, Cambridge, MA, pp. 1093–1109.
54a. Singer, W. (1980), *Verh. Dtsch. Zool. Ges.* **268.**
55. Bear, M. F., and Daniels, J. D. (1983), *J. Neurosci.* **3,** 407.
56. Bear, M. F., Paradiso, M. A., Schwartz, M., Nelson, S. B., Carnes, K. M., and Daniels, J. (1983), *Nature* **302,** 245.
57. Sejnowski, T. J. (1977), *J. Math. Biol.* **4,** 303.
58. Sejnowski, T. J. (1977), *J. Theor. Biol.* **69,** 385.
59. Bienenstock, E. (1980), *A Theory of Development of Neuronal Selectivity,* Ph.D. Thesis, Division of Applied Mathematics and Center for Neural Science, Cooper, L. N., Thesis Supervisor.
60. Von der Malsburg, C. (1973), *Kybernetic* **14,** 85.
61. Perez, R., Glass, L., and Shaler, R. J. (1975), *J. Math. Biol.* **1,** 275.
61a. Cooper, L. N., Munro, P. W., and Scofield, C. L. (1985), *Synaptic Modification, Neuron Selectivity, and Nervous System Organization* (Levy, W., Lehmkuhle, S., and Anderson, J. A., ed.) Erlbaum, Hillsdale, NJ.
62. Fregnac, Y., and Bienenstock, E. (1981), Specific Functional Modifications of Individual Cortical Neurons, Triggered by Vision and Passive Eye Movement in Immobilized Kittens, in *Pathophysiology of the Visual System: Documentta Ophthalmolgica* (Maffei, L., ed.) vol. 30, pp. 100–108 W. Junk, the Hague.
63. Scofield, C. L., and Cooper, L. N. (1985), Development and Properties of Neural Networks, in *Contemporary Physics,* vol. 26, **2,** pp. 125–145.

A Physicist's View of Vision

DONALD GLASER
Physics Department, University of California, Berkeley, CA

Visual mechanisms of animals are extremely complex, but, as a physicist, I shall treat them perhaps a bit simplistically. I shall start with two central questions: What is the function of our visual system, and what are the mechanisms by which it works? A little bit of natural history will be useful to begin to answer the first question. In the jungles of South America there is a strange species of jumping spider. The female has on her posterior a red mark in the form of a "V." The retina of the male is also in the form of a "V." When the male, wandering through the forest, notices an object of the right dimensions, and recognizes a brilliant red V, he knows he has found a mate. If the V is lacking, he knows it is only an opportunity for a good meal. His visual system is specially designed to fit his needs.

Frogs have in their retinas a mechanism that is very sensitive to small black dots in motion. They do not react in the presence of black dots of greater dimension, nor of miniscule black dots that are stationary; they respond only to black dots in motion. This phenomenon can be easily verified. When one inserts an electrode into the frog's nervous system at a precise point located behind the retina, one can record strong electrical discharges only when little black dots in motion are introduced into the frog's field of vision. Thus, the frog is well equipped for catching flies.

Neither the spider nor the frog has a realistic three-dimensional representation of the world in its brain. Nor does it need such knowledge. By these and many other well-known examples, we are led to the following questions: What fraction of information from the world around us do we ourselves take in to use in our own visual system? And what is our visual system designed to do with this information? Most of us believe that we have an accurate three-dimensional representation of the real world so we know where our bodies are with respect to other objects and have a concept of the world that is in agreement with physical reality. Does our visual system do that, and by what strategy does it operate?

One of the first steps in answering these questions is to measure what the visual system actually does by testing the vision of many people. That is the occupation of psychophysicists, who measure human abilities to detect motion, depth, color, and other aspects of real scenes. Another approach is to measure the anatomy and physiology of the visual system and other parts of the nervous system with which it interacts. The final step is to seek a theoretical explanation that explains the actual performance of our visual system in terms of connections among neurons and their firing patterns in response to various stimuli. But how can we build this bridge between the biology and the psychology of the visual system?

Perhaps it would be useful to recount a sobering conversation with an expert in the design of very-large-scale integrated circuits for use in computers and various automated machines. Nowadays circuit chips may contain thousands or hundreds of thousands of components. This man said that if someone gave him an integrated chip—even a modest one with perhaps a thousand components—he would not be able to tell its purpose or how it works on the basis of any kind of input or output measurements he could make, nor would an "anatomical dissection" of the chip be of any help. But if he were told ahead of time that the chip came from a hand calculator or a microwave oven or from the ignition control system of a car, then he could readily find out everything about its construction and function. Of course our visual system is enormously more complicated than such a chip, and, therefore, it is especially important to emphasize that we have no chance of understanding it unless we have hypotheses for what its functions are and some overall theory about how it works.

In his recent book, "*Vision*," David Marr proposes that studies of the visual system must begin with a "computational theory" that specifies what information is available in the image, how the information can be described for use by the visual system, and what result the visual system has to produce from that information. In other words, what does the image have to tell us so that we can detect motion, perceive depth, or name colors correctly?

A second step in vision is what Marr calls the algorithm. It is a rule by which computations are carried out mathematically. Usually, there are a number of different, equally good, ways to get the right answer using modern computers. The difficult question is, which of the possible rules does the visual system really use.

The last group of questions are, how does the brain actually carry out the calculation? Where in the brain is the primary information entered, where are the computations done, and where is the answer stored?

Although I have listed the three steps needed for understanding the visual system, beginning with theory, then algorithm, and, finally, neuronal hardware, I do not mean to imply that that is how the research process will always proceed. Sometimes clues gained from electrophysiology and neuroanatomy may give ideas about the purpose of a particular structure. Or maybe speculations about purpose will lead us to look for certain structures and certain patterns of neuronal firings. As the work proceeds, clues and ideas and facts from all three parts of the scientific procedure will be used in any combination that seems fruitful.

In the hands of a skilled artist, a few strokes of the pen can give a vivid image of a running horse, recall a familiar face, or describe a perching bird. What machinery in the brain could possibly use that apparently tiny amount of information to generate such elaborate and emotionally important results? The human eye contains a lens and, on its back surface, a retina containing a hundred million receptor elements. There are rods that are particularly useful at low light levels and cones that are especially sensitive to red, green, or blue. The retina also contains four other kinds of cells connected in a complicated network that add, subtract, multiply, and, in many ways, manipulate the primary image seen by the rods and cones, to finally send information to the brain via the optic nerve, which contains only a million fibers. The message contains information about movement, contour, contrast, color, depth, and other aspects of the scene. Most of these cells are analog elements that express their response in a smoothly varying voltage. Although the retina contains only five main types of cells, they are connected to each other in such a way that fairly complicated calculations can be carried out. Indeed, the retina is often thought of as a small independent brain unto itself. It is important to notice that the cells in the retina generate continuously varying voltages in response to light and do not produce chains of discrete spikes like those observed in many nerve fibers. The retina is then a sort of analog computer in which light-induced voltages are multiplied, added, subtracted, differentiated, and combined in complex ways. The final result of these calculations is to produce in the output cells, called retinal ganglion cells, a chain of spikes that propagate along the optic nerve to the brain. These signals do not simply report light intensities received at the eye, but have already calculated enough to be able to report to the brain the existence of colors, edges, motions, and other abstract characteristics of the images. In frogs, spiders, and many other lower animals, these messages already contain information about mates, food, and danger, which need very little further calculation for the animal to take some important action. By carrying out all of these calculations, the retina is able to reduce the information it receives on its hundred million rods and cones to a different kind of information that can be transmitted on only a million fibers. These fibers communicate to the brain the existence of important image features much more abstract than a simple list of which rods and cones are being illuminated and which ones are not.

Ingenious experiments have shown that not long after the signals leave the eye, they are received by the primary visual cortex, which is located on the rear surface of the brain. By sophisticated experimental techniques, it is possible to see that there is a copy of the retinal image spread out on this part of the brain, with distortions that emphasize the very detailed information collected from the fovea, the finest-grained part of the retina. From this "primary map," a number of calculations are carried out that send their results to other regions of the brain that specialize in the analysis of motion, color, stereo depth, and, probably, other features of the visual image. This part of the brain can be thought of as a collection of semi-independent computers connected to each other in a hierarchy, beginning with the eye at the bottom and proceeding upward from computer to computer, extracting more and more abstract information from the original image. At some very high levels in this hierarchy of computers, the brain may begin to recognize specific objects or situations and command that actions be taken.

Vision scientists have the fascinating task of trying to learn how each of these computers works, what information it receives, and what kind of result it sends out to the other computers in the network. Psychologists and psychophysicists work on these problems by measuring the response of human subjects to a variety of different pictures that move or flicker or have special relationships to earlier pictures. Physiologists study these problems by inserting electrodes into different parts of the visual system of cats, monkeys, and rabbits to listen in on the signals that these miniature computers are sending to each other and generating within themselves. Mathematicians and physicists try to invent logical mathematical schemes by which such computers might be imagined to do their work, and computer scientists construct computers that work by these principles to see if they are able to perform some of the tasks of the visual system. Anatomists carefully dissect the brains of many kinds of animals to find out exactly where the fibers go that connect one part with another and how the cells that are in a small area are related to each other. Neurochemists study the secretions of cells by which they influence their neighbors and the properties of cell membranes that are the targets of these neurochemicals. From the work of all of these diverse scientists comes a steady stream of new facts, ideas, speculations, and suggestions that are gradually building a picture of how the visual system works and describing in some detail exactly what our eyes tell us about the world outside. It is a fascinating study in which progress is now quite rapid, and it is also a very promising and exciting beginning in understanding how the rest of our brain works.

Section IV TECHNOLOGICAL DEVELOPMENT

Scanning in Medicine and Other Fields

A. M. CORMACK

Physics Department, Tufts University, Medford, MA 02155

Besides conventional imaging methods using optics, electrooptics, and X-rays, a number of unconventional methods have been developed over the past 80 years. Although the medical applications have received great publicity recently, the others have received very little because they have been widely spread in geography and time. All I can do is make two "sections" through all of this material and show that, in the sense of this meeting, they intersect in medicine.

The problem that I will discuss here is a mathematical one, called Radon's problem, which I will explain presently. What I want to do first is to take a number of applications of Radon's problem (in order of increasing size) that produce images in the ordinary sense in ordinary three-dimensional space. Then I will go through the history of the problem and show that there are more general ways of looking at it. Finally, I will talk briefly about some further generalizations of the problem.

What is Radon's problem? It is illustrated in Fig. 1. A function, f, is defined in a domain, D, of the plane. L is some straight line intersecting D, and g_L is the line of intergral of f along L. The problem is to determine f from point to point in D if g is known for some or all of the lines, such as L, which intersect D. (If you don't like line integrals, think of g_L as being the average value of f along L.) Radon's problem has a solution (in fact, many different forms of solutions), but this need not concern us. All that concerns us is that we can calculate from a set of measured quantities (the g_L) the values of f.

As I said, I was going to start by going through images in order of size. I am going to start with the work of Klug and his coworkers on human wart virus, the smallest object I know of to which tomographic methods have been applied. (I am sure that we are all distressed that Aaron Klug was unable to be here today, and I am particularly distressed because I am going to start with an application of his, both as a tribute to him as one of our most recent Nobel Prize winners and as a person

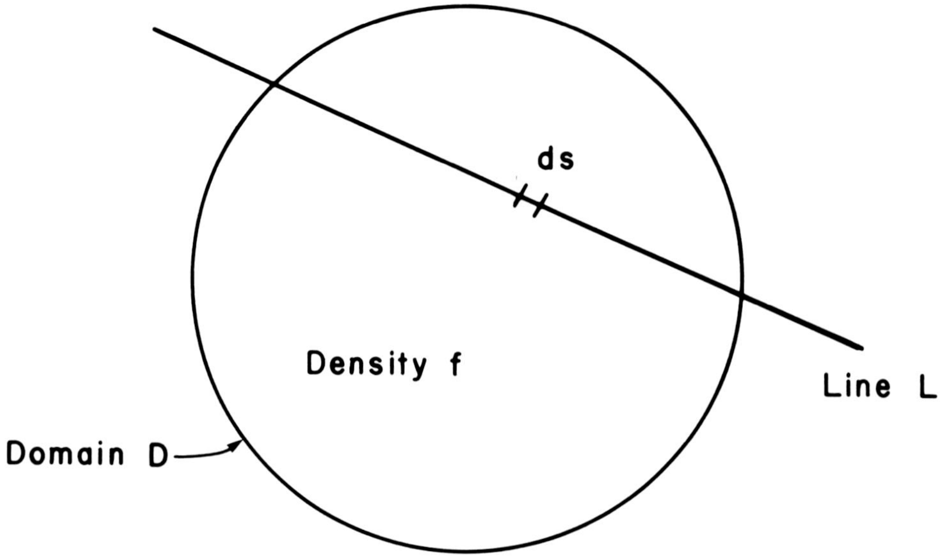

Fig. 1. Radon's problem. Given measured values of g for many lines like L, can one infer how f varies from point to point in the domain D?

$$g_L = \int_L f ds = A \text{ quantity measured along } L$$

whose work I have known and respected since we were both in Cape Town in 1946.) This application of Radon's problem is illutrated in Fig. 2, kindly provided by Dr. Klug. The object at the top is the one whose structure we wish to determine. The straight lines intersecting it represent the beam of an electron microscope intersecting the object from various directions and giving rise to the three projections indicated in the middle. (Another way of stating Radon's problem is: "How do we determine an object from its projections?") Klug uses Fourier methods to solve the mathematical problem and arrives, finally, at the structure of the object shown at the bottom.

Going up several orders of magnitude in size, we come to the scale of human beings, in whom the application of Radon's problem is to Computer Assisted Tomography (CAT). A fine beam of X-rays is fired along lines such as L in Fig. 1. The intensities of the incident and the emergent X-rays are measured, and g_L represents the average value of the absorption coefficient along L. From many measurements of g one can find f, in this case the absorptive coefficient of the tissue, and come up with a "CAT scan" of a section through the body. An early CAT scan of the head is shown in Fig. 3, and a scanner that produces such an image is shown in Fig. 4. This device obtains about a million values of g in about 10 s and from these produces the picture. It costs about a million dollars. For fun, in Fig. 5 I show a picture of my CAT scanner of 1963 with which I obtained about 250 values of g in 2.5 d, and that cost about a hundred dollars!

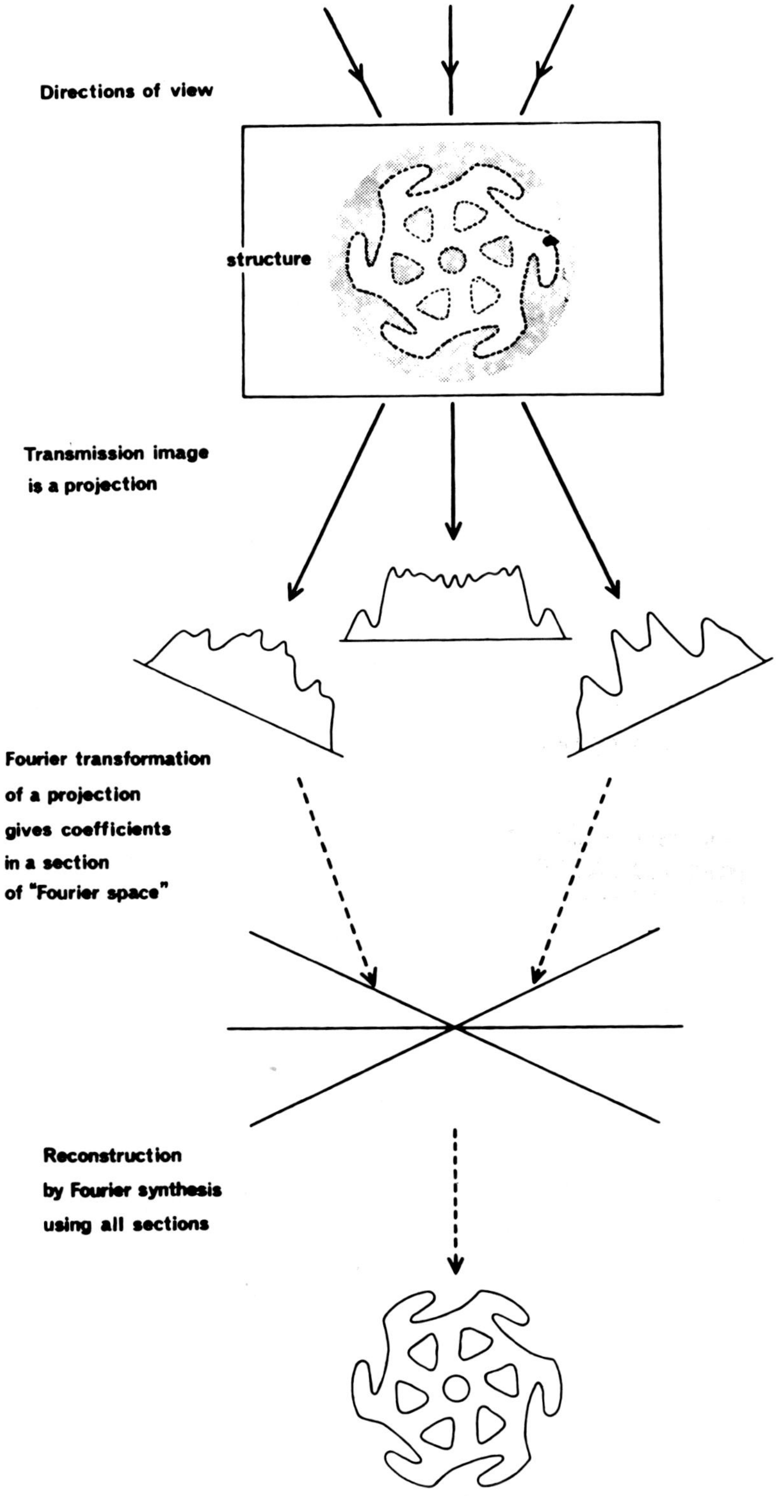

Fig. 2. Application of Radon's problem.

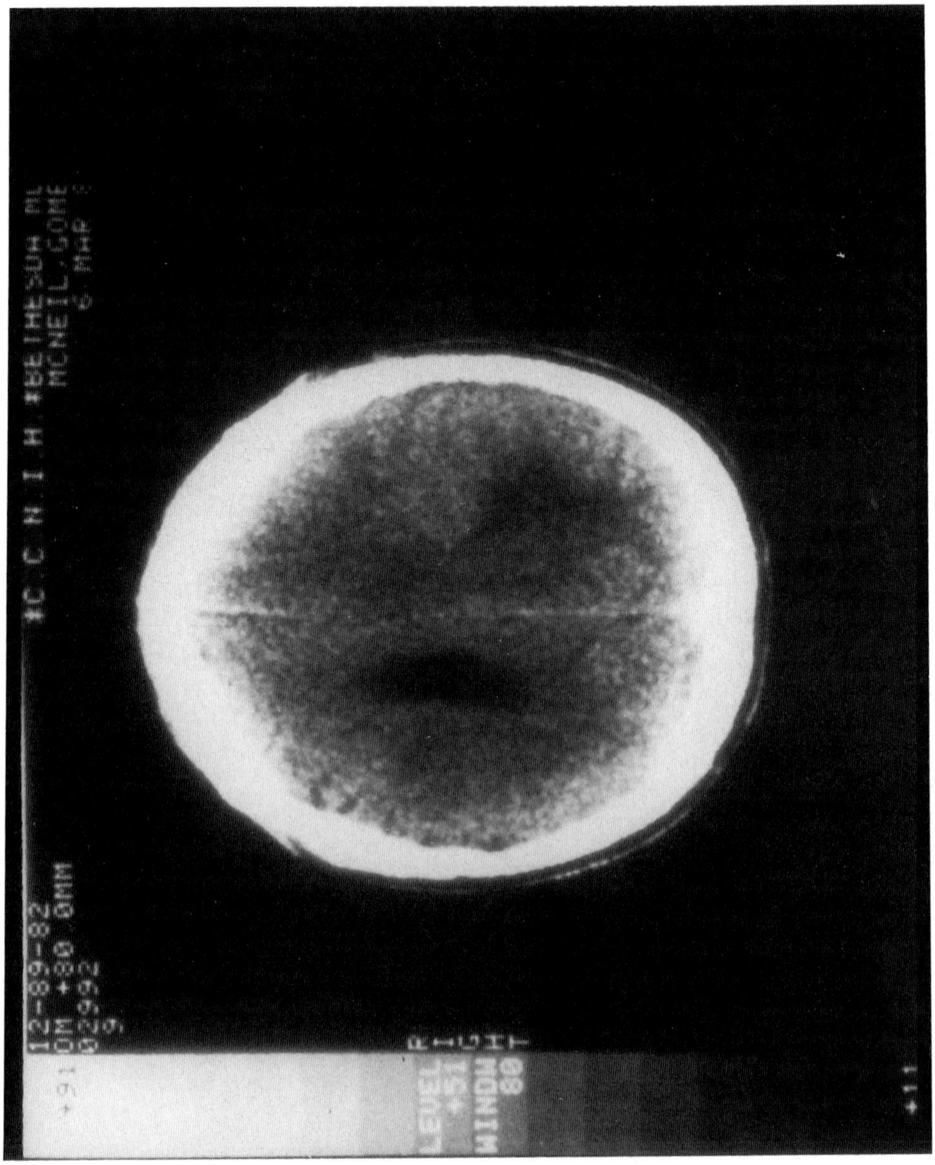

Fig. 3. Early CAT scan of a human head.

Fig. 4. A scanner.

Fig. 5. 1963 Model CAT-Scanner. Cost ~ $100.00.

The next application, on the same scale, involves a different type of physics, as illustrated in Fig. 6. Certain radioactive materials emit positrons, which are identical to electrons, except that they are positively rather than negatively charged. If a positron comes near an electron they will be attracted together, and after a short time they annihilate, producing two photons that go off in exactly opposite directions. This provides the basis for Positron Emission Tomography or PET scanning. Suppose that you have a distribution of some positron-emitting isotope in the head, say, as shown in Fig. 7. Suppose further that on either side of the head you have two detectors arranged so that they will give an output pulse only if they detect particles simultaneously or, as they say, in coincidence. Then, if a positron is annihilated in the hatched region in the figure and if one photon goes into one detector, the other photon must go into the other detector. Hence the rate at which the coincidence counter-registers coincidences is proportional to the rate at which positrons are being annihilated in the hatched region.

Now of course, most photons would miss the detectors if there were only two detectors, so in practice many more detectors are used, and in some research-type machines the number of detectors being contemplated is several thousand, almost completely enclosing the patient.

Fluorine-19 is a positron-emitting isotope that can be used to label a form of glucose; this in turn can be injected into a patient, and some of this labeled glucose will be metabolized in the brain. Figure 8 shows the results of PET scans of a patient's brain at different times after the injection of the glucose. What we see here is the rate of metabolism of glucose in the brain, as a function of time.

If one takes a normal person and puts him or her into a PET scanner, then by, say, scratching the back of their hand or whispering in one of their ears, one sees the response in those parts of the brain that are known to deal with stimuli from those regions. I think that no new information has yet been obtained about the functioning of the brain by this method, and I mention it only to point out two things: One, it shows the great sensitivity of PET scanning to brain function, and two, that this sensitivity to external stimuli provides a new source of noise over and above that provided by limited counting statistics.

Another positron-emitting isotope, is ^{11}C, and when you think of the enormous number of compounds of biological interest that contain carbon, and can therefore be labeled with ^{11}C, you will see that the potentialities of this method are enormous. It does not seem that PET scanning will ever be as widely used a diagnostic tool as CAT scanning because the isotopes that I have mentioned have half-lives on the order of 10 min, so you also need a cyclotron to produce those isotopes on the spot. This makes the whole operation rather expensive. However, one can imagine that a few PET scanners with thousands of detectors each, and the associated cyclotrons, would be invaluable for research purposes.

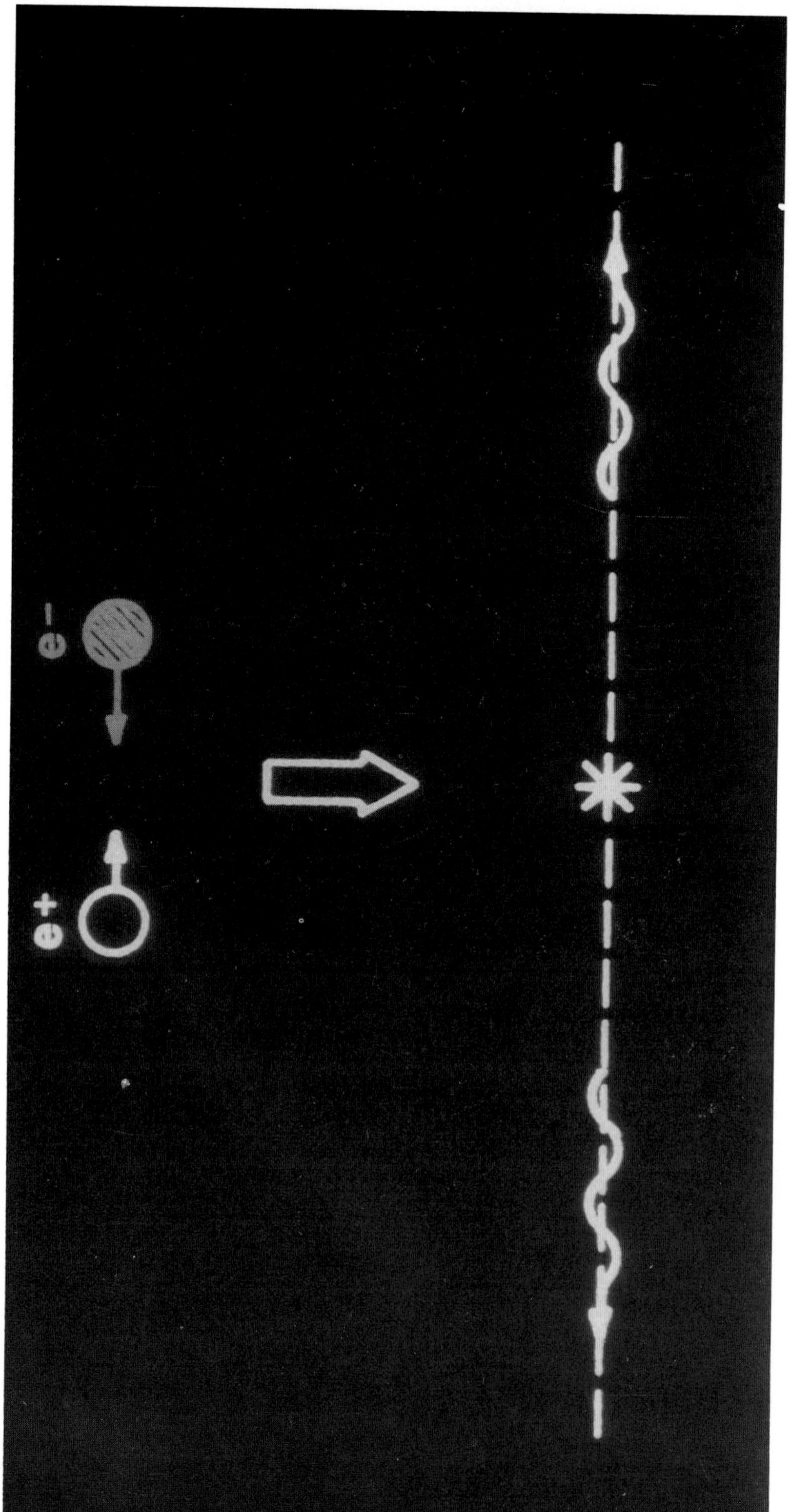

Fig. 6. Electron–positron annihilation into two photons.

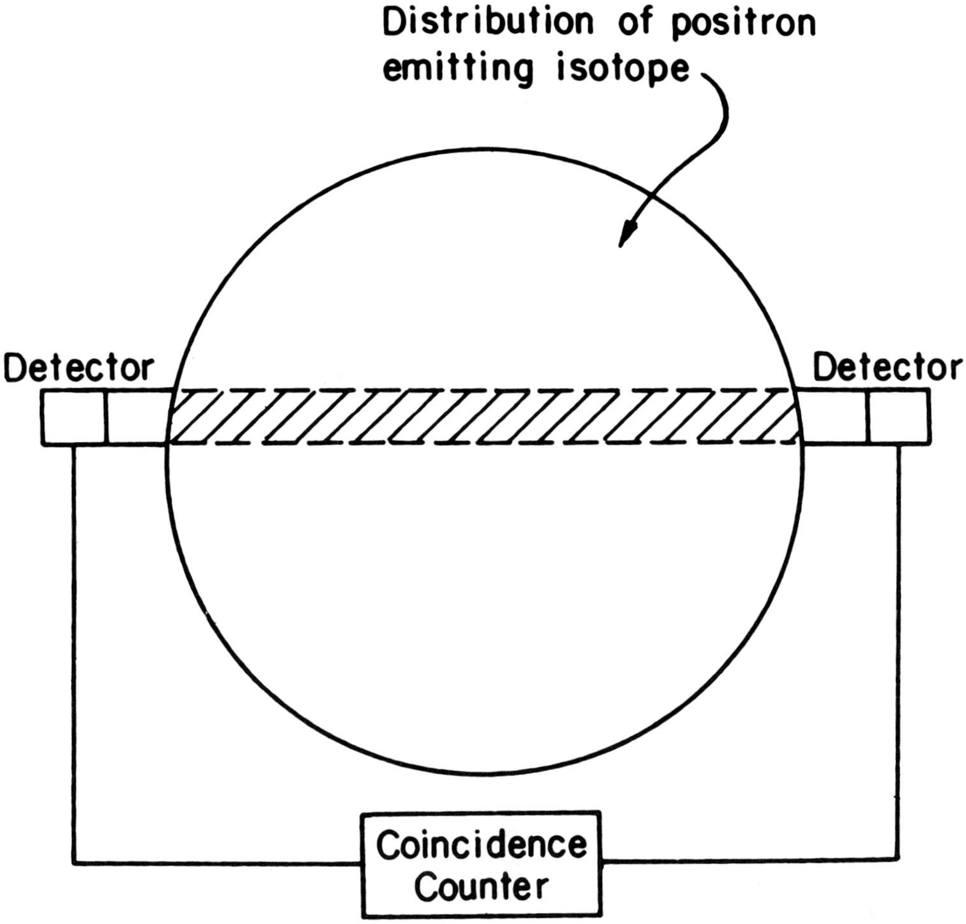

Fig. 7. Distribution of some positron-emitting isotopes in human head.

As my next application of Radon's problem I was going to use Nuclear Magnetic Resonance (NMR) scanning (or MRI, Magnetic Resonance Imaging, as some prefer to call it because "nuclear" has become a dirty word). However, because Paul Lauterbur is going to devote an entire article to this subject, I will skip to an application of Radon's problem to something a few orders of magnitude greater than the scale of human beings, namely, the ocean.

Suppose that you have a source and detector of sound immersed in water, suppose also that there is a temperature variation (and therefore a variation of the speed of sound) in the water between the source and the detector. The time that it takes a sound pulse to travel from the source to the detector will be the line integral of the inverse of the sound velocity along a line from the source to the detector. If a number of sources and detectors are placed in the plane around the region of ocean of interest,

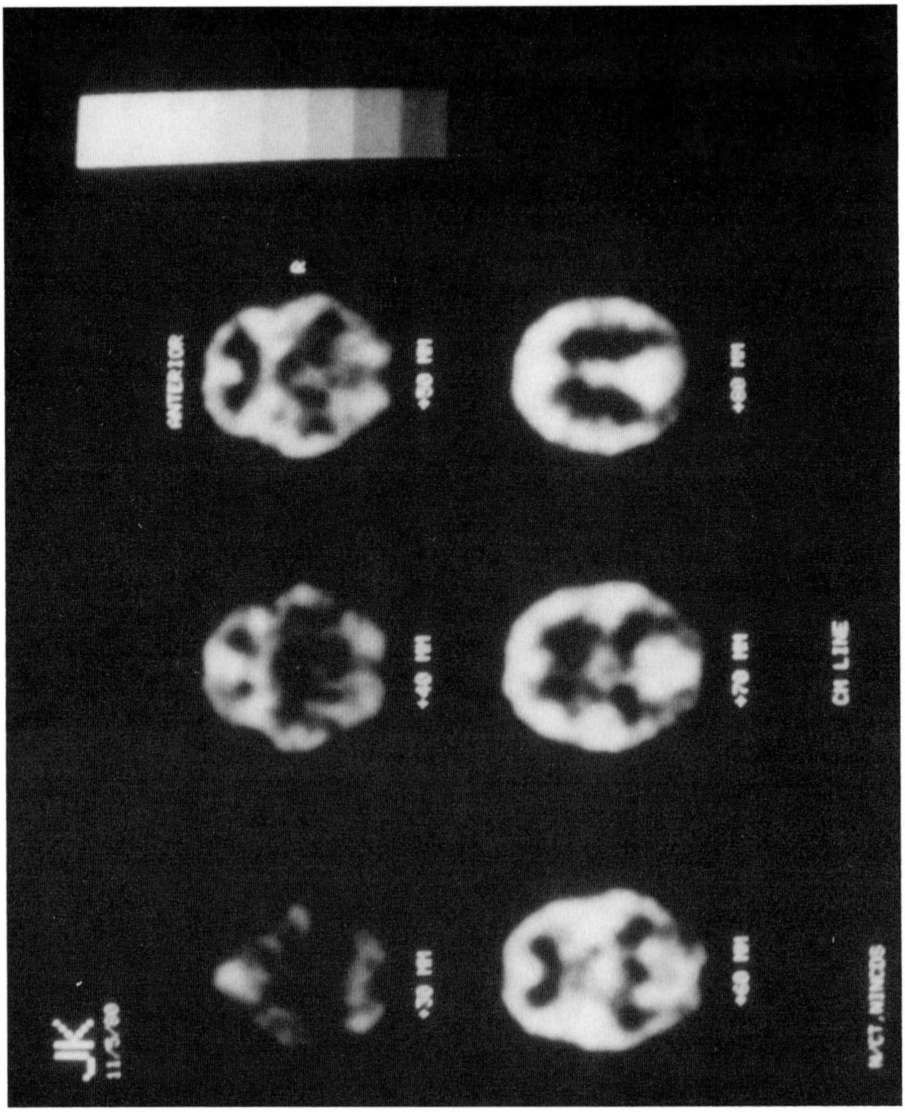

Fig. 8. Results of PET scans of human brain.

these will provide line integrals of the inverse of the sound velocity along lines intersecting the region, and we are back to Radon again. We can find the sound velocity from point to point in the region and, hence, the variations in the temperature of the water. (This is a gross oversimplification of very neat oceanographic problem.) Figure 9 (kindly provided by Dr. Carl Wunsch) illlustrates some results. The four squares represent 300 × 300 km regions of the ocean. The upper two squares represent two temperature distributions found by conventional methods, i.e., towing a thermometer behind a ship that traverses the squares many times. This of course takes a long time, and on completing the survey one would

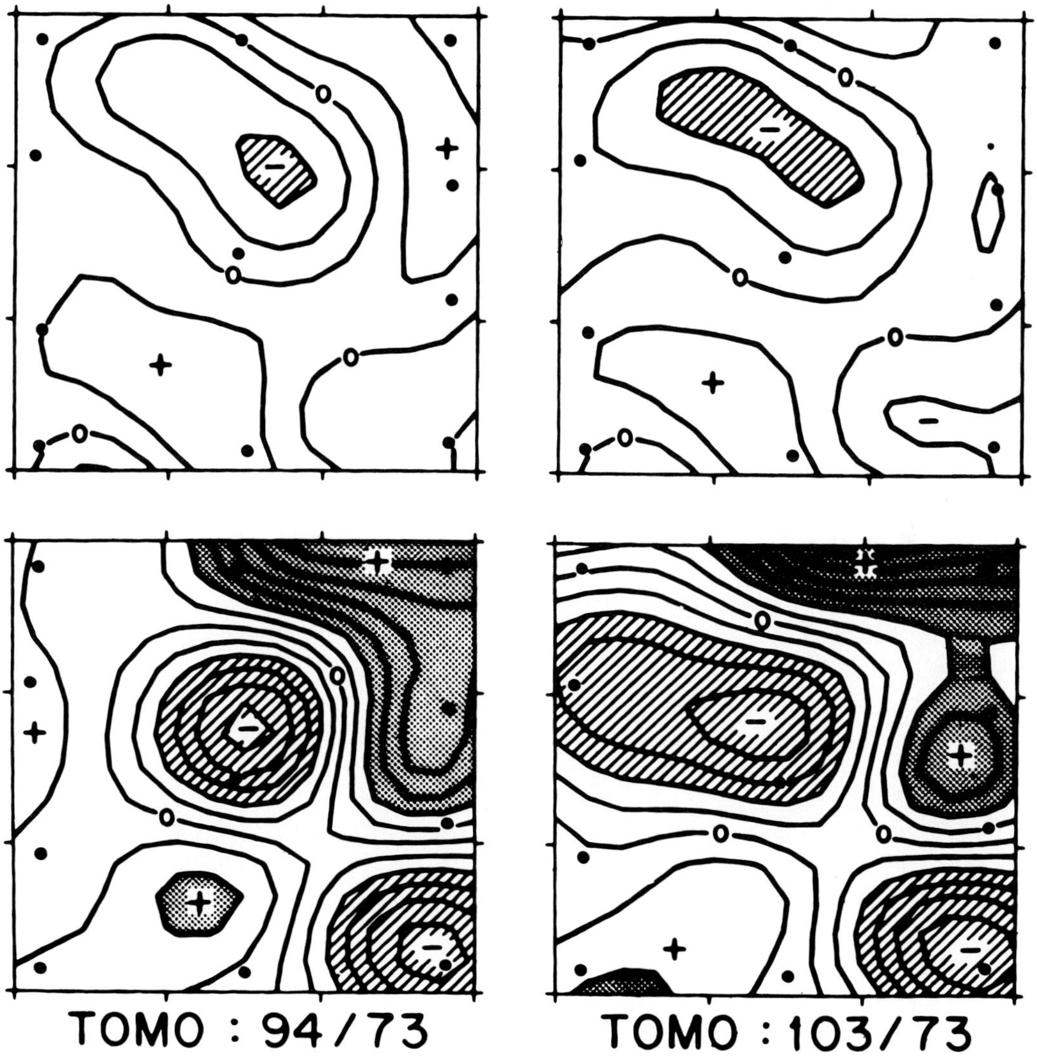

Fig. 9. Another application of Radon's problems.

wonder whether the temperatures measured at the beginning of the survey had changed or not. The lower two squares represent a tomographic survey of the same regions in a much shorter time. Indeed, to obtain the data shown in the lower two squares by conventional means in the time it took to obtain the data by tomographic means would have required a ship with a speed of 3000 knots (5500 km/h).

In my last example of this pass through the material, I would like to increase the scale by a few orders of magnitude and go to radioastronomy in the nineteen fifties. In those days there were a large number of circular radar antennas left over from World War II. If a number of these were placed along a line on the ground, and all were pointed in the same direction, and all were connected together electrically, they would behave together somewhat like a long antenna with a parabolic section perpendicular to its length. Such an antenna has good resolution in one direction and no resolution in a perpendicular direction. Pointed at an astronomical object, as Bracewell pointed his array at the moon, the output was the microwave emission at some wavelength integrated along a strip or line across the moon. Radon's problem again! But there was a snag. The antenna array was fixed to the ground and could not be moved around to integrate the microwave emission along an arbitrary line across the moon. All that could be obtained was the integrals along the limited number of lines provided by the relative motion of the Earth and the moon (this is the first appearance of the "limited number of views" problem in CAT scanning). What would be more natural than to increase the number of views by setting out another array of antennas along a line at right angles to the line of the first array? This was done, but it was done at about the same time as general methods of aperture synthesis were being developed, and it was immediately realized that this new array (a Mills' cross) provided a pencil beam that could look at points (rather than averages over lines) on the moon's surface. So Radon's problem in its "pure" form had a rather short history in radioastronomy, but it is illustrated in Fig. 10, which shows a microwave image of the moon at 9.1 cm. I thank Dr. Ronald Bracewell for pointing out this bit of astronomical history to me, and if I may add just a little more history, I'll point out that Bracewell was thinking about this line- or strip-integral problem in astronomy (in Australia) at about the same time that I first thought about the problem (in Cape Town) as it applied to CAT scanning.

Well, I have completed my first pass through the material and gone, in order of size, from viruses to the moon. All the applications of Radon's problem have resulted in ordinary images in ordinary two-dimensional space. Now I will turn to a historical treatment that will show certain generalizations of the problem itself and applications of these that produce more generalized images.

When I wrote my first article on CAT scanning in 1963, I had never heard of Radon's problem, but I felt sure that it was probably a standard part of the nineteenth century mathematical repertoire—I imagined that

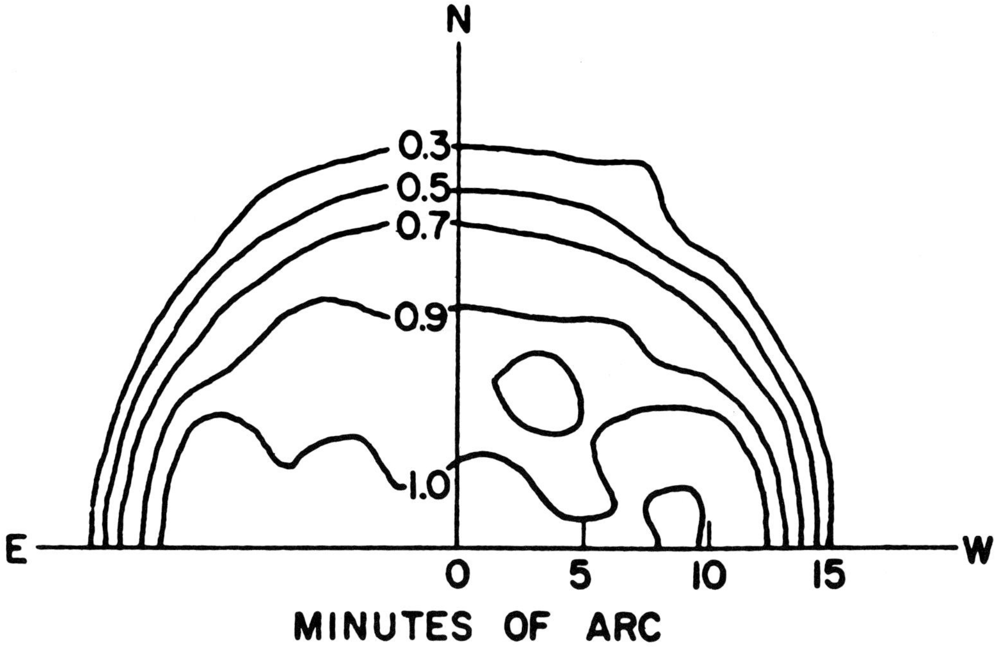

Fig. 10. Moon brightness at 9.1 cm reconstructed from fan-beam scans to a particular interval of 1′.28.

a Riemann or Cauchy might have dashed it off in a weak moment. Hence, I delayed publication while I enquired of mathematicians on three continents whether they knew of the problem or not. I published only after I had been assured that they did not know of it. However, in the last five or ten years I have come across or been informed of work that stretches back to the beginning of this century and may go back into the nineteenth century.

If, instead of a function *f* defined in the domain of the plane, one has a function defined in a domain *D* of three-dimensional space, and if one replaces line integrals in the plane, with integrals over planes in three-dimensions that intersect *D*, one has a natural generalization of Radon's problem: Given the integrals (or averages) of *f* over planes intersecting *D*, can one infer the values of *f* from point to point in *D*? This is illustrated in Fig. 11. The answer is again yes and the solution turns out to be simpler than the solution in two dimensions. This problem was addressed and solved by the great physicist H. A. Lorentz prior to 1905. Since he never published his result we don't know exactly when he did the work, and the only reason we know that he did it is that in 1905 a student of Lorentz stated the problem and its solution and attributed both to Lorentz.

Radon published the work that gave the problem his name in 1917, and he dealt with the problem for any number of dimensions. He pointed out that the distinction between the nature of the solutions in

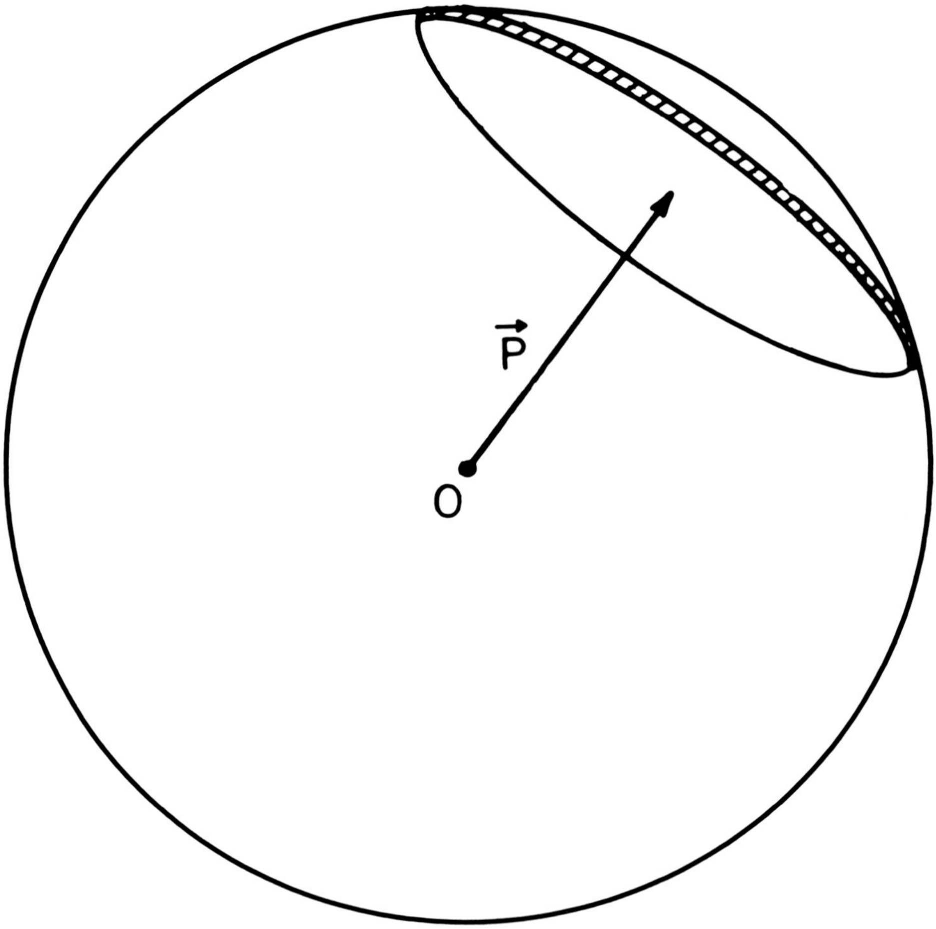

Fig. 11. A natural generation of Radon's problem.

two- and three-dimensions holds generally between spaces with an even number and spaces with an odd number of dimensions.

At the suggestion of Paul Ehrenfest, George Uhlenbeck generalized Lorentz's original result to *n*-dimensions in 1925, both being ignorant of Radon's work. Uhlenbeck was the first to use Fourier methods in dealing with the problem. In 1936 Cramer and Wold came up against the problem in statistics in which it can be stated: "How do you determine a statistical distribution from its marginal distributions (projections)?"

Also in 1936 we have what is, I believe, the first practical application of the problem, this time in astronomy. There are many stars in the neighborhood of the sun, and astronomers would like to know the distribution of velocities of these stars relative to the sun, say. For all of these stars we can measure their radial velocities, i.e., their velocities along the line of sight, by measuring the Doppler-shifts of their spectra. But for only a handful of stars can we measure their proper motions, i.e., their velocities perpendicular to the line of sight. Hence, for the vast majority of stars the problem is to infer the actual distribution of velocities from measurements of their radial velocities only. This problem was posed by Eddington and solved by Ambartsumian. A little thought will convince you that this is just Radon's problem again, not in ordinary three-dimensional space, but in three-dimensional *velocity–space* in which we would naturally express the distribution of velocities. Some of Ambartsumian's results are shown in Fig. 12 for a group of about 400 B-type stars. If one draws an arrow from the origin to any of the numbers of Fig. 12, that arrow represents the velocity of that number of stars. (For simplicity of presentation, Ambartsumian projected his three-dimensional results onto the galactic plane.) Figure 12 shows that you *can* do computed-

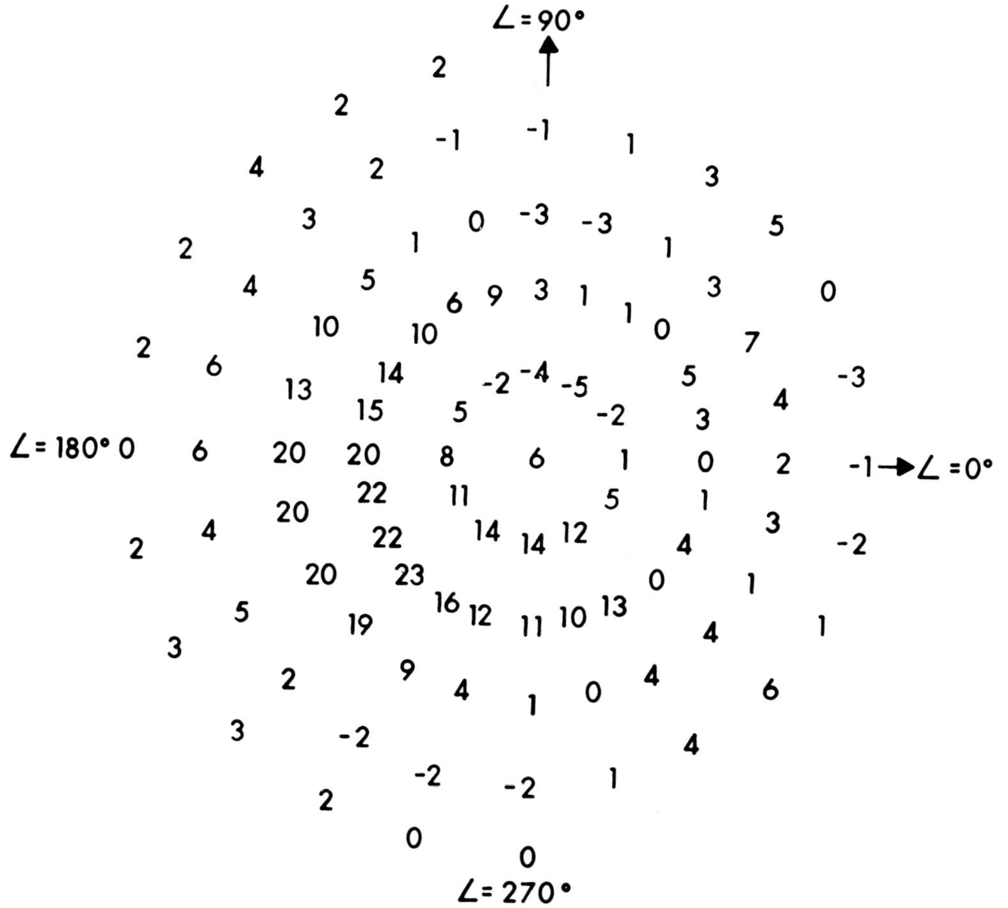

Fig. 12. Some of Ambartsumian's results for a group of about 400 B-type stars.

assisted tomography without computers, for there were no computers in 1936 and Ambartsumian used traditional (and or course, very slow) methods of numerical analysis.

As I mentioned earlier, my two passes through all this material were going to intersect in medicine, and in fact they do in NMR scanning. This can be viewed as a set of two-dimensional Radon problems involving sums along lines or as a three-dimensional problem involving sums over planes. Both methods have been used, but I shall not go into this because Paul Lauterbur will doubtless give more details.

What I mentioned earlier about positron–electron annihilation is true, but it is not the whole truth. In Fig. 6 I was illustrating what happened when a positron and an electron annihilate if they both start from rest, and therefore, their total linear momentum is zero. In this case, the resulting photons must fly off in exactly opposite directions so that their total linear momentum will also be zero. In general, the electron and the positron will both be moving before they annihilate and so together will have a resultant linear momentum that is not zero. To conserve momentum, the resulting photons must have the same amount of momentum, so they go off not along the same straight line, but along lines that have a small angle θ between them, as shown in Fig. 13. This angle θ is very small and can be completely ignored for PET scanning, but it is of great

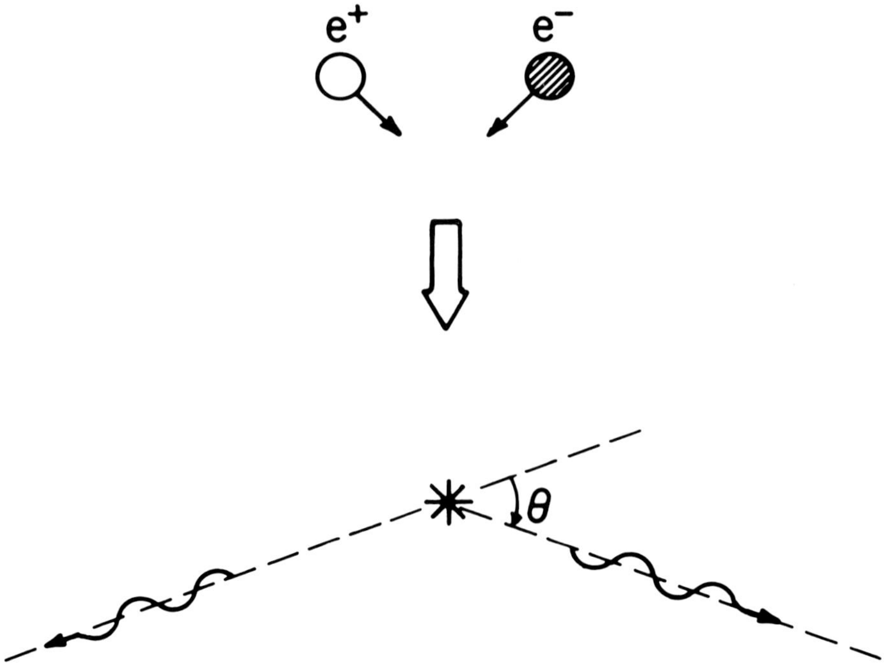

Fig. 13. Photons emitting along lines that have a small angle θ between them.

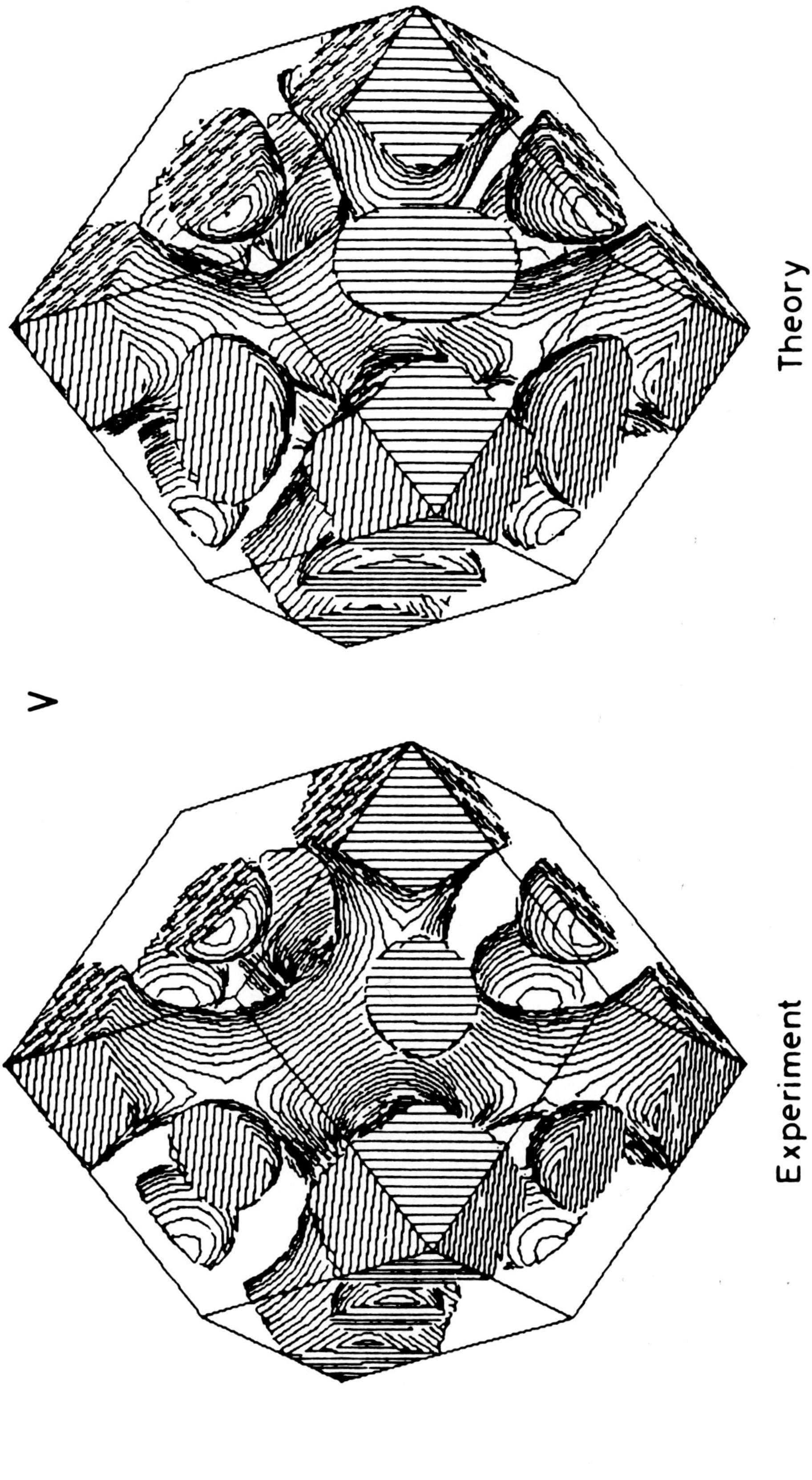

Fig. 14. The Fermi surface.

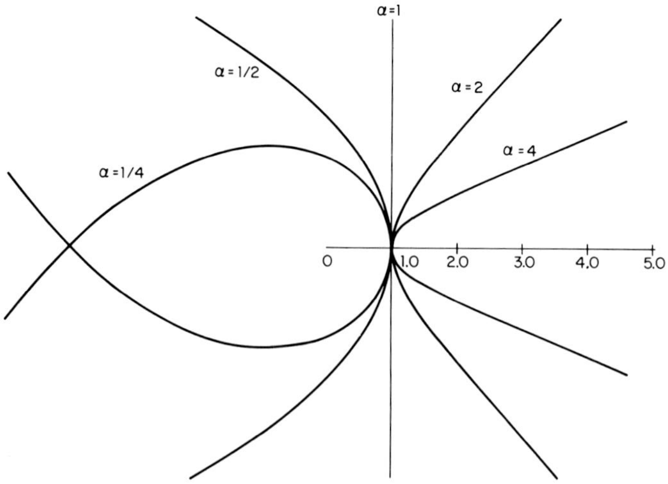

Fig. 15. The curves $(r/p)^\alpha \cos \alpha\,\theta = 1$ for $\alpha = 1/4, 1/2, 1, 2, 4$.

importance in solid state physics. By studying the distribution of this angle one may draw conclusions, using Radon's results, about the momentum distribution of the electrons in the solid, and hence, about the so-called Fermi surface of this solid. The Fermi surface is a surface in the three-dimensional momentum space of the electrons in the solid. So Radon's problem in this case gives us an image (of the Fermi surface) in this space. This is illustrated in Fig. 14, which was kindly provided by Dr. A. A. Manuel of the University of Geneva. The left-hand side shows the Fermi surface of vanadium, found directly from positron positron–electron annihilation experiments, the right-hand side shows the Fermi surface deduced by long and tedious bound-structure calculations. The agreement is striking.

What I have tried to show is the following. Suppose we wish to determine a function of n independent variables, and it doesn't matter what they are. They may be ordinary space variables, velocity or momentum components, or some statistical variables. Suppose, further, that what we can measure is some linear combination of these variables. Then this linear combination can be thought of as defining an $(n-1)$-dimensional plane in an n-dimensional space, and the solution of Radon's problem allows us to determine the original function from measurements of the linear combinations.

What about curves other than straight lines and surfaces other than planes? Over 20 years ago I discovered solutions to Radon's problem in which one integrates over circles through the origin rather than straight lines, and recently E. T. Quinto and I have given a much more detailed account of $(n - 1)$-dimensional spheres through the origin in an n-dimensional space.

In two dimensions I have recently considered two families of curves for which Radon's problem can be solved, one family containing the orig-

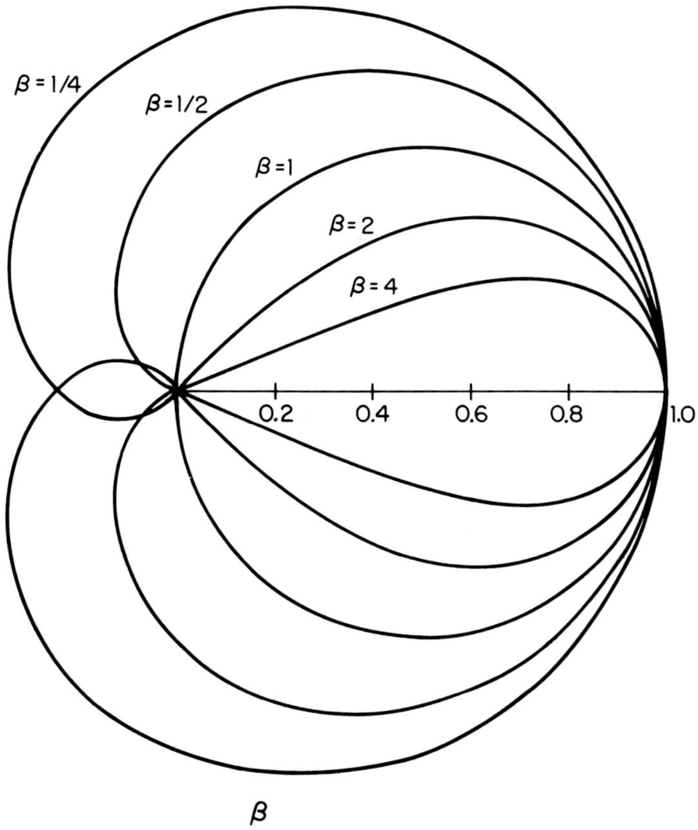

Fig. 16. The curves $(p/r)^{\beta} \cos \beta \theta = 1$ for $\beta = 1/4, 1/2, 1, 2, 4$.

inal straight lines, the other the circles that I have just mentioned. These are shown in Figs. 15 and 16.

In Fig. 15 we have typical curves of the family, which I have called α-curves. $\alpha = 1$ Represents the straight lines of the original Radon problem, $\alpha = 2$ represents a family of rectangular hyperbolae, $\alpha = 1/2$ represents a family of parabolae. In Fig. 16, $\beta = 1$ represents the circles I have mentioned, $\beta = 2$ is a family of Bernoullian lemniscates, $\beta = 1/2$ represents a family of cardioids, and so on. Further generalizations to very abstract spaces have been made by Helgason and others.

I hope that I have succeeded in demonstrating that Radon's problem represents a very powerful sampling technique with a wide variety of applications.

Genetic Engineering

A New Biotechnology

HARLYN O. HALVORSON

Rosenstiel Basic Medical Sciences Research Center, Brandeis University, Waltham, MA 02254 USA

THE DIVERSITY OF MICROBES

These are the reflections of a microbiologist watching the role of microorganisms in the dramatic revolution in biotechnology. It started in the 17th century with the remarkable Dutch merchant, Antony van Leeuwenhoek, who, while peering through his crude microscope, observed "very little animalcules." He was the first to observe this teeming milieu of microorganisms that surround us. It was not until the latter part of the 19th century that the role of "animalcules" in medicine and industry and their chemical basis entered the modern age. Louis Pasteur began a series of discoveries that can best be described as the "Pasteur century." The rapid development shortly after of the Delft school of microbiology, under Beijerinck and Kluyver and later extended to the United States by van Niel, quickly demonstrated that through enrichment techniques specific organisms could be isolated that carry out a desired set of chemical conversions. Microbiologists soon identified and learned to isolate nitrogen fixing bacteria, ammonia oxidizing bacteria, cellulose degrading bacteria, and so on. The first half of this century was a continual revelation of new and remarkable "animalcules," supporting the view of Beijerinck that for every organic molecule found on the earth there is some microorganism that can make a living on it. In the last few years we have discovered bacteria that have established a food chain in the depths of the Pacific off the South American coast by oxidizing sulfur released through vents in the ocean floor. These unusual microorganisms represent a group that utilizes an energy source that is not dependent upon energy dirived from the sun through photosynthesis.

ENZYME REGULATION

Early in the Pasteur century it was recognized that the enzymatic properties of cells could be manipulated, and therefore utilized, for human benefit. In 1882, Wortmann observed that microorganisms would produce amylase only when grown on starch, and 17 years later Duclaux found that *Aspergillus* would produce protease only when grown on a medium containing protein, such as milk. However, it was not clear from these early experiments whether changes in enzymes were because of a Darwinian selection or an inherent capacity in each cell in the population. Dienert (*1*) showed that adaptation in yeast to galactose fermentation could occur in the absence of cell division. Many years later, Benzer (*2*) in Monod's laboratory showed that adaptation occurred in each and every cell in the population of *Escherichia coli*, a conclusion directly demonstrated later by Rotman (*3*). From subsequent investigations from the Pasteur Institute, Jacques Monod and colleagues (*4*) went on to clarify the process of adaptation of β-galactosidase as a negative regulation, identify

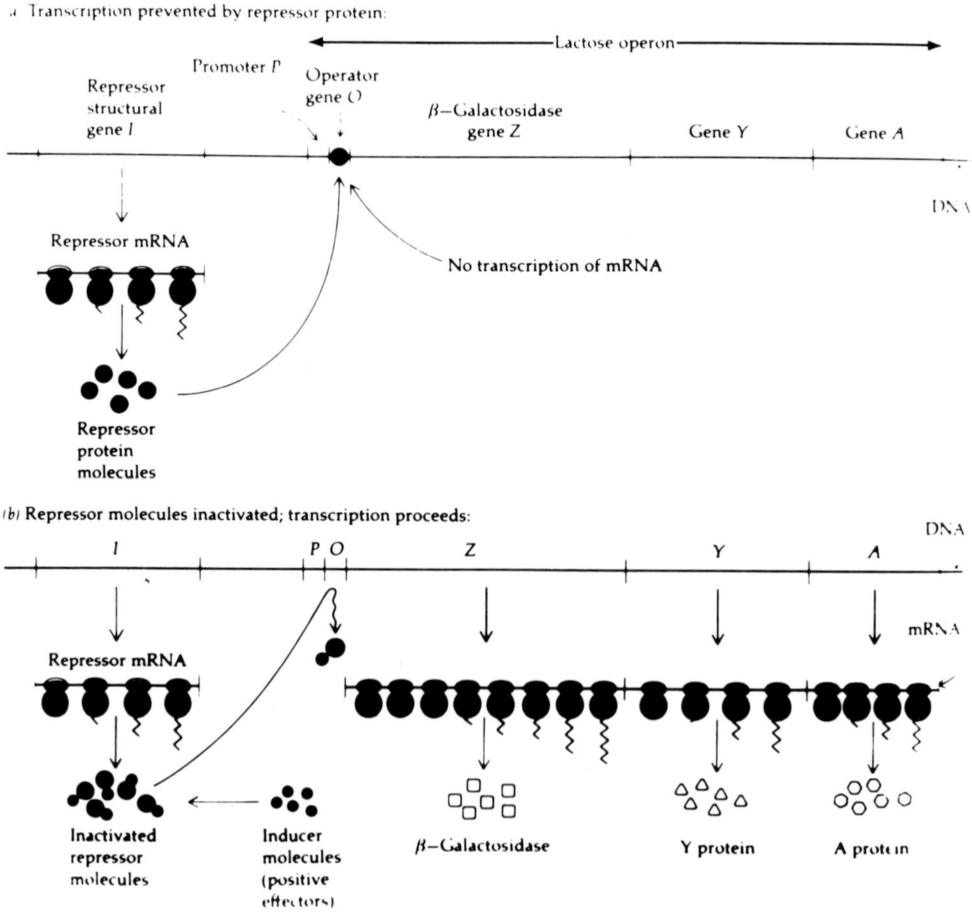

Fig. 1. Mechanism of regulation of β-galactosidase in *E. coli*.

the genes involved, and set the framework for modern manipulation of enzyme expression through studies of regulation at the genetic level.

Figure 1 summarizes the mechanism of β-galactosidase regulation and the function of a bacterial operon in which a number of genes can be subjected to the control of a common regulatory gene. Through mutation at either the i or o locus, this entire operon could be turned on or off. Under totally derepressed conditions, by appropriate selection of mutants, 7% of the dry weight of the *E. coli* cell produces β-galactosidase from a single gene copy.

Although positive regulatory mechanisms were later uncovered in bacteria, the pioneering studies of the β-galactosidase system showed us that the known efficiency of microbial cells could be understood and manipulated by placing several enzymes with related functions in a pathway under common regulatory control. The rate of enzyme synthesis could be governed by both cytoplasmic regulatory signals and chromosomal sites adjacent to the gene that recognizes these signals. The era of molecular regulation was quickly recognized.

During the 1940s and 1950s, through the efforts of Tatum, Lederberg, and others, the genetics of bacteria was placed on a solid fundamental basis that provided the opportunity to examine genetically biosynthetic pathways. As shown in Fig. 2, by selecting mutants that re-

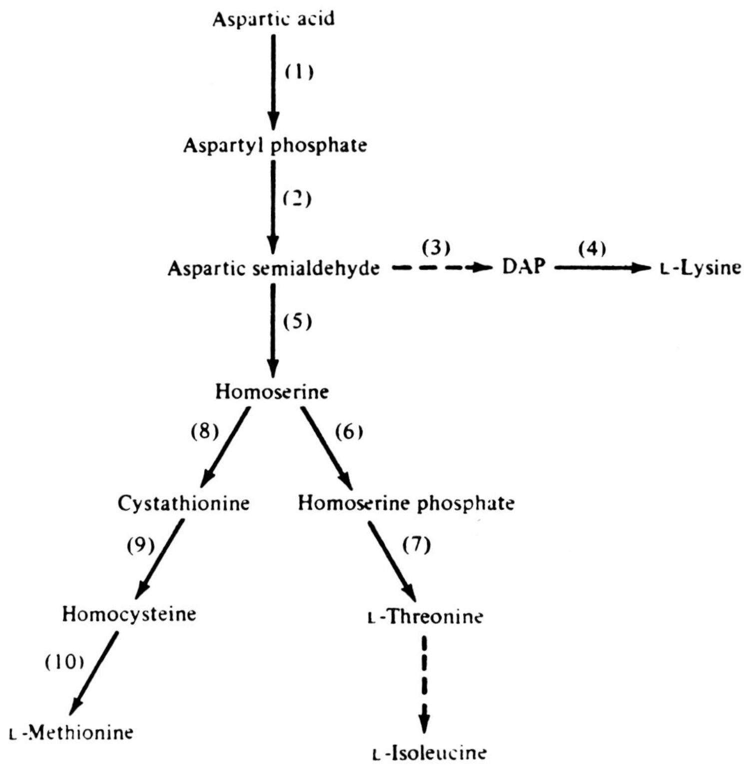

Fig. 2. Metabolic pathway of the aspartic acid family of amino acids.

quire the presence of the end product in a pathway for growth, one can isolate not only mutants blocked in various stages of that pathway, but also mutants that will excrete intermediates. This figure shows the pathway for the conversion of aspartic acid to lysine, methionine, and isoleucine, all members that derive their carbon skeleton from aspartic acid. A mutation in this microorganism to homoserine dependency, that is, a mutation to eliminate reaction 5-homoserine dehydrogenase, will divert into the lysine branch all of the carbon that enters the aspartic acid pathway. The same result may be obtained if the culture is mutated to methionine and threonine dependence by blocking reactions eight and six, respectively. When the vitamin biotin is lacking, threonine auxotrophs produce both homoserine and lysine. The list of mutants and amino acids that are produced by biosynthetic defects is lengthy and demonstrates the versatility of mutation as a tool to develop amino acid producers. The Japanese were the first to isolate microorganisms that release large quantities of glutamic acid into the mediium and to recognize the potential of such mutants in producing other amino acids. Furthermore, mutations in genes controlling regulatory functions have added to our capabilities of manipulating a biosynthetic pathway to produce numerous compounds of commercial interest. In addition, antibiotics, important intermediates in metabolism, and fine chemicals, have all been produced by selection and mutation in microorganisms.

During the past 30 years there has been extensive manipulation of the biosynthetic pathways for antibiotic synthesis. At present we now have over 150 antibiotic compounds on the market out of the more than 7000 naturally occurring and 30,000 semisynthetic, prederived antibiotics. Although most of these were identified empirically, an increasing number of cases was predicted from knowledge of the pathways of biosynthesis in which direct selection is based on our knowledge of regulation in bacteria. An example of this is the selection for analog resistant mutants. This strategy has proven very successful in selecting for mutants of *Penicillium* and *Acremonium* that overproduce the β-lactam antibiotics, penicillin and cephalosporin (*5*). The biosynthetic pathways for these two antibiotics involve the amino acids α-amino adipic acid, cysteine, and valine. Analogs of these amino acids can be used to select for strains with improved cephalosporin C productivity. Using analogs has increased the probability of finding such overproducing strains. Among this list are analogs of methionine. Although methionine is not a precursor of cephalosporin C biosynthesis, it is a required nutrient for fermentation of this organism. Strains were selected that overproduce cephalosporin C, but the mechanism remains unknown.

RECOMBINANT DNA TECHNOLOGY

By the end of the 1960s we could manipulate many pathways to produce cells under given controlled culture conditions, yielding products of industrial interest, particularly in inexpensive nutrient medium. However, only a modest interest was evident in seeking biological sources of products. The technology advances that arose through fermentation expertise, regulation, and genetic selection have peaked. We are now losing our trained experts in fermentation technology, and even support for studying the regulation of new and diverse pathways is beginning to wane. Although the genetics in new and diverse groups of organisms, such as *Bacillus* and *Streptomyces* began to emerge, there was no parallel rise in interest in using these for new technology. By the early 1970s the stage was set for an unexpected marriage between biology and technology. The beginnings were steeped in basic research, and the implications were thus unrecognized. Restriction enzymes were characterized that were capable of recognizing specific sites on the DNA molecule and cleaving them. These enzymes had long been suspected on physiological grounds. DNA of specific bacterial strains had unique restriction sites for cleavage. A number of these enzymes are shown in Fig. 3. The bacterial restriction endonucleases recognize specific sequences within double-stranded DNA. Large numbers of these enzymes have been isolated and are used in cloning. They cut DNA within or near a particular recognition

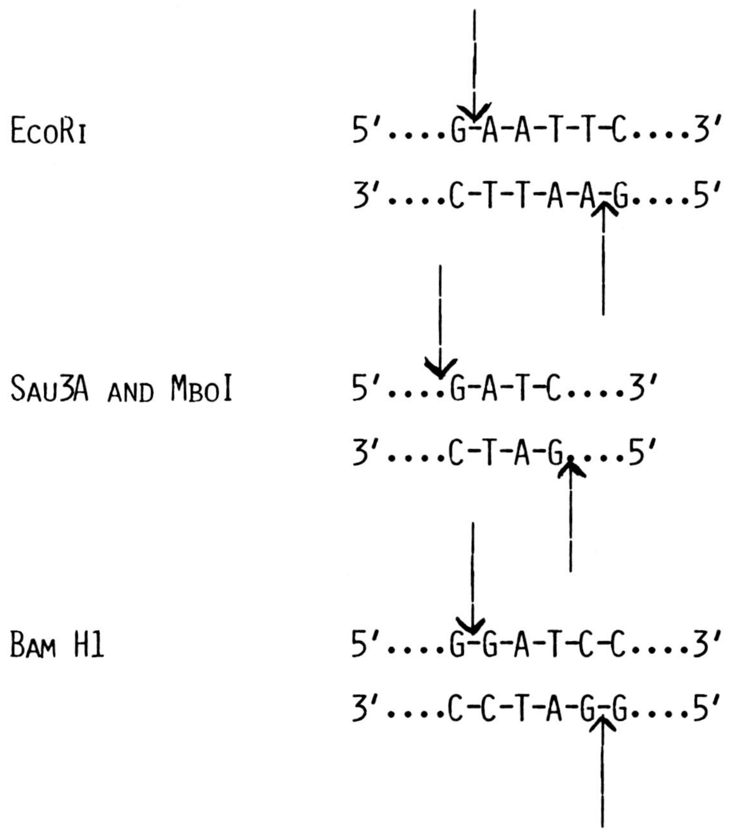

Fig. 3. Examples of restriction enzymes.

sequence, typically 4–6 nucleotides with a twofold axis of symmetry. The enzyme *Eco*RI, for example, recognizes the first hexanucleotide sequence. It does not cleave at the axis of dyad symmetry, but at positions four nucleotides apart. This yields fragments of DNA with protruding cohesive 5′-termini. Thus, any DNA molecule containing such recognition sites can be joined to any others to form novel recombinant molecules.

Several different enzymes cleave within the same target sequence. Some of these recognize tetranucleotide sequences. *Mbo*I and *Sau*3A recognize GATC, whereas *Bam*HI recognizes GGATCC. Depending upon the source, a library of enzymes is available that is capable of cleaving DNA molecules at specific sites. The further discovery that one could seal these breaks by enzymes called ligases gave a tool to the molecular biologist to manipulate the genome. That is, by specific cleavages one could open up a segment of vector DNA and with ligase, in the presence of cloned segments of DNA, anneal the molecule to include a desirable insert. The stage of molecular cloning was reached. The process was incredibly simple and involved selecting a system in which DNA could be inserted into a given cell. Among the mechanisms that were known or discovered were viruses and plasmids. The latter are extrachromosomal genetic elements found in a variety of bacterial species. They are, in general, double-stranded, circular DNA molecules that contain 1000–200,000 bases. These plasmids contain genes coding for enzymes that are often advantageous to the bacterial host.

Through transformation, plasmids can be transferred to bacteria under conditions that render the cells temporarily permeable to small DNA molecules, thus permitting a very simple selection of those bacteria that have been transformed. The recipient bacterial cell now has a second chromosome with a new array of genes. The ability to insert genes that are homologous to the chromosomes provides a very simple mechanism for inserting new genetic elements. This is shown in Fig. 4. The plasmid is cleaved with a restriction endonuclease, in this case *Bam*HI, located in the tetracycline-resistant gene in the plasmid. At the same time, the DNA to be inserted is digested with the same enzyme. After ligating the two DNAs at the appropriate concentrations, the ligation mixture is used to transform ampicillin sensitive *E. coli* to ampicillin resistance. Some of the colonies that grow in the presence of ampicillin will contain recombinant plasmids, others will contain plasmid DNA that has been recirculized during ligation without insertion of foreign DNA. To discriminate between the two kinds of transformants, a number of colonies are streaked on plates containing ampicillin or tetracycline. The colonies that survive and grow in the presence of tetracycline contain plasmids with an active tetracycline-resistant gene. Such plasmds are unlikely to carry inserts of foreign DNA. The colonies that grow only in the presence of ampicillin contain plasmids with inactive tetracyline-resistant genes. These plasmids derived by insertion inactivation are likely to contain foreign DNA sequences.

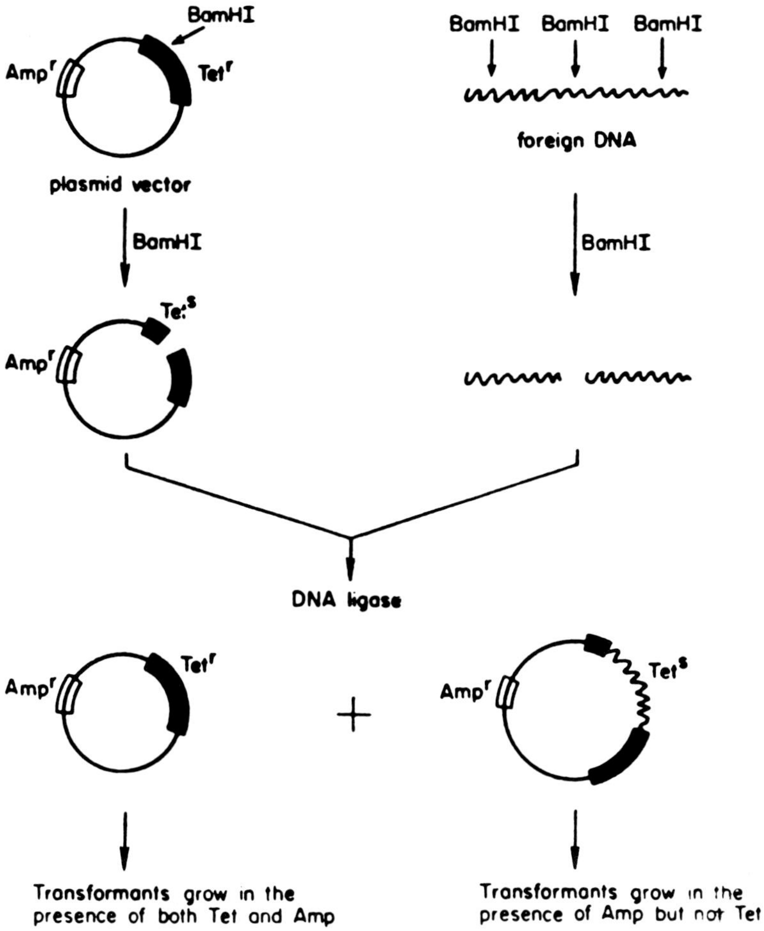

Fig. 4. Insertion of genes into plasmids.

HOST–VECTOR SYSTEMS

This methodology has been widely applied. In selecting a commercial application, a number of factors should be considered in choosing a host–vector system. First, the host should be inexpensive to grow; a tough and resistant strain is desirable to withstand high hydrostatic pressure or lysis after death. In the selection of a vector there are three major considerations. The vector must be able to replicate in the host and produce multicopies, increasing the yield of functions that are contained within it. The relaxed colicin family in bacteria is one example and the 2μ circles in yeast, another. The vectors should contain several markers that can be used for selection. The first one serves for the maintenance of the plasmid, either antibiotics in bacteria of histidine or leucine auxotrophy in yeast. The second marker is used for screening the insertion of the foreign DNA, as seen earlier in insertion inactivation. Finally, the marker should have a promoter that has high efficiency and can be regulated by repression or derepression. Examples are β-galactosidase in *E. coli* and glycolytic enzymes and acid phosphatase in yeast. For promoting the expression of either a bacterial or eukaryotic gene in *E. coli,* a number of well defined *E. coli* promoters have been used. These include the promoters of the *lac* operon of *E. coli* and the ampicillin-resistant gene of pBR322 that codes for the enzyme β-lactamase. In addition, efforts have been made to bring the translation of transcripts under the control of translation-initiation signals on the plasmid. An example used by Goeddel and colleagues (*6*) is summarized in Fig. 5. They brought the expression of the eukaryotic human growth hormone in *E. coli* under the control of both a promoter and a ribosome binding site. The structural gene for the human gene hormone was fused to a chemically synthesized DNA fragment encoding for the *N*-terminal amino acids of this protein. The fused segment was inserted into a plasmid carrying the *lac* promoter and the β-galactosidase ribosome binding sequence. The hybrid plasmid contained the proper spatial arrangement between the regulatory elements, and the AUG starting codon specified the production of the 186,000 monomers of the growth hormone/*E. coli* cell.

It is often desirable to clone in another organism. *Bacillus subtilis* is an obligate aerobe that sporulates and has been widely used in the fermentation industry for the manufacture of enzymes. The development of a *B. subtilis* cloning system has been hindered by the absence of suitable vector replicons. Ehrlich (*7*) isolated plasmids from *Staphococcus aureus* that code for resistance to kanamycin and chloramphenicol and could be transformed and cloned in *B. subtilis.* Figure 6 show the construction of an *E. coli–B. subtilis* hybrid replicon (*8*). Plasmic pC194 has a single *Hin*dIII site that can be used as a *Hin*dIII cloning vector in *B. subtilis.* The

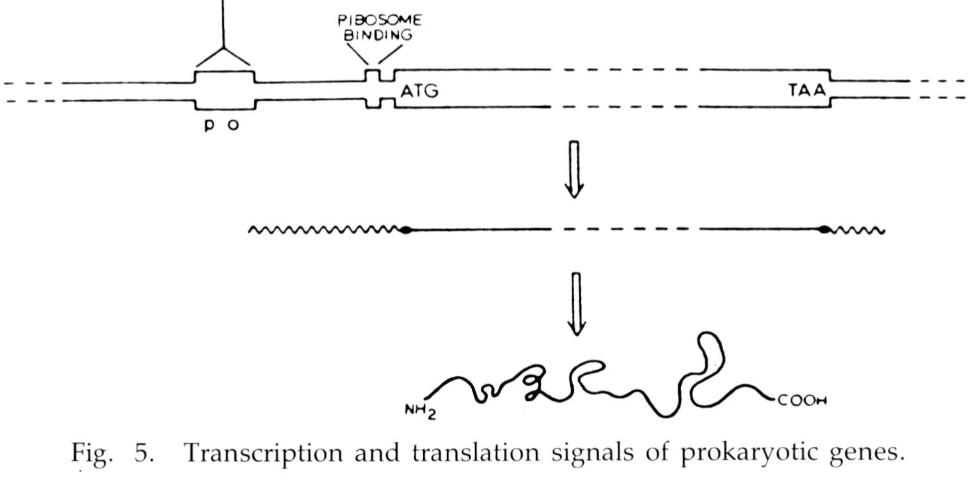

Fig. 5. Transcription and translation signals of prokaryotic genes.

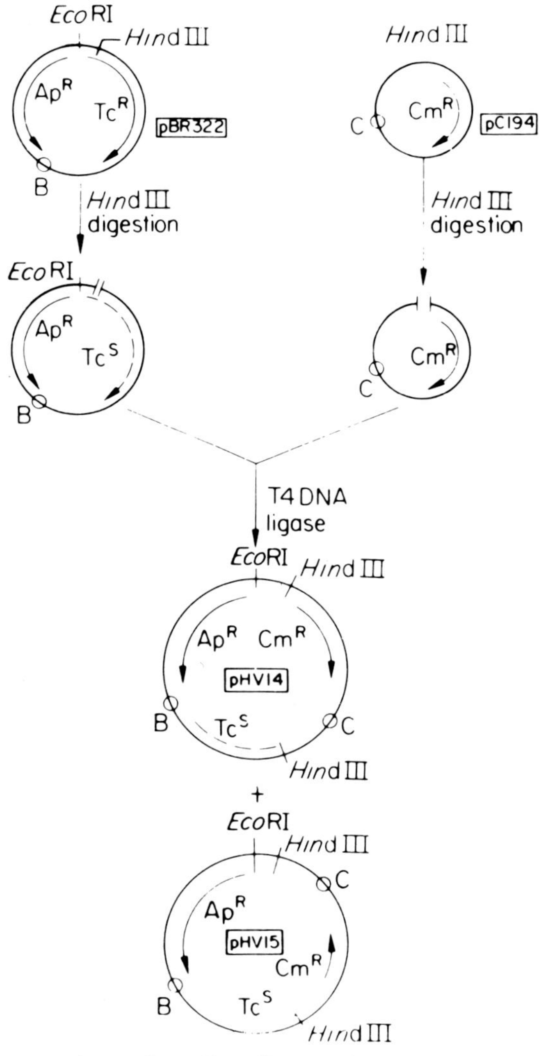

Fig. 6. Construction of an *E. coli–B. subtilis* hybrid replicon. O_B and O_C represent the origins of replication of pBR322 and pC 194, respectively.

related pBR322 can be used also as a *Hin*dIII vector in *E. coli*. Thus, by ligating these two, a hybrid plasmid was constructed that can serve as a vector and be expressed in both *E. coli* and *B. subtilis*. This mixture was used to transform *E. coli* to chloramphenicol resistance (Cm^r). These transformants were obtained and, as expected, were tetracycline-sensitive because of the insertion of the pC194, but were still ampicillin resistant.

Because of its longstanding industrial importance, there was a great deal of enthusiasm in cloning in *Saccharomyces cerevisiae*. However, not until the demonstration of transformation in yeast by Hinnen and colleagues in 1978 (*9*) was it possible to introduce cloned segments into yeast. They achieved the transformation process by incubating DNA with yeast spheroplasts, and then allowed these cells to regenerate their cell walls in a stabilizing medium. When this was accomplished, shuttle plasmids were constructed from naturally occurring 2μ circular DNA. In addition, segments of yeast chromosomal DNA that provided replicator functions were included in the plasmid. Insertion of centromere DNA sequences helped to stabilize the replicating fragments at high copy number.

PROTEIN SECRETION

The attractiveness of gene splicing for producing biological materials of commercial interest is greatly enhanced if the product can be secreted from the cell. Protein secretion in bacteria is a well-known phenomenon and coupling of the secretion mechanism to control a given protein has already been used. As mentioned earlier, yeast is attractive as an industrial microorganism because it does not lyse in a stationary-phase culture or in cultures that are grown in fermentors to high cell densities. The growth medium contains only a few major protein species, representing less than 0.5% of the total yeast protein. Thus, secretion control in yeast holds a great deal of promise. Interest in protein secretion from *S. cerevisiae* has been primarily focused on invertase and acid phosphatase enzymes, both of which are secreted into the periplasmic space. Novick and Schekman (*10*) examined the secretion pathway in temperature-sensitive secretion mutants and recognized that secretion occurs through organelles. Perlman et al. (*11*) and Carlson and Botstein (*12*) demonstrated that the enzymes are expressed as a precursor having hydrophobic amino-terminal signal sequences that are subsequently removed during the secretion process. *Saccharomyces cerevisiae* carries a single invertase structural gene, *SUC2*, that produces distinct invertase mRNA and polypeptides with a secreted and cytoplasmis form of the enzyme. The two translation polypeptides, at 60 and 62 kdaltons, correspond to the nonglycosylated cytoplasmic form of invertase and the precursor of glycosylated secreted invertase, respectively. The polypeptide map and DNA sequencing of the cloned invertase gene are shown in Fig. 7. The

```
                                                                     β   |
MET LEU LEU GLN ALA PHE LEU PHE LEU LEU ALA GLY PHE ALA ALA LYS ILE SER ALA SER MET THR ASN GLU
ATG CTT TTG CAA GCT TTC CTT TTC CTT TTG GCT GGT TTT GCA GCC AAA ATA TCT GCA TCA ATG ACA AAC GAA
1

THR SER ASP ARG PRO LEU VAL HIS PHE THR PRO ASN LYS GLY TRP MET ASN ASP PRO ASN GLY LEU TRP TYR
ACT AGC GAT AGA CCT TTG GTC CAC TTC ACA CCC AAC AAG GGC TGG ATG AAT GAC CCA AAT GGG TTG TGG TAC
                                 100

ASP GLU LYS ASP ALA LYS TRP HIS LEU TYR PHE GLN TYR ASN PRO ASN ASP THR VAL TRP GLY THR PRO LEU
GAT GAA AAA GAT GCC AAA TGG CAT CTG TAC TTT CAA TAC AAC CCA AAT GAC ACC GTA TGG GGT ACG CCA TTG
                                                                        200
```

Fig. 7. Yeast invertase sequence.

amino acid sequence from cell-free translation of the cytoplasmic mRNA contains a 19-amino-acid sequence—the signal sequence—that is absent from the purified enzyme. The amino acid sequence from the cytoplasmic enzyme, whether determined by cell-free translation of the mRNA or analysis of the purified enzyme, both lack the 19-amino-acid signal sequence. Although encoded by the same structural gene, the in vivo localization of invertase to both the secreted and cytoplasm appears to be a function of the presence of a signal sequence in P62 and its absence in p60. The features of the signal sequence in invertase are common to those of other secretory proteins. Bedouelle and colleagues (*13*) and Emr and colleagues (*14*) have shown that with the well-characterized maltose-binding protein and the lambda-phage receptor protein in *E. coli*, point mutation causing substitutions of a single charged amino acid for hydrophobic residues within the signal sequence prevents secretion. Substitution of polar amino-acid analogs in leucine-rich signals prevent both in vivo and in vitro secretion as shown by Hortin and Boime (*15*). To test the possibility suggested by Blobel (*16*) and others (*17*) that the protein confirmation of the signal sequence may control insertion to the lipid bilayer, Perlman and I compared the sequences of 39 different secretory proteins from both prokaryotic and euakaryotic cells. Since the signal sequence of a presecretory protein could encounter an aqueous, proteinaceous, or lipid environment as it spans the membrane lipid bilayer, the hydrophobic amino acids within the signal sequence might form intramolecular hydrogen bonds with other transmembrane proteins or intramolecular bonds in a lipid environment. Based on probability calculation from sequence and X-ray structural data by Chou and Fassman (*18*), the model in Fig. 8 was constructed for a secreting protein (*19*). Of the 39 proteins, although varying in length and amino acid composition, the labile peptides share a hydrophobic core of approximately 12 amino acids, the distance needed to span the lipid bilayer. A positively charged residue, lysine or arginine, usually precedes the hydrophobic core. Core termination is defined by the occurrence of a charged residue, a sequence of residues that may induce a β turn in the polypeptide or an interruption in potential α-helix or β-extended strand structure.

Furthermore, we propose the existence of a signal peptide recognition sequence, A-X-B, with the preferred cleavage site located after the sixth amino acid following the core sequence. The orientation of the carboxyl end of the polypeptide toward the lumen of the endosplasmic reticulum, coupled with the predicted β turn typically found between the core and the cleavage site, implies reverse hairpin insertion of the signal sequence, as proposed by Steiner and colleagues (*17*) in 1980.

Thus, the overall yeast pathway appears to start with a translation of the preprotein at the endoplasmic reticulum where the presequence (amino-terminal signal peptide) is removed after passage through the membrane and is then transported to the Golgi, where vesicle intermediates are formed. These vesicles then move to the bud of the plasmid

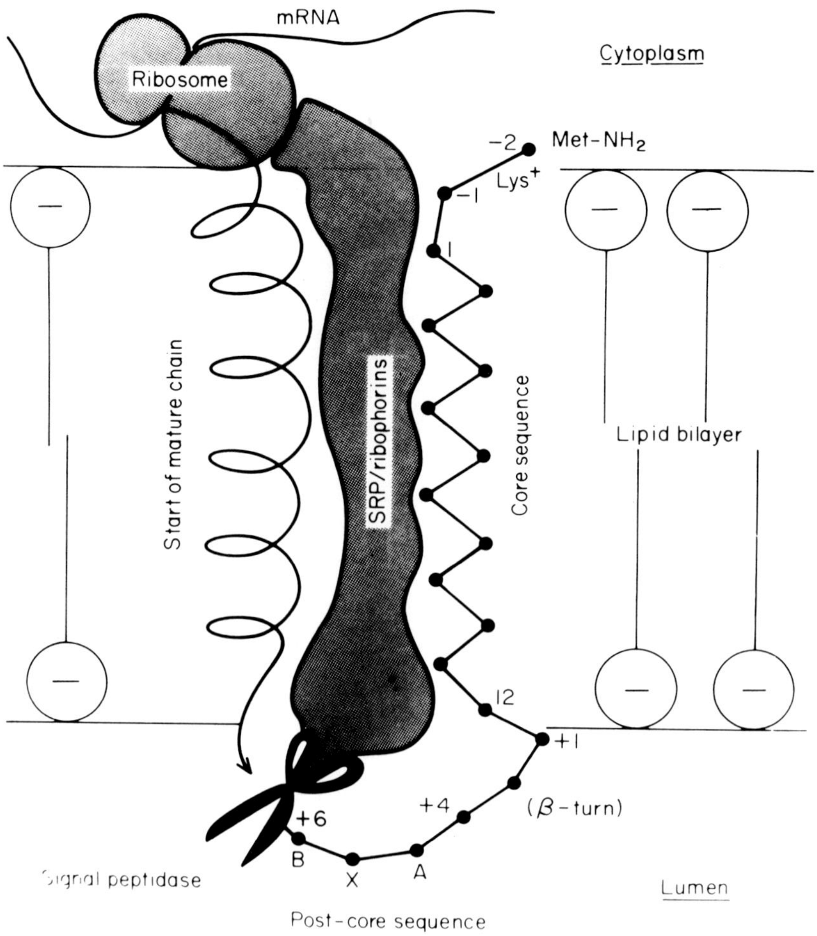

Fig. 8. Proposed orientation of a typical nascent secretory polypeptide with a 20-amino-acid signal sequence at the time of signal peptidase cleavage.

membranes, where fusion occurs, with the subsequent release of the protein from the cell. Hitzeman and colleagues (*20*) have recently employed plasmids that allow expression of heterologous genes in yeast. They constructed a portable yeast promoter from a highly expressed yeast glycolytic gene encoding 3-phosphoglycerate kinase. When the gene for human interferon-α_1, containing the secretion signal, was inserted in this plasmid, the interferon gene was transcribed, translated, and processed to yield secreted interferon. The amount of interferon secreted varies depending on the growth stage of the cells, with maximal secretion occurring at the stationary phase in shake flasks and at high cell density. Thirty percent of the interferon is present in the medium.

The second promising and well-studied secreted enzyme in yeast is that of the derepressible acid phosphatase. The current regulatory model for acid phosphatase in *S. cerevisiae* (Fig. 9) is based upon the extensive

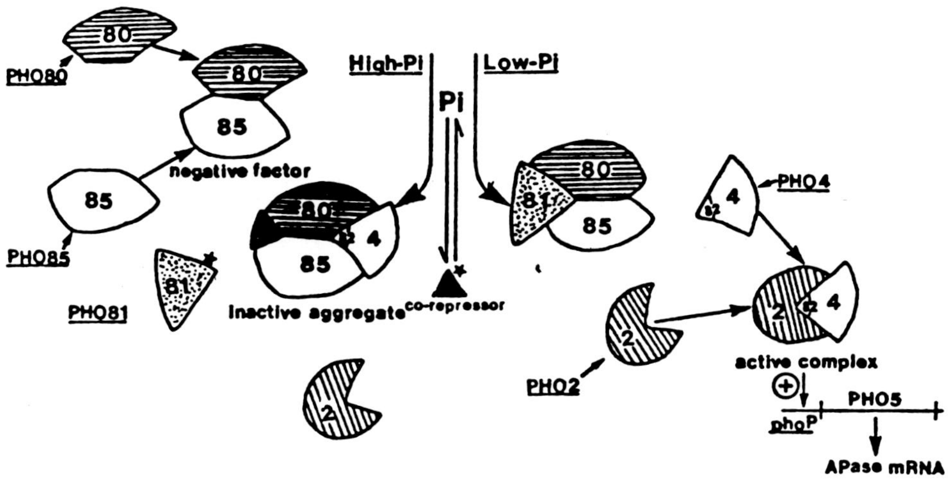

Fig. 9. Regulatory model for APase regulation in *S. cerevisiae*. *Alternatively, corepressor could function by binding to *PHO81* to mediate binding of *PHO81* to the inactive aggregate.

studies by Toh-e and colleagues (*21*) in Japan, complemented by the efforts of Bostian et al. (*22*) and of Kramer and Anderson (*23*). The phosphatase genetic system is a dispersed gene system consisting of numerous structural and regulatory genes scattered throughout the genome. This regulatory model involves a simultaneous functioning of the regulatory factors by direct molecular interaction. A few molecules of the regulatory factors are produced constitutively. Under repressed conditions, the *PHO*4 gene product aggregates with a complex of the *PHO*80 and *PHO*85 gene products at a protein domain identified by *PHO*82 and is unable to activate transcription of *PHO*5. Under derepressed conditions, the *PHO*81 products bind to this aggregate and are regulated by the aporepressor, which is either a phosphate or a metabolite, and the *PHO*4 product is released. It binds, in combination with *PHO*2 protein, to an operator of the structural gene and activates transcription. Bostian, in collaboration with Kramer and Trill (*24,25*), sequenced clones of this gene, as shown in Fig. 10. The *PHO*3 gene product, produced in high-phosphate, and the *PHO*5 gene product, expressed in low-phosphate, are both preceeded by a signal or leader sequence. These are cleaved at a common glycine residue at position 17. Canonical eukaryotic-promoter consensus sequence and regions of conserved homology between these genes are both shown to the left of the leader sequence. The boundaries for regulatory sequence is upstream of the *PHO*5 gene that is contained within the *Bam*HI restriction site, 554 base pairs upstream of the initiator codon. Removal of the UAS fragment from —276 to —554 leads to the loss of expression of the *PHO*5 gene. This is analogous to the upstream activation sites (UAS) observed for the *CYC*1 and *GAL*10 genes in yeast and other higher eukaryotic genes.

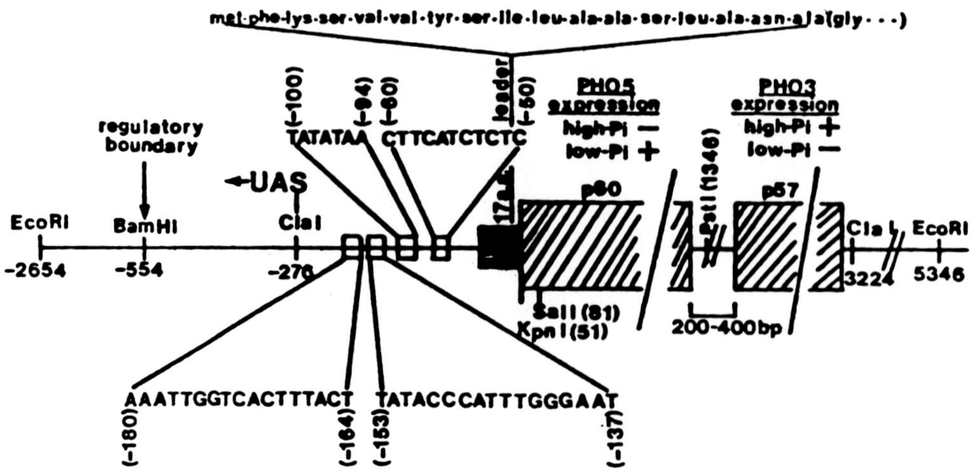

Fig. 10. Schematic model for *PHO5/PHO3* gene organization.

From a technical point of view, the acid phosphatase system is very attractive. The enzymes in phosphate metabolism are regulated by external growth concentrations of inorganic phosphate. Under nonlimiting conditions they are present in basal amounts. However, we found that when cells are starved for inorganic phosphate, these enzymes are derepressed, internal reserves of polyphosphate are depleted, mRNA, polypeptides, and active enzymes are synthesized, and the overall metabolism becomes dependent upon exocellular acid phosphatase (*26*). When inorganic phosphate is replenished, the enzymes are repressed through a coordinate pathway.

Protein Engineering

Genetic engineering techniques have developed to the point where we can now synthesize DNA fragments that code for precise amino-acid sequences. From X-ray crystallography one can determine the likely structure that a given amino acid would have in a secretory protein. Thus, we can design not only leader sequences, but starting with the known crystal structure for the final protein, modify the gene to alter that structure in a highly predictable fashion. These include active sites, binding sites for regulatory effector molecules, and other domains of the molecule itself. Although this technology is in its infancy, genetic engineering companies now employ protein crystallographers. As pointed out recently by Ulmer (*27*), if enzymes are to be more widely used as industrial catalysts, we need to develop methods to tailor their properties to the process of interest. Molecular studies of enzymes isolated from bacteria that are thermostable, salt tolerant, mutants resistant to analogs or to allosteric regulation, or strains that grow at high or low pH will all contribute clues on structure and function. Structural and conformational studies of the purified enzymes should overcome some of the difficulties in incremental conventional mutagenesis. As pointed out by Ulmer, an interesting prospect is to modify the most abundant protein on earth, ribulose-1,5-biphosphate carboxylase. This enzyme is responsible for carbon dioxide fixation in photosynthetic organisms. The enzyme uses molecular oxygen as a substrate, and this results in photorespiration in plants. Approximately 50% of the fixed carbon is lost in this process, and, therefore, there is a great deal of interest in methods for eliminating or reducing the activity of this enzyme. The gene has been cloned, and its structure is under current investigation.

APPLICATION IN AGRICULTURE

The technological revolution is clearly evident from the fact that we can now genetically alter proteins, such as immunotoxins that would function as pharmacological reagents, antibodies that would serve as diagnostic reagents, short synthetic peptides that would elicit antibody response, insecticides and various microbial enzymes that would function as immobilized catalysts. This revolution is particularly significant in agriculture. Two recent developments have made it possible to manipulate the genome of plants. It is now possible to conduct genetic transfer in plants through intraspecific protoplast fusion. Protoplasts of sexually incompatible species have been fused to give rise to somatic cell hybrids. In such hybrids, partial elimination of parental chromosomes is common. Such a combination occurs in the potato–tomato somatic hybrid, for which numerous phenotypes and carrier types are encountered. Also, plant transformation vectors have been discovered. Among these, *Agrobacterium tumefaciens,* a soil bacterium that incites crown gall disease in a variety of dicotyledonous plants, has proved a successful plant vector (*28*). Tumor-inducing plasmids insert their genome into the nuclear DNA of a host plant, leading to transformed cells. A number of opportunities exist for genetic engineering in plants (*29*). First, seeds of legumes in cereal grains directly provide humans with approximately 70% of the dietary protein requirement. Thus, increasing seed proteins is a highly desirable goal. Second, utilizable nitrogen needs of agricultural crops can be supplied by genetically engineering the ability to reduce N_2 directly to ammonia through nitrogen fixation. Surveys of legumes suggest that enhancement of nitrogen fixation efficiency may be obtainable. Third, the ultimate value of plants lies in their ability to convert solar energy into stored chemical reserves through photosynthesis. Methods for the transformation of this apparatus are now available, and there are hopeful prospects of increasing the efficiency of the electron flow through photosystems I and II, raising the level of light saturation and enabling more efficient light harvesting. Fourth, a large proportion of total world crop production is lost through pest or pathogen damage. This can be diminished by increasing our understanding of pathogenicity, providing early detections of infection, or engineering resistant crop varieties. Finally, we can improve cultivation practices in which crops are subjected to a variety of environmental extremes. Adaptation of plants to stress, such as water deficiency or high temperatures, involve specialized plant morphology, i.e., leaf surface area and the presence of fewer stomatal openings, promoting greater plant water retention. Although the molecular respones that are induced by stress are unknown, heat-shock proteins have been identified and can be studied. Eventually, mechanisms for engineering resistance in varieties will be found.

The advent of the miracle drugs, which for the past 40 yr has provided us with a defense against infections, has focused attention on the secondary metabolites of microorganisms. Although industry concentrated on antibiotics, many other potential applications remain. Demain, a microbiologist long interested in industrial microbes, recently reminded us (*30*) that Streptomyces produces a narrow spectrum antibiotic, monensin, active against coccidiosis, one of the major economic diseases of poultry. *Bacillus thurengiensis* produces a crystal protein active against mosquitoes. Low-molecular-weight microbial metabolies active against insects include inhibitors of respiration, protein synthesis, and membrane-active agents. A nucleoside analog, nikkomycin, is now being used as an agriculture antifungal agent that functions as an inhibitor of chitin synthesis. Pharmacologically, there are many examples; to cite but one, cyclosporin A, an antifungal, antibiotic produced by *Tolypocladium inflatum,* is used as an immunosuppressive agent in human organ transplants. Vasodilators have been isolated from *Streptomyces aureofaciens.* Demain enumerated (*30*) a number of other enzymes that appear to be involved in disease processes. From the studies of Umezawa and colleagues (*31*), a number of effective enzyme inhibitors have been isolated from microbial broths. These include inhibitors of 3-hydroxy-3-methylglutaryl-CoA reductase, the rate-limiting enzyme of chloresterol synthesis, dopamine-β hydroxylase, influencing hypotensive activity in animals, intestinal glycosidases affecting carbohydrate-dependent diseases, such as diabetes, and pancreatic esterases in delayed-type hypersensitivity. A particularly promising area has been protease. A number of potent protease inhibitors have been isolated from *Streptomyces* broths. Best known of these are leupeptin, antipain, chymostatin, elastinal, bestatin, and pepsadin. Their importance is indicated by the fact that elastinal and antipain are active ingredients in chemical mutagenesis. Pepstatin inhibits focus formation by murine sarcoma virus on mouse cells. Elasnin, a fungus metabolite, inhibits the enzyme elastase, which is involved in chronic obstructive lung diseases, such as emphysema. Bestatin, a product of *Streptomyces olivoreticuli,* is a specific inhibitor of aminopeptidase B and leucine aminopeptidase is believed to be involved in the immune response. These represent promising targets for future exploitation.

As the late Sol Spiegelman predicted—the future is depressingly bright. In this biotechnology revolution, the scarcest resource that we have is not energy, but training and information.

ACKNOWLEDGMENTS

This work was supported by the US Public Health Service Grant AI 10610 and by a NSF grant PCM-802217.

REFERENCES

1. Dienert, F. (1900), *Ann. Inst. Pasteur* **14,** 139.
2. Benzer, S. (1953), *Biochim, Biophys. Acta* **11,** 383.
3. Rotman, B. (1973), *Fluorescence Techniques in Cell Biology.* Thaer, A. A. and Sernetz, M. eds.), Springer-Verlag, NY, p. 333.
4. Monod, J., and Jacob, F. (1962), *Cold Spring Harbor Symp. Quant. Biol.* **26,** 249.
5. Vournakis, J. N., and Elander, R. P. (1983), *Science* **219,** 703.
6. Goeddel, D. V., Shepard, H. M., Yelverton, E., Leung, D., and Crea, R. (1980), *Nucleic Acids Res.* **8,** 4057.
7. Ehrlich, S. D. (1977), *Proc. Natl. Acad. Sci. USA* **74,** 1680.
8. Old, R. W., and Primrose, S. B. (1981), *Principles of Gene Manipulation, an Introduction to Genetic Engineering,* University of California, Berkeley, CA.
9. Hinnen, A., Hicks, J. B., and Fink, G. R. (1978), *Proc. Natl. Acad. Sci. USA* **75,** 1929.
10. Novick, P., and Schekman, R. (1979), *Proc. Natl. Acad. Sci. USA* **76,** 1858.
11. Perlman, D., Halvorson, H. O., and Cannon, L. E. (1982), *Proc. Natl. Acad. Sci. USA* **79,** 781.
12. Carlson, M., and Botstein, D. (1982), *Cell* **28,** 145.
13. Bedovelle, H., Bassford, P. J., Fowler, A. V., Beckwith, J., and Hofnung, M. (1980), *Nature* **285,** 78.
14. Emr, S. D., Hall, M. N., and Silhavy, T. J. (1980), *J. Cell Biol.* **86,** 701.
15. Hortin, G., and Boime, I. (1980), *Proc. Natl. Acad. Sci. USA* **77,** 1356.
16. Blobel, G., and Dobberstein, B. (1975), *J. Cell Biol.* **67,** 835.
17. Steiner, D. F., Quinn, P. S., Chan, S. J., Marsh, J., and Tagu, H. S. (1980), *Ann. NY Acad. Sci.* **343,** 1.
18. Chou, P. Y., and Fasman, G. D. (1978), *Adv. Enzymol.* **47,** 45.
19. Perlman, D., and Halvorson, H. O. (1983), *J. Mol. Biol.* **167,** 391.
20. Hitzeman, R. A., Leung, D. W., Perry, L. J., Kohr, W. J., Levine, H. L. and Goeddel, D. V. (1983), *Science,* **219,** 620.
21. Toh-e, A., Inouye, S., and Oshima, Y. (1981), *J. Bacteriol.* **145,** 221.
22. Bostian, K. A., Lemire, J. A., and Halvorson, H. O. (1983), *Mol. Cell Biol.* **3,** 839.
23. Kramer, R. A., and Anderson, N. (1980), *Proc. Natl. Acad. Sci. USA* **77,** 6541.
24. Thill, G. P., Kramer, R. A., Turner, K., and Bostian, K. A. (1983), *Mol. Cell Biol.* **3,** 570.
25. Anderson, N., Thill, G. P., and Kramer, K. A. (1983), *Mol. Cell Biol.* **3,** 562.
26. Bostian, K. A., Lemire, J. A., and Halvorson, H. O. (1983), *Mol. Cell Biol.* **3,** 839.
27. Ulmer, K. M. (1983), *Science* **219,** 666.
28. Ream, L. W., and Gordon, M. P. (1982), *Science* **218,** 854.
29. Barton K. A., and Brill, W. J. (1983), *Science* **219,** 671.
30. Demain, A. L. (1983), *Science* **219,** 709.
31. Umezawa, H., Takeuchi, T., Nitta, K., Okami, Y., Yamamoto, T., and Yamaoka, S. (1953), *J. Antibiot.* Ser. A **6,** 101.

Renewable Fuels and Materials

Oil From Plants

MELVIN CALVIN

Department of Chemistry and Lawrence Berkeley Laboratory, University of California, Berkeley, California 94720

INTRODUCTION

Growing oil, one of the clearest examples of renewable fuel, means just that. Up until now, we have been mining oil, the ancient photosynthetic product of several hundred million years ago. The question now most frequently arising is about new sources of ancient photosynthetically produced materials, and there is no longer a tenable positive answer to this question. Other alternatives must be found.

An immediate need for alternate energy sources is emphasized by the history of productivity of drilling rates for US domestic petroleum from 1945 to the present. It has been demonstrated that the amount of oil found/ft of well drilled is falling, from 35 barrels of oil/ft drilled in 1945 to less than half that amount in 1975; the rate of discovery is falling. Also, the energy cost of drilling and extracting oil is rising (*1*). The productivity is falling, the cost of obtaining the oil is rising in terms of energy, and somewhere near the year 2000 the energy cost of finding and extracting a barrel of oil will exceed the energy content of that barrel of oil. When that occurs, even if there is oil in the ground, the energy cost of extraction will prohibit any further use of the oil because it will cost more, in energy terms, to extract it than it will be worth.

CARBON DIOXIDE PROBLEM

One alternative that has been suggested is to use coal (also a product of ancient photosynthesis), huge supplies of which are available in the United States, the Soviet Union, Europe, and China, but this also has some problems. About 20–30 yr ago, we were admonished to restrict the use of coal for environmental reasons, such as the destruction of the land by strip mining, destruction of the miners by deep mining, and destruction of the people by the effluents of burning coal that consist of acid rain and unburned hydrocarbons, which are carcinogens. We transformed our coal-burning power plants into clean (gas) or low-sulfur (oil) burning power plants to eliminate or reduce environmental hazards. As a result of the oil embargo in 1973, there has been a return to the idea of using coal for power plants, especially by indirect combustion (coal conversions to gas or liquid hydrocarbons that would be useful) rather than by older combustion methods. When we suggest that coal in any form can be used as an alternative to burning oil, there will be an environmental constraint to prevent the use of coal power over a very extended period.

The combustion of fossil carbon in any form, but particularly coal, has created a problem, particularly over the last 100 yr, but in an accelerated form in the last 20 yr. This is a result of the combustion of fossil carbon, and it does not matter in which form the fossil carbon is burned. Carbon that has been stored in the ground for several hundred million years and is suddenly used, generates excess carbon dioxide. It is not possible to get the heat/energy values from the combustion of fossil carbon (oil or coal) without the production of CO_2. As a result, there has been a 7% increase in the CO_2 levels in the atmosphere over the past 20 yr, even by just burning mostly oil, which is not just carbon, but carbon and hydrogen; oil has approximately two atoms of hydrogen for every atom of carbon, and the heat comes from the burning of hydrogen as well as the burning of carbon. When coal is burned, however, there is less than one atom of hydrogen for one atom of carbon, with the result that roughly twice as much carbon dioxide/million Btu is created than from the burning of oil.

It is possible to observe the annual increase in the CO_2 levels in the atmosphere from the data in Fig. 1, which is taken from a station at the top of Mauna Loa in Hawaii. This site was chosen because of its distance from urban and natural disturbances, with the result that the data obtained is more truly representative of the actual atmospheric situation since it is not contaminated with other pollutants. (There are other stations throughout the world where information about CO_2 levels is obtained. The suggestion was made some years ago that it would be important to establish a similar station in the tropical forests in the Amazon to obtain the vertical CO_2 profile so that the effect of the forest itself on this increase could be observed.) The annual changes (a rise in winter and a fall in summer) of the CO_2 levels in the atmosphere are clearly observable. Notice that the fall in CO_2 concentration in the summer is never as great as the rise in the winter, with the result that the net CO_2 level has been rising constantly since 1958, when the station on Mauna Loa was established, from 315 ppm to over 330 ppm. We can actually extrapolate backward by another type of measurement; the ^{13}C content of tree rings that were laid down in 1860. From the ^{13}C deficiency we can estimate the

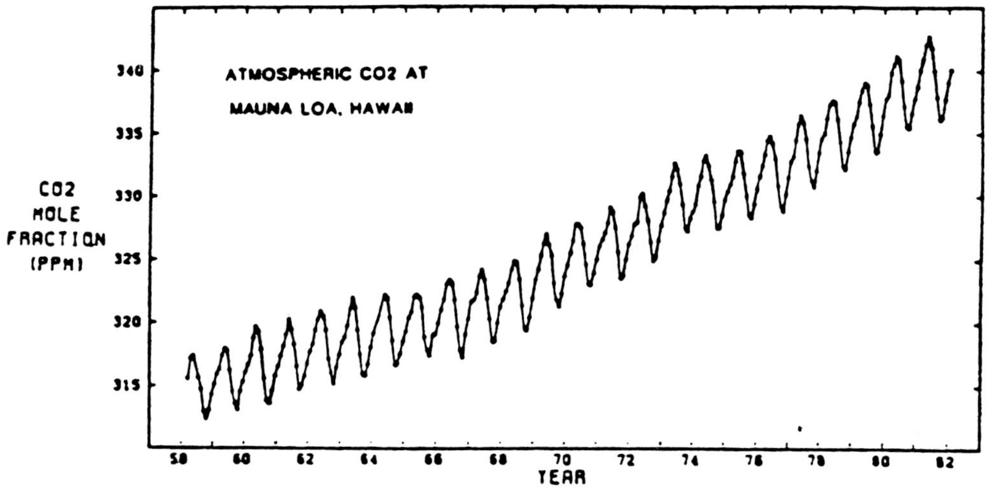

Fig. 1. CO_2 concentration at Mauna Loa, Hi (Data from C.D. Keeling, et. al., Scripps Institute of Oceanography)

total carbon dioxide that was in the atmosphere in 1860; it turns out to be about 290 ppm for that time. So, from 1860 to 1980, the rise has been about 15% in CO_2 concentration, and of that 15%, one-half has been in the period of 1958–1981. There has been a slight decrease in worldwide carbon-based fuel usage since the oil embargo period, but that is a relatively small perturbation on the overall effect of the rise of the CO_2 concentration.

Why is the CO_2 level important? Carbon dioxide is a peculiar gas. It is transparent to visible light, and approximately 99% of the visible light is converted to heat when it strikes the surface of the earth after passing through the atmospheric CO_2 blanket. This heat is reirradiated back into space, but the carbon dioxide is opaque to infrared light and absorbs it and rereflects some of it back down to the earth's surface. The CO_2 blanket thus acts as a one-way valve, letting heat into the surface of the earth and not allowing heat to escape again into space, with a result that the earth's surface temperature rises (2,3).

Effect of Rising World Temperature

Let us discuss for a moment what the rising temperature of the earth may be expected to be and what the possible economic and social costs might be as a result. Estimates have been made by various means and the data obtained show the rising temperatures as a result of synthetic fuel use, which might be as great as 4°C from what it is today, a very large change in the global average temperature. Even if we now use nonfossil replacement fuels, with fast growth we still have the inertia of the CO_2 rise that has already begun and will continue to do so, as will the temperature rise. Synfuels and nonfossil replacement fuels with no growth still show an increase in the projections for global temperature. There is no way to stop the temperature rise that has already begun, and it will take 50–100 yr before the temperature rise levels off or even begins to fall again. We can be sure that if such temperature rises take place, there will be profound effects on agriculture on a global scale, on human distribution, and on human societies, which will have to adjust to a relatively rapid change.

Several years ago the National Academy of Sciences organized a study on the effects of the carbon dioxide concentration increase in the atmosphere, and in 1982 they updated their findings (*2*). The 1982 report indicated that the increase in temperature of 3°C postulated in 1979, which was predicted to accompany a doubling of atmospheric CO_2 concentration, needs "no substantial revision" at this time. The potential of increasing concentrations of atmospheric CO_2 to produce climatic effects is substantial, and the scientific uncertainties that surround the nature and extent of these factors require extensive worldwide monitoring of temperature, ocean cycling, and other factors.

Is it possible to detect any consequence of the 15% rise in the CO_2 level that has occurred in the last 100 yr? Can we extract this rise from the "noise" of the annual fluctuations in time and in space of the temperature and weather in general? A number of efforts have been made to find "early warning signals" large enough to be believable, but not so serious as to cause agricultural, economic, and social damage. If we wait until the evidence is unambiguously clear, until the temperature actually rises unambiguously, it will be too late to alter our living patterns enough to depart from the inertial curve. We need some "early warning signals" to inform us about whether the temperature rise is happening and the order of magnitude of its occurrence (*4,5*). I have collected three or four different "signals," which are independent observations, but, when taken together, I believe you will agree that they are unambiguous evidence that the heating up effect is already beginning to show. The simplest is the temperature history of some locality; the temperature change data for the Northern Hemisphere from 1880 to 1981 are shown in Fig. 2, indicating that 1981 was a peak year for warm weather, however, there are tremendous fluctuations in this data. The rise in temperature from 1880 to 1980 is substantial, even allowing for the large "noise" level from year to year. This is not too good an early warning "signal" because the annual fluctuations are so great it is hard to distinguish a trend. Evidence has also been accumulating that the ocean surface temperature is increasing as a result of increasing atmospheric CO_2 concentration, and an extrapolation has been made as to the time dependent response of the ocean temperature to a possible instantaneous doubling of CO_2 concentration (Fig. 3) (*2*).

A more sophisticated piece of evidence concerns the polar ice cap. If the global temperature is indeed rising, one would expect that the polar ice caps would show some consequence of that increase since they are an

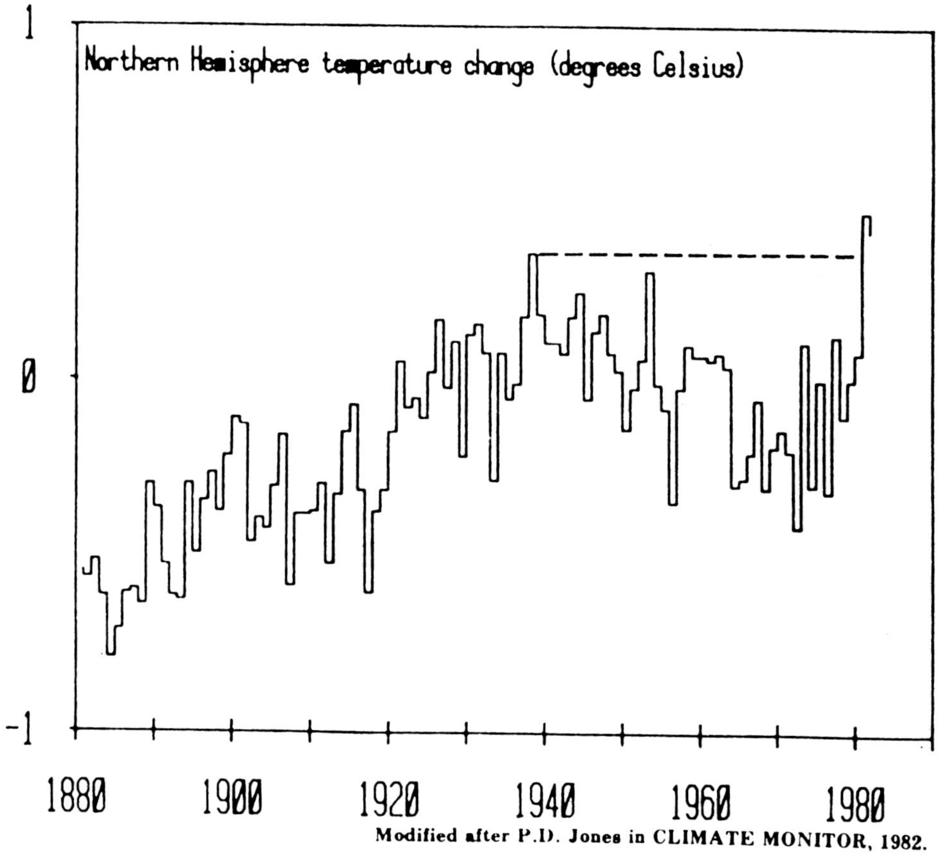

Fig. 2. 1981 sets modern heat record.

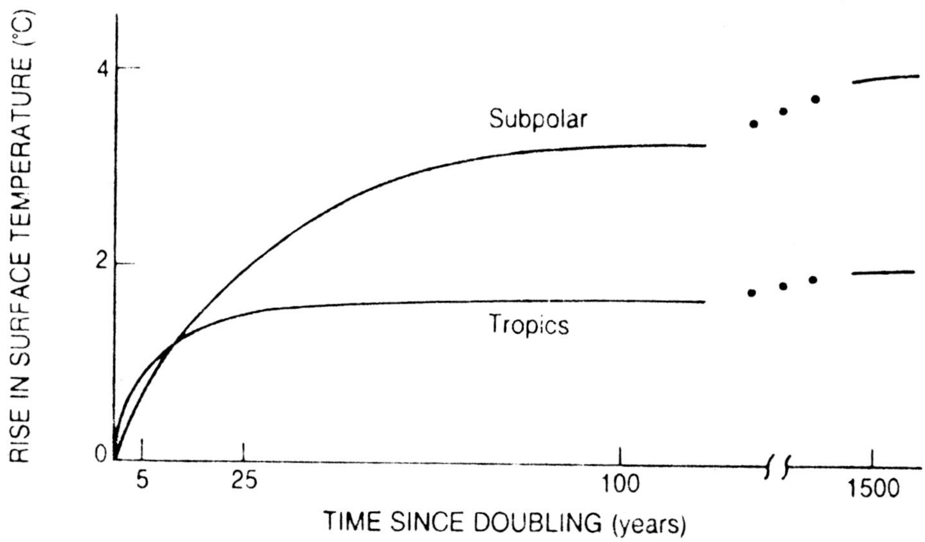

Fig. 3. Time-dependent response of ocean surface temperature to doubling of atmospheric CO_2 (National Academy of Sciences, 1982).

accumulation of many years. The accumulation of ice on the polar ice caps is an integrated effect unlike the fluctuations in annual temperature. (Temperature measurements are point measurements, whereas the amount of polar ice is an integrated measurement of the amount of heat coming through and leaving the earth's surface, and it smooths out the "noise" level.) Last year there was a satellite photograph of the South Polar ice cap, and that photograph, together with surface measurements by the US and Soviet navies has provided evidence that the South Polar ice cap has decreased in the last 100 yr. This data indicates that the South Polar ice cap has been diminishing, and in the last 7 yr it has fallen from about 12–13 million km^2 to approximately 10 million km^2, in other words, 2 million km^2 of ice have disappeared (melted) from the South Polar ice cap. From temperature distribution data on the ice cap, it is possible to state that the thickness of the remaining ice cap has also decreased (*6*). One can calculate how much ice has melted, and, therefore, how much "new" water has come into the oceans. We can then deduce that the sea level should have risen, and risen more rapidly in the last 20 yr than before. The global mean sea level trend based on tide gage measurements indicates that from 1800 to 1940 the rise was about 1 mm/yr and from 1940 to 1980 the rise has been 2–3 mm/yr. That, I think, is unambiguous evidence, together with the satellite data and the temperature information for the Northern Hemisphere, that the heating effect (greenhouse effect) of the CO_2 is indeed occurring and will have its consequence on agricultural patterns throughout the globe in various ways. Unfortunately, we do not as yet have good enough meteorological models to predict microclimatic changes, (*7*), i.e., to predict where the heat will be greatest, where the rainfall will be increased/decreased, and where the agricultural patterns of the globe will change during the relatively brief period of time (20–30 yr) that may be available.

To give some clue as to what the surface change in the sea level might become, we can examine a period about 50,000 yr ago, when the temperature was higher and there was less polar ice. We find evidence, at least on the East Coast of the United States, that is unambiguous. It has been determined that most of Florida was totally under water, or marshland, and a good bit of the East Coast was flooded or marshy from New Jersey, south to Florida. This is what happened about 50,000 yr ago when the Antarctic ice was less than it is today. We currently are going in that direction once more, with the decreasing South Polar ice cap, increasing global temperatures, and other results of increased atmospheric CO_2 concentration.

What is the alternative? We cannot burn coal much longer. If we do, we are headed toward major temperature changes. Therefore, the coal-burning alternative is only of limited use, for a short time. I believe that coal should not be used at all for environmental reasons, and that the temperature change effects alone can and should limit the use of coal.

PLANTS AS ENERGY SOURCES

We are left with only one alternative, and that is some way of harvesting the sun on an annual basis and not simply using the sun's harvest that has been stored in the ground for several hundred million years. As a first alternative, we must learn to harvest the sun annually, which means growing plants that will produce the kinds of materials we need (*8,8a–d*). We already are doing that today on a good scale, in the form of forestry activities, but these are minor compared to what must be done to meet the fuel and material needs of the next generation. The fact that we can use green plants as a source of almost all the necessary fuel and materials is established through the photosynthetic carbon cycle itself.

Sugar Cane–Puerto Rico

In some countries, energy agricultural practices have already begun. About 10 yr ago, just before the oil embargo, the Puerto Rican government constructed a very large power plant on the south coast, intended to be fueled by oil. That became impractical, but Puerto Rico has an alternative to oil. It is possible to grow sugar cane there on a large scale, and research activities have led the Puerto Rican gvernment to encourage the planting of a special type of sugar cane, called "energy" cane. Normally, sugar cane is bred to reduce the fiber and raise the sugar content, so the processing of the cane for sugar can be economic. However, in Puerto Rico they realized that the fiber from the cane could be used to run the power plant. They therefore returned to an examination of some old clones that were poor in sugar and rich in fiber. These were planted experimentally a few years ago and the Puerto Ricans have now developed a type of "energy" cane (Fig. 4) (*9*). The "energy" cane contains only about 8–9% sugar. Actually, the sugar produced/acre from the energy cane is about the same as in the standard (*Gran cultura*) cane, which is about 14–15%, but it contains about three times as much fiber (bagasse), which is the important part of the cane for fueling the power plants. By utilizing energy cane, Puerto Rico now has both sugar and fiber (fuel), which can be fed into the power plants.

Sugar Cane—Brazil

When we first began looking for energy alternatives back in 1974, I knew there existed plants that were commercially producing hydrocarbons on a large scale, namely, the *Hevea* rubber trees, found mostly in Malaysia and Indonesia, and that these trees were native to the Amazonian forest in Brazil and had been transferred from Brazil through Kew Gardens in London to the southeast Asian plantations, where they grow

Fig. 4. Energy cane, Puerto Rico (Photo Gene Elle Calvin)

today. I thought if *Hevea* is native to Brazil, belonging to the family Euphorbiaceae and making hydrocarbons, might there not be other members of that family, or even of that genus, that would produce hydrocarbons of lower molecular weight, which might be useful as crude oil.

It is in Brazil on a trip in 1974 that we learned about the sugar cane industry and how it can be a self-contained, energy-efficient process. Brazil grows more sugar cane than any country in the world, and in 1974 they produced 10 million t of raw sugar and 400 million L of fermentation alcohol from the molasses of that sugar. As we traveled around Brazil looking for other species of Euphorbiaceae, we saw mostly sugar cane, and we visited mill after mill, finding that the Brazilians were harvesting the cane under conditions of maximum sucrose production.

If the Brazilians wanted to use alcohol as a possible source of liquid fuels and materials, the cane should be harvested for maximum fermentables rather than maximum sucrose. It was easy to determine that if the Brazilians waited for the sugar cane to ripen (for the glucose and fructose phosphate to dimerize), some of the fermentables would be lost because the dimerization is an energy-consuming process. However, if they harvested the sugar cane at the point of maximum fermentables, that is, a little earlier, they should have a higher alcohol yield. Indeed, the Brazilians now harvest a few weeks earlier and get 10–15% more alcohol/ha than they do when the sugar cane is grown for sucrose production only.

In 1975 the Brazilian government made a decision to go into the alcohol-for-fuel program, what they call the Proalcohol program, and this encouraged the growing of cane for fermentables, which meant modifying the growing procedures, modifying the mill equipment to use the sugar juice directly, and so on. The sugar mills themselves became self-supporting in energy in the sense that they had enough bagasse from the cane to generate the steam to run the crushers, run the fermentation process, distill the alcohol, and fuel the trucks that collected the cane. The products from these mills were excess electricity and alcohol.

The Brazilians are now reaching the stage in which they can use the excess electricity from the sugar cane operation to produce nitrogen fertilizer, which is required for the soil for cane plantations. The potassium and phosphorus in the ash from the furnaces are available, but the nitrogen is gone from the soil, so it is necessary to find a way to replenish it. The excess electricity from burning the bagasse can be used to electrolyze water, make hydrogen, and build small ammonia generators. The sugar plantations of Brazil, therefore, will be the first completely self-contained "energy plantations," which can not only produce sugar, but also liquid fuels (alcohol) and petrochemicals from the alcohol that comes from the fermentation process.

HYDROCARBON-PRODUCING PLANTS

Our trips to Brazil and elsewhere (Malaysia, Australia, Spain) were taken for the purpose of finding plants that could produce hydrocarbons of low molecular weight as alternative fuel and materials sources (*8,8a–e*). We continued to look for species in the family Euphorbiaceae and we found a very large number of the genus *Euphorbia*, having roughly 2000 species, all of which produce a latex (similar to *Hevea* latex) containing one-third hydrocarbon, one-fifth protein, and the balance water. Not only are there 2000 different species of *Euphorbia* growing all over the world and under all kinds of conditions and growth habits (tiny plants to large trees), but these growth habits and circumstances can be selected for the kind of land and climate that is available (*10*).

One of the best candidates for hydrocarbon production is *Euphorbia lathyris* (commonly called the "gopher plant"), and plantations of this species have developed in various parts of the world, specifically in the southwest of the United States (*11*), California (*8a*), and Spain (Fig. 5). There are currently plantations of *E. lathyris* in other Mediterranean countries, Africa, the Canary Islands, and Australia (*12*). The reason for our choice of *E. lathyris* was its universal distribution throughout the United States, growing as far north as Montana, in the Southwest, and even on the East Coast. It requires only 12–15 in of water annually and can grow in land that is not suitable for food production. The reason that *E. lathyris* might be one of the best candidates for an alternate energy crop comes from several factors. First, it is possible to grow this plant using irrigation on semiarid land not suitable for food or fiber production. Second, the plant grows to size for harvest in a 5–7 mo period.

Fig. 5. *Euphorbia lathyris,* Spain

Third, the extraction proces is standard for the chemical industry. Also, in addition to the oil, the plant contains a substantial quantity of sugars, fermentable to alcohol. It seems that a yield of 6–10 barrels of oil/acre/yr is achievable using wild seed. Agronomic and genetic improvements in the plant, of course, would increase that yield.

Because *E. lathyris* has been studied more than any other candidate for hydrocarbon production, data is available for many of the factors that are required for a crop of this type, such as genetic variability, irrigation requirements, process development, and analyses of the final products (*10–13*). There is no question that in spite of its widespread distribution, *E. lathyris* may not be the best candidate for arid lands since the water requirement is too high and the resulting products not economically competitive with crude oil. Efforts should be made to learn whether specialty chemicals from *E. lathyris* could be used for feedstock development, in addition to those fractions that are similar to crude oil, thereby making growing *E. lathyris* more economically competitive and useful (*13*).

Another species, *Euphorbia tirucalli,* a perennial that is common in Brazil, Africa, and Israel and also grows prolifically in the southwest of the United States, has been cultivated for its oil producing quality, but it requires more water than *E. lathyris.* At the present time, *E. tirucalli* planatations are being developed in Okinawa and Thailand (Fig. 6), the latter effort actually under the direction of a Japanese chemical company.

There is another plant family, the Asclepiadaceae (milkweeds), that is widely distributed throughout the world. We have seen great areas of some of these plants (*Calotropis procera*) in Puerto Rico, and they are also being developed in the United States by a private firm, Native Plants, Inc., as a possible candidate for liquid fuel production (*14*). Just recently I have learned from Australia of the existence of plantations of *C. procera* (Fig. 7) there as well. Several thousand hectares in Australia have been planted with this species, and the effort is being sponsored by the Australian government to determine whether this plant would be a suitable candidate for fuel and materials production (*15*). The chemistry of *Calotropis* is similar to that of *E. lathyris,* and the extraction process that has been developed for *Euphorbia* and the various transformations can be used with the milweeds.

The processing sequence to recover tepernoids and sugars from *E. lathyris* (Fig. 8) is almost the same for *E. tirucalli* and *C. procera* (*16–18*). This extraction procedure was worked out in the laboratory and has not been used on a pilot plant scale. The process sheet is calculated for 1000 dry t/d of *E. lathyris* material, which would yield 80 t of crude oil and 200 t of fermentable sugars that could produce 100 t of alcohol. About 500 t of bagasse is used to run the process, with a resulting 200 t of bagasse that could be used to distill the alcohol. The fermentation alcohol, which is a byproduct of processing the dried material from *E. lathyris,* is, of course, a starting point for an entire petrochemical industry.

The whole process is self-contained, as in the sugar cane situation, but the economics of processing *E. lathyris* and the products recovered from that effort are more conjectural. Some time ago (1980) I made calculations for a small miniplant, and the price that we would have to receive for the oil and alcohol to make the process economically feasible would

Fig. 6. *Euphorbia tirucalli*, Thailand. (Yuko Chem. Co., Osaka, Japan).

Fig. 7. *Calotropis procera,* Australia (Lyall Williams)

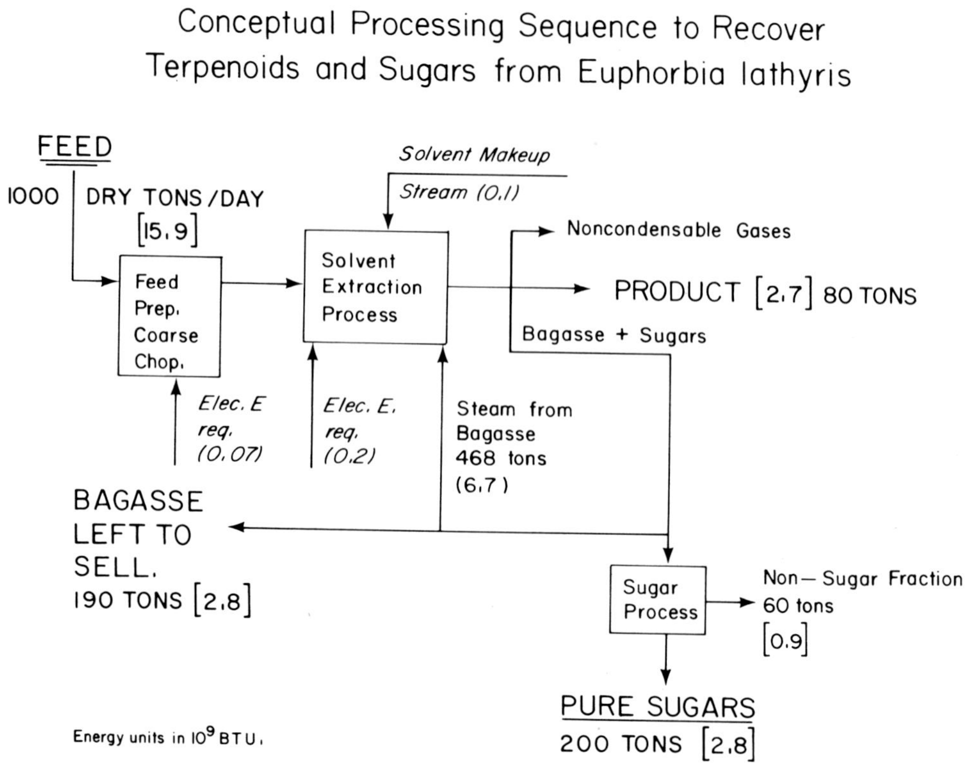

Fig. 8. Conceptual processing sequence of recover terpenoids and sugars from *Euphorbia lathyris*.

be $40/barrel for the oil and $50/barrel for the alcohol. This type of return would make the process economical, including capital costs for construction of the extraction plant and payments to the farmer for growing the *E. lathyris,* which would be approximately $500–300/acre.

Studies of *E. laythris* have shown that, although it is possible to crack the material from this plant, much as crude oil, it might be more economic to try and determine whether this species and its products could contain material more useful for chemical feedstock production, i.e., more specialty chemicals that would not compete with the cheaper crude oil and that would have a more unique situation in world markets (*13*).

The crude oil from the *E. lathyris* has been converted, using special zeolite catalysts developed by the Mobil Corporation, to the usual suite of products, which are similar to those from the standard catalytic cracking of petroleum (*19*). These include olefins, paraffin, aromatics, and nonaromatics. This information confirms the desirability of the products of *E. lathyris* as possible raw materials to substitute for crude oil. Since these results were obtained, as a result of additional data on processing and cracking the material from *E. laythris,* it now appears that a price of $100/barrel for the oil would be more realistic. This is still within reason, considering that it is only three times more than the 1985 price/barrel for crude from OPEC.

Isoprenoid Formation in "Oil" from Plants and Trees

The chemicals in the plant that constitute the black oil are mostly triterpenes (C_{30}), sterols, and sterol esters. It is for this reason that the cracking pattern makes such desirable products. The biosynthetic route to the terpenes in the plant has not been worked out in great detail, but the route in *E. lathyris* is probably similar to that for rubber biosynthesis, except that the end products from the *E. lathyris* are lower-molecular-weight compounds. The biosynthetic route is from sugar via the glycolytic cycle to pyruvate, which is then built up to mevalonic acid, and goes on to give isopentenylpyrophosphate (IPP). The IPP further polymerizes into a variety of isoprenoids. Normally, in the *E. lathyris* the material goes on through the isoprenoid biosynthetic pathway to squalene (C_{30}), which is then folded up to make the C_{30} terpenoid alcohols that constitute the greater percentage of the oil. The actual route is: Sugar to pyruvate to acetyl phosphate to mevalonic acid and on to dimethylallylpyrophosphate (DMAP) and (IPP). The isomeric C_5 compound (DMAP) drops off a pyrophosphate to create a carbonium ion that takes an electron pair from the double bond of IPP to form a C_5–C_5 link, the resulting carbonium ion then drops off the proton to give an identical structure (allylpyrophosphate structure) that can go on to add another C_5 to make a C_{15} compound, then add another C_5 piece to make a C_{20} compound. This process can continue on to polyisoprenes, essentially to rubber.

However, if the plant contains a cyclase enzyme, the carbonium ion can take an internal double bond and close at a cyclic C_{10} compound instead of adding another C_5 group. In the cases of *E. lathyris* and *E. tirucalli*, two C_{15} compounds come together to form squalene, a C_{30} open chain with no pyrophosphate, which then folds up to form the triterpenes, closed C_{30} compounds, mostly sterols.

Would it be possible to improve the yield of the C_{15} compounds in the plants (which are the desired substances) if the process could be stopped at the C_{15} step? This would mean that no cracking operation would have to be performed on the materials since C_{15} is essentially diesel, and the C_{15} cyclic sesquiterpenes can be used directly as diesel fuel without further processing.

Hydrocarbon-Producing Trees

In our search for hydrocarbon-producing plants, we encountered a tree in Brazil that can be tapped and produces sesquiterpenes (diesel oil) (*8f–h*). This tree belongs to the Leguminosae family, and there are several species growing in Brazil. We found one, *Copaifera langsdorfii*, growing in the Botanical Garden in Rio de Janeiro, but that particular species does not produce as much diesel-like fuel as others. In the Ducke Forest in Manaus, we observed the *C. multijuga,* which produces approximately 20 L of sesquiterpene fuel material/tree in 24 h every 6 m. The tree is harvested by drilling a hole in the trunk about 3 ft from the gound, the hole is about 2 cm in diameter and goes into the heart wood of the tree. A pipe is inserted in the hole and the oil drains out of the pipe into a bucket (Fig. 9). After the material is "harvested," the hole is plugged with a bung, and 6 mo later the tree will produce another 20 L from the same hole. The oil comes not from the cambium, as does the rubber latex in *Hevea,* but from the 1–2 mm diameter vertical pores in the heart wood.

The material from the *C. multijuga* is a sesquiterpene that can be put directly into an automobile without any further processing or refining. An experimental plantation of *C. multijuga* has been started at the Amazon research institute, INPA, in Manaus, to show that the sesquiterpenes were "clean" enough to be used directly in automobile engines. They are also studying agronomic practices in connection with this species, with the hope of improving the yield, or perhaps of being able to use more than one hole in each tree for diesel fuel production.

The oil from the trees of the *Copaifera* species, called "copaiba oil," has been analyzed by gas chromatography, and three main products are copene, bergamotene, and carophyllene, all C_{30} compounds and all of them cyclic compounds with C_{15} (*20*).

The biosynthetic method by which the diesel oil from the *Copaifera* is made is the same as that used by the *E. lathyris* (Fig. 10), up to the C_{15} step. The *Copaifera* cyclizes the C_{15} farnesyl pyrophosphate, that is, drops the phosphorus off to give the cyclic C_{15} compounds. One type of enzyme is responsible for the difference in the two end products of *C. multijuga* and *E. lathyris,* and that is the farnesyl pyrophosphate cyclase enzyme. With *E. lathyris,* this compound is dimerized, whereas with the *Copaifera,* the material is cyclized, with many cyclic C_{15} products as a result.

Another family, the Pittosporaceae, contains trees that produce fruit (nuts) that are rich in terpenes. The terpenes, rather than seed oils (glycerides), seem more desirable as fuel candidates. One of the first members of the Pittosporaceae family that came to my attention is native to the Philippines and produces a fruit (nut) that is rich enough in terpenes to be used for illumination. This is *Pittosporum resiniferum,* and the oil from the "petroleum nut" has been analyzed and found to contain a simple mixture of terpenes (*21,22*).

With the *P. resiniferum,* we can assume an average oil yield of 10% and a maximum yield of 18 kg of fruits/tree/yr at a high density planting (2 × 2 m). A maximum oil yield/acre/yr would be 10 barrels. With smaller fruits the yield would be approximately 6.5 barrels/acre/yr. The figure of 6.5 barrels/acre/yr is the same as for *E. lathyris,* and the 10 barrels/acre/yr is the same as the yield for palm oil, which is the best oil yield so far obtained from seed crops. Therefore, the *P. resiniferum* appears to be a very

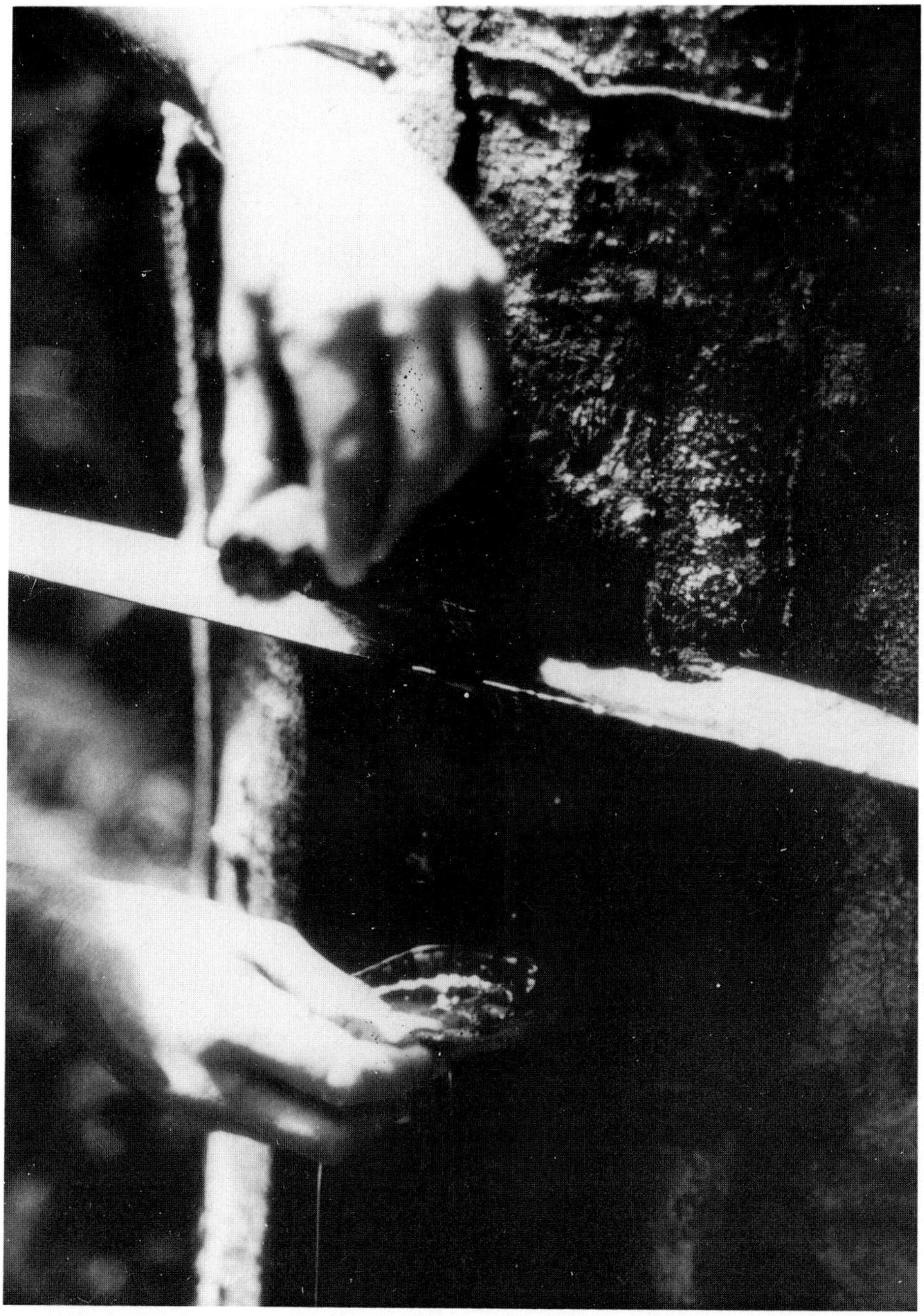

Fig. 9. *Copaifera multijuga*, Manaus, Brazil, showing flow of oil from bung (Ancona).

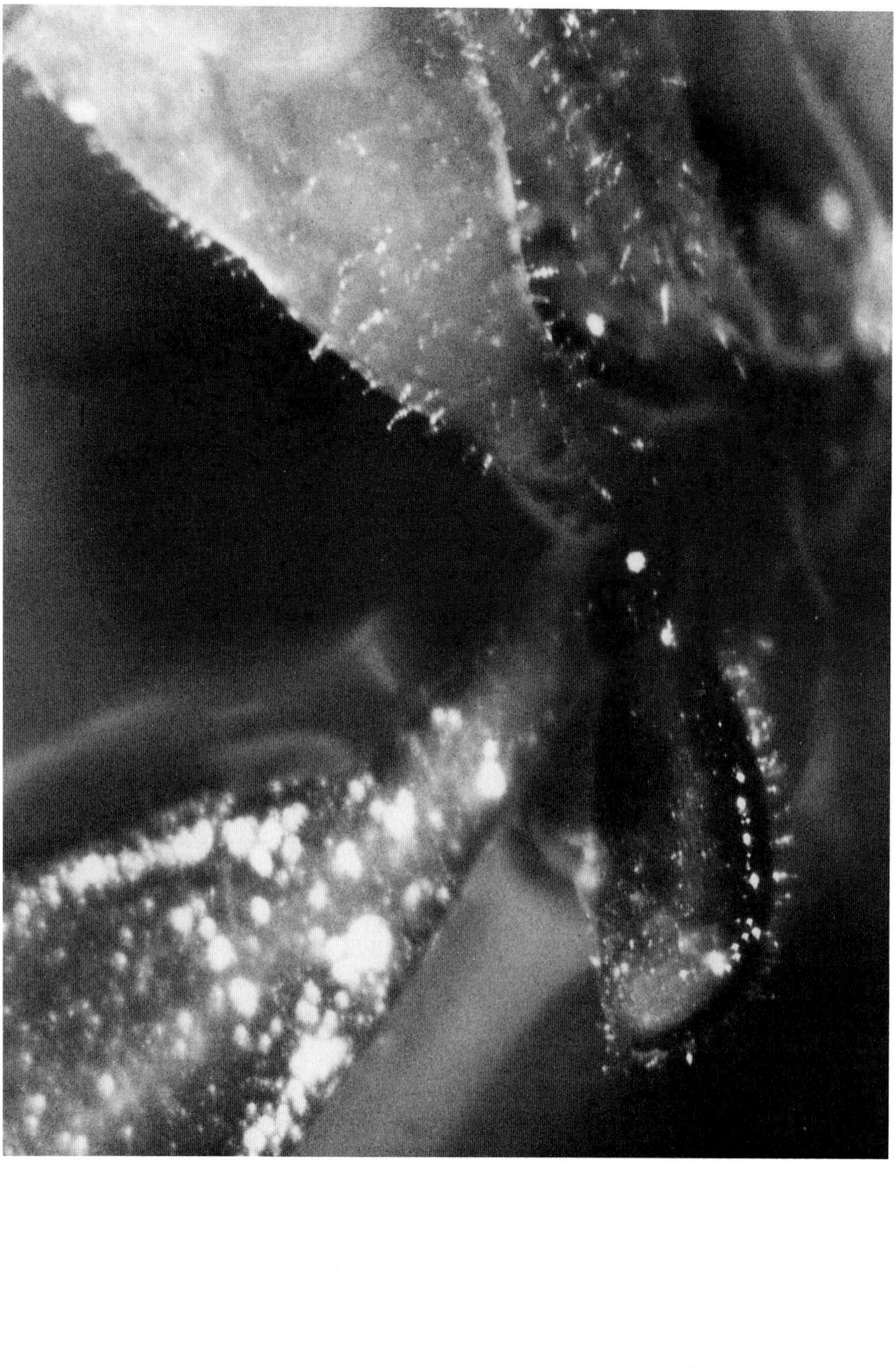

Fig. 10. *Euphorbia lathyris* shoot from protoplast callus. (Redenbaugh and Kawaguchi, Plant Genetics, Inc.)

desirable fuel crop if high-yielding trees can be selected and planted at high density.

We then examined the fruit of a species of *Pittosporum* that is native to California (*P. undulatum*). When the fruits were extracted they were found to have the same terpenes (C_{10} compounds) as the ones native to the Philippines (*21*). Because this genus does grow prolifically in the Philippines, and even in such places as California, these trees may turn out to be good candidates for fuel (terpene) production.

A comparison of biosynthetic routes might be of interest at this point. In the case of *E. lathyris*, the sequence is all the way to C_{15} compounds, followed by dimerization to C_{30} materials. In the *Copaifera* (diesel tree), the sequence goes all the way to C_{15}, followed by cyclization to the sesquiterpenes. For the *Pittosporum*, however, the route is to the C_{10} compounds, followed by cyclization to create monoterpenes in the fruit.

GENETIC ENGINEERING

The question now arises as to whether it is possible to transfer the gene for the production of sesquiterpenes from the *C. multijuga* (or other *Copaifera* species) to such plants as *E. lathyris (8f,h)*. It is not yet possible to grow *Copaifera* in the United States, and, therefore, gene transfer seems to be a method of using the characteristics of sesquiterpene production in that species in another plant species, which can then be grown on an annual basis under conditions that exist in the United States and elsewhere. There is only one enzyme, farnesyl pyrophosphate cyclase, involved in moving the C_{15} instead of having the compounds go all the way to C_{30}, as is the normal route in the *Euphorbia*. In other words, a single gene transplant from a donor cell of the *Copaifera* to the acceptor cell of a plant, such as *E. lathyris*, would be required. However, it is necessary to find a donor cell that has the genes for the desired enzyme, take out the messenger RNA, make a copy of the DNA, insert the copy DNA into a plasmid, clone the plasmid in *Escherichia coli*, and then, by means of the plasmids or other vector, insert the gene into a selected plant. Eventually, the piece of genetic information can be integrated into the nuclear gene of the transformed cell, which has been done with bacteria, but not yet accomplished with higher plants.

The regeneration of whole plants has not yet been accomplished through genetic engineering, although some plants, such as tobacco and carrot, have been regenerated from single cells. To try and accomplish whole plant regeneration by means of genetic engineering techniques, it is necessary to have the two plant species available. The *Copaifera*, since it does not grow in the United States, must be cultivated in greenhouse conditions. We managed to obtain some seeds of various species of *Copaifera* (*C. langsdorfii*, *C. lucens*, *C. multijuga*, *C. officinalis*) and have tried to germinate them without too much success. We do have, however, some plant materials that hopefully will be able to provide us with the donor cells we need to place the enzymatic information into the *E. lathyris* for transformation.

However, to accomplish gene insertion into the acceptor plant, it is necessary to prepare a plant tissue culture. We have been able to take the callus tissue from *E. lathyris* and separate it into protoplasts. The protoplasts do not reproduce very well because they tend to aggregate into new callus tissue. We have, however, separated single protoplasts from the callus, and they do divide and form new callus in only a few hours. We have also regenerated whole plantlets from single protoplasts through such a callus.

One alternative method to introducing new genetic material into plant protoplasts would be to use the technique of cell fusion, that is, take a cell that has the desired gene and fuse it with a cell that does not, then select the plant fused cells (*23*). Eventually, there should be a cell into which the desired gene has been integrated. We have used a cell selection system and measured the process by flow cytometry. We took two sets of protoplasts from *E. lathyris*, staining one of them with rhodamine and the other with fluorescein, both of which fluoresce in different regions of the spectrum. The two cell lines were mixed and a fusion agent added so that the fused cells can be selected. This selection is done via flow cytometry, a method that passes the cells in front of a laser beam (*23*). Two photomultipliers on each side of the laser beam on the flow cytometer examine light of different wavelengths. As the cells go by, a charge is placed on them only if they have both types of fluorescence. That charge allows separation of the charged from uncharged (fused from unfused) cells.

The next step in the process is to regenerate plants from leaf protoplasts, which can be done (Fig. 10). Whole plants have just been regenerated in this way (*24*). The next step will be to examine the regenerated shoots to see if any of them make cyclic C_{15} compounds, at which point we can begin to optimize the process and its steps to produce the new plant material, which will be able to synthesize the sesquiterpene compounds for fuel and materials use.

CONCLUSION

The use of plants for fuel is not a new idea. Two different efforts were made over 40 yr ago in this direction. The first was an attempt by the Italians in Ethiopia to use a member of the Euphorbiaceae family, *E. Abyssinica*, as a source of fuels. These plants grew prolifically throughout Ethiopia, and the Italians built an extraction plant for the oils from these plants. However, time did not permit the development of this idea, and the Italian abandoned the effort when they left Ethiopia in 1938. They had hoped to use this plant species, which contained oil, as the basis of a petroleum industry (*25*). In addition, the French in Morocco in 1940 used another species of Euphorbia, *E. resiniferum*, and succeeded in obtaining oil from wild plants. Their yield was 3 metric t of oil/ha from wild plants (*26*); they did not continue this effort, however.

It is clear that the idea of using plants to create hydrocarbon-like materials as a substitute for petroleum will become more important, especially in some of the less developed areas of the world that have land unsuitable for food production. Various efforts have been made—in Spain, Okinawa, Thailand, Australia—toward this end, and attempts are underway to improve agronomic yields, develop small scale extraction plants, and learn more about the composition of the oil itself, using species of plants that heretofore have not been used for large-scale cultivation. In the United States, efforts along these lines in the southwest and California have shown that it would be possible to produce approximately six barrels of oil/acre/yr on semiarid land.

What is now needed is an effort on the part of the agricultural community to commit itself to an "energy" agriculture that would have long term benefits for our country. However, it is my feeling that we will have to be shown by others that this type of development is feasible before we begin to use this most important resource, the annually renewable green plant, for our own benefit.

ACKNOWLEDGMENT

The work described in this paper was sponsored, in part, by the Assistant Secretary for Conservation and Renewable Energy, Office of Renewable Energy, Biomass Energy Technologies Division of the US Department of Energy under Contract No. DE-AC03-76SF00098.

REFERENCES

1. Hall, C. S., and Cleveland, C. J. (1981), *Science* **211,** 576.
2. Smagorinsky, J. (ed.) (1982), *Carbon Dioxide and Climate: A Second Assessment,* National Academy of Sciences, Washington, DC.
3a. Macdonald, G. J. (ed.) (1982), *The Long-Term Impacts of Increasing Atmospheric Carbon Dioxide Levels,* Ballinger, Cambridge, MA; and *Carbon Dioxide and Climate: The Greenhouse Effect* (1982), Hearing before the Subcommittee on Natural Resources, Agriculture Research, and Environment and Subcommittee on Investigations and Oversight of the Committee on Science and Technology of the U.S. House of Representatives, Washington, DC.
4. Hansen, J., Johnson, D., Lacis, A., Lebedeff, S., Lee, P., Rind, D., and Russell, C. (1981) *Science* **213,** 957.
5. Bryan, K., Komro, F. G., Manabe, S., and Spelman, M. J. (1982), *Science* **125,** 57.
6. Kukla, G., and Gavin, J. (1981), *Science* **124,** 497.
7. Schlesinger, M. E., (1982), *CO_2-Induced Climatic Warming: A Review of Model Research and Prospectus for First Detectability.* Oregon State University, Climatic Research Institute, Corvallis, Report No. 36.

8. Calvin M. (1976), *Photochem. Photobiol.* **23,** 425; (1977), *Energy Res.* **1,** 299; (1978), *Chem. Eng. News* **50(12),** 30; and (1980), *Die Naturwiss.* **67,** 525.
8a. Calvin, M. (1979), *Bioscience* **29,** 533.
8b. Calvin, M., Nemethy, E. K., Redenbaugh, K., and Otvos, J. W. (1982), *Experientia* **38,** 18.
8c. Calvin, M. (1981), *Proc. 1st. Asian and Pacific Chemistry Congress,* Singapore National Institute of Chemistry (1982), pp. 1–15.
8d. Calvin, M. (1983), *Science* 219, 24.
8e. Nemethy, E. K. *CRC Crit. Rev. Plant Sci.* **2,** 117.
9. Alexander, A. G., (1980), *The Energy Cane Concept for Molasses and Boiler Fuel,* Symp. on "Fuels and Feedstocks from Tropical Biomass", San Juan, Puerto Rico.
10. McLaughlin, S. P., Kingsolver, B. E., and Hoffman, J. J., (1983), *Econ. Bot.,* **37** 150.
11. Johnson, J. D., and Hinman, C. W. (1978), *Science* **208,** 460.
12. Coffey, S. D., and Halloran, G. M. (1981), *Euphorbia: Perspective and Problems, Proc. Natl. Conf. on Fuels from Crops,* Melbourne, Australia; and Stewart, G. A., Hawker, J. S., Nix, H. A., Rawlins, W. H. M., and Williams, L. R. (1982), *The Potential for Hydrocarbon Fuels from Crops in Australia,* CSIRO, Melbourne, Australia.
13. Kingsolver, B. E. (1982), *Biomass* **2,** 281.
14. Adams, R. P. (1982), *Production of Liquid Fuels and Chemical Feedstocks from Milkweed,* Institute of Gas Technology, Miami, FL.
15. Carruthers, I. B., Griffiths, D. J., Home, V. and Williams, L. R. (1984), *Biomass,* 4, 275.
16. Nemethy, E. K., Otvos, J. W., and Calvin, M. (1979). *J. Am. Oil Chem. Soc.* **56,** 957; (1981), *Pure Appl. Chem.* **53,** 1101; and (1981), Natural Production of High Energy Liquid Fuels from Plants, in *Fuels from Biomass,* (Klass, D. L., and Emert, G. H., eds.), Ann Arbor Science, Ann Arbor, MI, pp. 405–419.
17. Adams, R. P., and McChesney, J. D. (1983), *Econ. Bot.* **37,** 207.
18. Buchanan, R. A., Otey, F. H., Russell, C. R., and Cull, I. M. (1978), *J. Am. Oil Chem. Soc.* **55,** 657.
19. Weisz, P. B., Haag, W. O., and Rodewald, P. G. (1979), *Science* **206,** 57.
20. Wenninger, J. A., Yates, R. L., and Dolinsky, M. (1967), *J. Am. Oil Chem. Soc.* **50,** 1304.
21. Noble, B. F. (1979), *Canopy* **4(4),** 6.
22. Nemethy, E. K., and Calvin, M. *Phytochemistry* (1982) **21,** 298.
23. Redenbaugh, K., Ruzin, S., Bassham, J. A., and Bartholomew, J. C. (1982), *Z. Pflanzenphysiol.* **107,** 65.
24. Redenbaugh, K. Plant Genetics, Inc., private communication.
25. Frick, G. A. (1938), *Cactus and Succulent J.* **10**(9), 60.
26. de Steinheil, P., (1941), *Rev. Gen du Caoutchouc.* **18(2),** 55.

NMR Imaging in Biomedicine

PAUL C. LAUTERBUR

Department of Chemistry, State University of New York, at Stony Brook, Stony Brook, NY 11794

The first successful nuclear magnetic resonance (NMR) experiments in bulk matter were carried out by Bloch and his collaborators at Stanford and by Purcell and his collaborators at Harvard, and were published in 1946. In the 38 yr since those first papers, NMR techniques have been widely applied in many areas of physics, chemistry, biochemistry, and biology. These applications are based almost entirely on the fact that the quantized energy levels available in a magnetic field to a nucleus with spin angular momentum and a magnetic moment are separated by an energy that is proportional to the strength of the magnetic field at the nucleus. This field is always slightly different from an applied extermal magnetic field because of perturbations by the nearby electrons and magnetic nuclei and by local and long-range magnetic susceptibility effects. These small differences result in shifts in the resonance frequency. Also, the fluctuations in the magnetic field at the nucleus caused by its motions and those of nearby nuclei and impaired electrons determine the transition rates between the nuclear Zeeman energy levels and the rates at which phase coherence of the detectable component of the nuclear magnetization are lost. Such changes in the relaxation times are useful because of their sensitivities to molecular motions and local environments.

In order to observe directly the changes in resonance frequencies corresponding to the small differences in magnetic field caused by chemical and physical structure, it is necessary to apply an external magnetic field that is quite uniform over the object, and also to study objects that are themselves homogeneous, so as to avoid the confusing superposition of many different spectra. Therefore, in order to investigate molecular and crystal structures in great detail, much effort has been devoted to making magnets with very homogeneous fields. The residual field inhomogeneities must often be dealt with in some formal way, and the practice was sometimes adopted of approximating their effect by introducing a relaxtion term, T_2^*, by analogy with T_2, the exponential transverse relaxation time that appears in the phenomenological Bloch equation that describes, for example, the behavior of the nuclear magnetization in simple liquids. This choice corresponds to assuming a Lorentzian line shape, independent of the structure of the sample or the exact nature of the nonuniformity of the external magnetic field. The equations thus become tractable, and the approximated inhomogeneity effects could be factored out and discarded. Unfortunately, this was a case of "throwing the baby out with the bath water," because it made it very difficult to imagine that the effects of the inhomogeneity of the applied magnetic field could potentially give rise to whole new fields of application for NMR. Nevertheless, there were some situations in which the interaction of a field inhomogeneity with the sample shape was noted and explicity taken into account.

The new applications of NMR depend on a very simple idea. If the shape of the applied magnetic field is known, the NMR signal from any object of known structure can be calculated by integrating the response over all surfaces of constant magnetic field. If signals are obtained for many different magnetic field distributions, it might be imagined that it should be possible to encode enough information on the spatial distribution of the signal sources to determine it uniquely by some mathematical inversion procedure. The result would be an image, but this would be inconsistent with familiar principles. The wavelengths associated with the frequencies commonly used in NMR experiments are of the order of meters or tens of meters. The laws of physical optics, as usually applied to image formation, permit resolution only down to approximately one wavelength, and it did not seem reasonable to be making images with resolution independent of wavelength. There is no contradiction, however. As a very simple example, imagine two light-emitting objects much less than one wavelength apart. If the frequencies emitted were different, a spectroscope would resolve the two lines, regardless of the separation of the sources, and if the frequencies depended in a known way on the locations of the sources their separation could be deduced. That is exactly what happens when an NMR experiment is carried out in a known linear magnetic field gradient. The nuclei are tuned to different Larmor frequencies at different locations, and the observed signal strength as a function of frequency gives a one-dimensional projection of the distribution of nuclear spins. In three dimensions, each frequency corresponds to a surface of constant field, and from the integrated signal over different sets of such surfaces a complete image of the object can be built up.

Such considerations, leading to the idea that NMR provides a new and general method for image formation, did not arise spontaneously. They were stimulated by my observations of experiments being carried out on rats by a group further investigating an observation made shortly before that some animal tumors had longer water proton NMR relaxation times than other selected tissues in the animal. Although the results were fascinating, it seemed likely that their practical importance would be quite limited if the measurements could be made only on tissue specimens removed from the animal or human. The problem was quite clear; was there a general method for distinguishing NMR signals in one part of an object from those elsewhere? The sensitivity pattern of an *rf* coil obviously gives some degree of localization, but it seemed unlikely to be enough. The use of inhomogeneities in the applied magnetic field seemed correct in principle, but the only method that came to mind immediately was the use of gradients of many orders, ideally resembling a complete set of spherical harmonics, to give signals from which a one-center approximation to an image could be constructed. This approach had two obvious faults; it would be experimentally cumbersome and inelegant, and the construction would not treat all points in the same way and would probably converge extremely slowly. A more attractive solution then presented itself. A set of linear gradients applied in different directions would generate a set of projections that must contain all of the necessary information, and the experiments would be absurdly simple. The remaining problem was to find or devise a mathematical method for deducing an image from its projections. Although neither I nor several colleagues knew of any such method, it seemed that an iterative method might be forced to converge on a satisfactory solution. Simple simulations by hand calculation, using 8×8 arrays and four projections, confirmed that such was indeed the case. A computer was then programmed to work with larger arrays, more realistic objects, and, eventually, experimental data. At that point, a local computer scientist came across an article describing iterative reconstruction algorithms. Not only was the idea not new, but the algorithm I was using was essentially that already known as "multiplicative ART." One difference was that my initial programs used a discrete approach, which avoided interpolations in the two-dimensional image by taking only points on the projections corresponding to the rows of points at 0, 45, 90, and 135°, giving a scheme that could be implemented efficiently in fixed point arithmetic on a slow computer with limited core memory.

With the necessary mathematical techniques in hand, an ordinary analytical NMR spectrometer was used to generate projection data. A large nearly linear field gradient was produced by turning the controls of the electric shim system on a Varian A-60 spectrometer so as to give the worst possible signal and then scanning slowly (in 500 s) over a 1-kHz frequency range. Instead of changing the gradient direction, the experimental sample was rotated to different orientations, an equivalent procedure if no anisotropic susceptibility effects or shifts are present. Scans on a phantom made from two 1-mm tubes containing H_2O, fixed inside a 5-mm tube containg a D_2O/H_2O mixture to minimize magnetic susceptibility effects while maintaining contrast, were hand-digitized and entered into the computer, giving crude but recognizable images. Subsequent experiments generated partially saturated signals that produced T_1 images and then, after various observations on plant and animal tissues and small organisms, gave an image of a 4-mm living clam (probably *Gemma gemma*).

Along with the development of the necessary mathematical techniques and their use to make images of small objects, I was concerned with the problems of increasing the scale of the experiments to whole-body size. The first problem was to calculate signal-to-noise ratios for small volumes in very large *rf* coils. The results were encouraging, as was a search of the literature for designs of suitable, large magnets. It seemed likely that quite practical whole-body magnets with rather uniform fields of about 1000 G (0.1 *T*) could be built, and that images with resolutions of several millimeters could be reconstructed. A design for such a sytem was even done and privately circulated at that time. At that point, all of the information from which the general medical potential of NMR imaging could be estimated was known.

As time went on, larger animals, such as mice, were imaged, malignant tumors were distinguished from surrounding tissue in vivo, and much larger magnets, gradient coil systems, and *rf* coils were constructed. New techniques were developed for combining spectra with images, obtaining true three-dimensional images from plane integrals, and studying flow and surgical specimens, and medical and commercial interest began to develop rapidly.

Other laboratories were also active, of course, with significant early contributions coming from groups working with Mansfield, Hinshaw, Ernst, and Hutchison. The field is now much too broad and complex to cover in a single short review, but many books and reviews are available for further study.

ACKNOWLEDGMENTS

Since this work began, almost 100 undergraduate students, graduate students, postdoctural research associates, collaborators, technicians, and secretaries have helped with it and have shared in the excitement of trying new things. Their assistance and enthusiasm have been invaluable. Financial support has been supplied at various stages of the work by the National Cancer Institute, the National Heart, Lung, and Blood Institute, and Petroleum Research Fund of the American Chemical Society, the State of New York, Bruker Physik AG, and General Electric Corporation, the Chevron Oil Field Research Corporation, and the National Science Foundation, who have the gratitude of all 100 of us.

Section V METHODOLOGICAL DEVELOPMENTS

LIFE AND PHYSICS

New Perspectives

ILYA PRIGOGINE

Université Libre de Bruxelles, Brussels, Belgium

INTRODUCTION

The idea for a symposium on "The New Frontiers at the Crossing Between Physical and Life Sciences" comes at a time when we are facing an entirely new situation in so far as these topics are concerned. I will focus my discussion here on the problem of the origins of life as it may be thought about using some newer conceptual tools of physics.

As we all know, life is an evolutionary process. We are thus supposed to adopt an evolutionary paradigm in order to cope with biological time. It is now clear that physics had to go a long way to design an adequate conceptual frame: Classical physics stressed a quite atemporal description of phenomena, and we are only in the first steps of the necessary reconceptualization.

It should be emphasized that in classical physics all processes could in principle be reduced to the same laws of motion as those formulated by Newton: Once initial conditions are known, the outcome of every situation could in principle be predicted with absolute certainty.

The two characteristics of this scheme that I would like to stress are as follows: The first corresponds to a deterministic description of nature: no place is left to any uncertainty. The second characteristic corresponds to a reversible description; Newton's laws do not imply any distinction between past and future. (*1*). This grandiose intellectual structure was not without problems. How to understand the position of life, of humankind, in a universe described as a giant automaton?

How does one reconcile the two attitudes? Are humans outside the universe they describe? The famous French biologist Jacques Monod had a way of making this contrast explicit by stating that the laws of biology were compatible with the laws of physics, but not included in them.

In the classical version we inherited from the 17th century, nature appeared as passive, as submitted to the creative impulse of humankind. The history of modern culture is a dramatic story indeed. There were moments when the program of classical science seemed near completion: A fundamental level, which would be the carrier of deterministic and reversible laws, seemed in sight. However, something went wrong. The scheme had to be enlarged, and the fundamental level, in the sense of the classical program, remained elusive.

Today, wherever we look we find evolution, diversification, and instabilities. A fundamental reconceptualization of science is going on. We have long known that we are living in a pluralistic world in which we find deterministic, as well as stochastic, phenomena; reversible, as well as irreversible processes. We observe deterministic phenomena, such as the frictionless pendulum or the trajectory of the moon around the earth; moreover, we know that the frictionless pendulum is also reversible. But other processes are irreversible, diffusion or many chemical reactions for instance: And we are obliged to acknowledge the existence of stochastic processes if we want to avoid the paradox of referring the variety of natural phenomena to a program printed at the moment of the Big Bang. What has changed since the beginning of this century is our evaluation of the relative importance of these four types of phenomena.

The artificial may be deterministic and reversible. The natural contains essential elements of randomness and irreversibility. This leads to a new vision of matter: No longer passive, as described in the mechanical world view, but associated with spontaneous activity. This change is so deep that I believe we can really speak about a new human dialog with nature.

Even at the start of this century, continuing the tradition of the classical research program, physicists were almost unanimous in admitting that the fundamental laws of the universe were deterministic and reversible. Processes that did not fit this scheme were supposed to be exceptions, even artifacts resulting from some apparent complexity, which had itself to be accounted for by invoking our ignorance or lack of control of the variables involved. Now that we are at the end of this century, more and more of us think that the fundamental laws of nature are irreversible and stochastic; that deterministic and reversible laws are applicable only in limited situations.

It is interesting to inquire how such a change could occur over a relatively short time span. It is the outcome of unexpected results obtained in quite different areas of physics and chemistry, such as the fields of elementary particles and cosmology, or the study of self-organization in far-from-equilibrium systems.

Who would have believed, fifty years ago, that most, perhaps all, elementary particles are unstable? That we would speak about the evolution of the universe as a whole? That far from equilibrium, molecules may communicate, to use anthropomorphic terms, as witnessed in the chemical clocks to which we shall later turn in more detail?

All of these unexpected discoveries have also had a drastic effect on our outlook on the relation between the "hard" and "soft" sciences. According to the classical view, there was a sharp distinction between simple systems, such as those studied by physics or chemistry, and complex systems, such as those studied in biology and human science. Indeed, one could not imagine a greater contrast than the one that exists between the simple models of classical dynamics or the simple behavior of a gas or a liquid and the complex processes we discover in the evolution of life or in the history of the human societies.

But now we find this gap being filled. Over the last decade, we have learned that in nonequilibrium conditions, simple materials, such as a gas or liquid, or simple chemical reactions can acquire complex behavior. This unexpected complexity leads to the appearance of broken temporal or spatial symmetry, to chaos, and to patterns of increasing diversity (2).

FAR FROM EQUILIBRIUM

In its evolution toward complexity, with all its connotations of irreversibility and stochasticity, thermodynamics has played an important role. It may be useful to contrast the description of a system in terms of classical dynamics with the thermodynamical one. In the former, we consider, typically, a given number of points interacting through some type of potential. A typical example would be the system formed by the sun and earth. Of course, there are also the other planets and stars, but they are treated as a kind of perturbation. In contrast, thermodynamics is based on the concept of entropy. Entropy has quite unique behavior. Its changes with time can be split into two additive terms: d_eS, corresponding to the flow of entropy that the system exchanges with the outside world; and d_iS, corresponding to the production of entropy inside the system. Because d_iS is always positive or zero, irreversibility can only create entropy, but not destroy it. For an isolated system, there is no entropy flow, and S changes only as a result of entropy production: Entropy increases monotonously as the system approaches equilibrium, corresponding to a maximum of entropy.

For a long time interest in thermodynamics was concentrated on isolated systems at equilibrium. Today, our interest is shifting to nonequilibrium systems interacting with their surroundings through entropy flow. Let us emphasize an essential difference with the dynamical description. In thermodynamics, we are dealing with embedded systems. The interaction with the outside world through entropy flow plays an essential role. This immediately brings us closer to situations like towns or living systems, which can only survive because of embedding in their respective environments.

From the start, the thermodynamical point of view is one of interaction, we could say a "holistic" one. But there is more: Suppose that in the gravitational example quoted above, we have some foreign body approaching the earth. This would lead to a deformation of the earth's trajectory that would then remain so deformed forever. This class of systems has no way of forgetting perturbations.

This is no longer the case when we include the effects of dissipation. A damped pendulum will reach a position of equilibrium, whatever the initial perturbation. Again, let us emphasize how much closer to life is the thermodynamic description. If I impose large oscillations on an undamped pendulum, its period slows down and remains so slowed forever. However, when I run, my heart beat increases; but when I rest, it returns to its normal rate. In thermodynamics, perturbations, may be forgotten. In the thermodynamic description that includes dissipation, we have attractors.

Without attractors, our world would be chaotic. No general rules could ever have been formulated. Every system would pose a problem in itself. We can now also understand in quite general terms what happens when we drive a system far from equilibrium. The attractor that dominated the behavior of the system near equilibrium may become unstable as a result of the flow of matter and energy that we direct at the system. Nonequilibrium becomes a source of order; new types of attractors, more complicated ones, may appear and give the system new remarkable and space–time properties.

Let us illustrate these general statements with two examples that are widely studied today. We first consider the so-called Bénard instability. It is a striking example of instability in a stationary state, giving rise to the phenomenon of spontaneous self-organization; the instability is a result of a vertical temperature gradient set up in a horizontal liquid layer. The lower face is maintained at a given temperature, higher than that of the upper. As a result of these boundary conditions, a permanent heat flux is set up, moving from bottom to top.

For small differences of temperature, heat can be conveyed by conduction, without any convection; but when the imposed temperature gradient reaches a threshold value, the stationary state (the fluid's state of "rest") becomes unstable: Convection arises, corresponding to the coherent motion of a huge number of molecules, increasing the rate of heat transfer. Under appropriate conditions, the convection produces a complex spatial organization in the system *(3)*.

There is another way of looking at the Bénard phenomenon. There are two elements involved: heat flow and gravitation. At equilibrium conditions, the gravitational force has hardly any effect on a thin layer on the order of 10 mm. In contrast, in the Bénard instability, gravitation gives rise to macroscopic structures. Nonequilibrium matter becomes much more sensitive to outer world conditions than matter at equilibrium. I like to say that at equilibrium, matter is blind; far from equilibrium, it may begin to see.

This can also be illustrated by the example of chemical oscillations. The Belousov–Zhabotinski reaction is a well-known chemical clock. The striking results obtained by Zhabotinski became available in the late 1960s. I cannot discuss here the detailed mechanism (*see* refs. 3,4). But I would like to point out some general features. Ideally speaking, we have a chemical reaction whose state we control through appropriate injection of chemical products and elimination of waste products.

Suppose that two of the intermediate components are formed by red and blue molecules, respectively, in comparable quantities. We would expect to observe some kind of blurred color with, perhaps occasionally, some flash of red or blue spots. This is, however, not what actually happens. Under appropriate conditions, we see—in sequence—the whole vessel become red, then blue, then red again; we have a chemical clock. In a sense, this violates all of our intuition about chemical reactions.

We were used to speaking of chemical reactions as being produced by molecules moving in a disordered fashion and colliding at random. In contrast, the existence of chemical clocks shows that far from being chaotic, the behavior of the intermediate species is highly coherent. In a sense, the molecules have to be able to "communicate" in order to synchronize their periodic changes of color. In other words, we are dealing here with new supermolecular scales, both in time and space, produced by chemical activity. In a stirred chemical reactor, we can only observe temporal behavior (which may be periodic or "chaotic"). In other cases, we may also see the appearance of space structures.

Another important concept is bifurcation. A system presenting a multiplicity of solutions may be characterized by a bifurcation diagram. We see that at critical points new types of solutions emerge. There is also another elment I would like to emphasize. Near a bifurcation point, the system has a choice between the two branches; we may therefore expect a stochastic element.

Quantum mechanics appears to us so revolutionary because it introduces a stochastic description in the microscopic world. The new developments we have summarized here suggest that this stochastic type of behavior may also appear in the macroscopic world.

CLIMATE AND CHIRALITY

The concepts that we outlined in the previous section are today diffused throughout a large domain, including biology, the social sciences, and climatology. One of the main problems we confront is to understand the large-scale regularities that shape our environment.

Let me give two examples of what I have in mind. Consider first the problem of climate. We know now that climate has fluctuated violently over the past. Climatic conditions that prevailed during the last two or three hundred million years were extremely different from those found at present. During these periods, with the exception of the quaternary era (which began about two millions years ago) there was practically no ice on the continents, and the sea level was higher than its present level by about 80 m. A striking feature of the quaternary era is the appearance of a series of glaciations, with an average periodicity of one hundred thousand years, on which is superimposed an important amount of noise. What is the source of these violent fluctuations that have obviously played an important role in our history? There is no indication that the intensity of solar energy may be responsible for these fluctuations.

Let us now consider a problem by which, on the contrary, we are stunned at the remarkable uniformity and lack of fluctuation. In chemistry, we deal often with chiral molecules. For example, in most amino acids, the central carbon is asymmetric: There are two configurations called the D and L isomers. The relation between the two is somewhat similar to left- and right-handed spirals. The curious fact is that amino acids in proteins are always in the L configuration.

A somewhat similar example is the case of particles and antiparticles. From the point of view of quantum theory, they play nearly identical roles. Still, the world is composed of particles, whereas antiparticles are produced in accelerators.

We now have the tools to begin to understand these two extreme situations: Climate is described by highly nonlinear equations, involving both the absorption and the emission of solar light. In short, this leads to two types of climate: a cold one and a moderate one. There are, of course, fluctuations around these two climates. However, these fluctuations taken alone would not be sufficient to tilt the climate from one type to another.

But, in addition, the eccentricity of the earth's orbit leads to a very slight periodocity in the solar influx. This has a major consequence, which was derived by two independent research groups *(5,6)*. Such a very small outside perturbation leads to amplification of the transition probability from one climate to the other.

We begin in this way to understand the glaciation periodicity of one hundred thousand years. The point that I want to emphasize here is that the internal noise may amplify minute external perturbations, leading to major changes in climate.

Let us turn to the second question: We have seen that matter under nonequilibrium conditions may become very sensitive to outside perturbations. This is especially true in the neighborhood of bifurcation points.

A very minute difference, which may exist in the stability of right- or left-handed molecules (leading for example, to differences in the energy of activation on the order of 10^{-15}) may suppress the stochasticity inscribed in a "symmetrical" bifurcation and lead to a preferential selection of one of the states *(7)*.

This may lead to applications in a wide range of fields. We may construct new types of highly sensitive switches. However, a price has to be paid; these switches are very slow. We have to maintain the bifurcation parameter near the bifurcation point for long periods (the exact value of which may vary from hours to minutes, according to the kinetic equations).

HUMAN BEINGS AND ENVIRONMENT

Living beings are undoubtedly the most complex and organized objects found in nature, in view of both their morphology and their functioning. They already serve as prototypes from which physical scientists can get the motivation and inspiration needed for an understanding of complexity. They are literally historical structures, since they have the ability to preserve the memory of forms and functions acquired in the past, during the long period of biological evolution.

Moreover, they function definitely under conditions far away from equilibirum. An organism as a whole receives continuous fluxes of energy and matter, it transforms into quite different waste products that are evacuated into the environment. These nonequilibrium conditions are observable at all scales of the organism. It is then legitimate to inquire whether some of the above features of biological systems can be attributed to transitions induced by nonequalibrium constraints and appropriate destabilizing mechanisms similar to chemical autocatalysis. This is probably one of the most fundamental questions that can be raised in science. No exhaustive answer can be claimed today, but one can mention some examples in which the connection between physicochemical self-organization and biological order is striking: The reciprocal relations between chemotaxis and amplification of inhomogeneity in the aggregation process of *Dictyostelium discoideum,* the accrasial amoeba; the threshold behavior of the immune system, induced by a cellular dynamics similar to classical equations of enzyme kinetics; adaptative phenomena in social insects, whose behavior presents highly probabilistic features at the level of the individual, and coherent patterns at the scale of the colony.

These examples, which I cannot develop here, are promising in the sense that they permit us to believe that in the future the gap between biology and physics will narrow as we design a new approach to complexity (*8*).

CONCLUSION

Is what we traditionally called fundamental physics the basic description of nature, or is it instead a simplified description of a reality to which we only now begin to gain some access through the study of complex processes? It would be presumptuous of us to try to answer yet, but the very fact that we can ask such questions shows that our perspectives have changed drastically since the time when, at least on our level, the laws of nature seemed to have reached their final formulation.

REFERENCES

1. Prigogine, I., and Stengers, I., (1979), *La nouvelle alliance,* Paris, Gallimard (1979). English translation *Order out of Chaos, Bantam Books, NY (1984).*
2. Prigogine, I., (1980), *From Being to Becoming. Time and Complexity in the Physical Sciences,* Freeman, CA.
3. Nicolis, G., and Prigogine I., (1977) *Self-Organization in Nonequilibrium Systems—From Dissipative Structures to Order through Fluctuations,* Wiley, NY.
4. P. de Kepper and I. Epstein, (1982) *Journal of the American Chemical Society,* **104,** 9, 2657.
5. Nicolis, C., (1982), *Stochastic Aspects of Climatic Transition-Responses to a Periodic Forcing,* Tellus **34,** 1–9. 6. Benzi, R., Parisi, G., Sutera, A. Vulpiani, A. (1982) Stochastic *Resonance in Climatic Change, Tellus* **34** 10–16.
7. Kondepudi, D. K., and Nelson, G. W., (1983) *Physical Review Letters, 50n,* n° **14** 1023–26.
8. Nicolis, G., and Prigogine, I., (1985) *Understanding Complexity,* to appear *Piper Vlg.*

Cell Differentiation and Malignancy

LEO SACHS

Department of Genetics, Weizmann Institute of Science, Rehovot 76100, Israel

ABSTRACT

An understanding of the mechanism that controls growth and differentiation in normal cells would seem to be an essential requirement to elucidate the origin and reversibility of malignancy. For this approach I have mainly used normal and leukemic blood cells, and in most studies have used myeloid blood cells as a model system. Our development of systems for the in vitro cloning and clonal differentiation of normal blood cells made it possible to study the controls that regulate growth (multiplication) and differentiation of these normal cells and the changes in these controls in leukemia. Experiments with normal blood cell precursors have shown that normal cells require different proteins to induce growth and differentiation. We have also shown that in normal myeloid precursors, growth-inducing protein induces both growth and production of differentiation-inducing protein so this ensures the coupling between growth and differentiation that occurs in normal development. The origin of malignancy involves uncoupling of growth and differentiation. This can be produced by changes from inducible to constitutive expression of specific genes that result in asynchrony in the coordination required for the normal developmental program. Normal myeloid precursors require an external source of growth-inducing protein for growth, and we have identified different types of leukemic cells. Some no longer require and other constitutively produce their own growth-inducing protein. But addition of the normal differentiation-inducing protein to these malignant cells still induces their normal differentiation, and the mature cells are then no longer malignant. Genetic changes that produce blocks in the ability to be induced to differentiate by the normal inducer occur in the evolution of leukemia. But even these cells can be induced to differentiate by other compounds, including low doses of compounds now being used in cancer therapy, that induce the differentiation program by other pathways. This differentiation of leukemic cells has been obtained in vitro and in vivo, and our in vivo results indicate that this may be a useful approach to therapy. In some tumors, such as sarcomas, reversion from a malignant to a nonmalignant phenotype can be a result of chromosome changes that suppress malignancy. But in myeloid leukemia, the stopping of growth in mature cells by induction of differentiation bypasses the genetic changes that produce the malignant phenotype. These conclusions can also be applied to other types of normal and malignant cells.

CLONING AND CLONAL DIFFERENTIATION OF NORMAL BLOOD CELLS IN CULTURE

The cloning and clonal differentiation of normal blood cells in culture made it possible to study the controls that regulate growth (multiplication) and differentiation of different blood cell types (*1–6*). We first showed (*7,8*), and this was later confirmed by others (*9*) that normal mouse myeloid precursor cells cultured with a feeder layer of other cell types can form clones of granulocytes and macrophages in culture (Fig. 1). We also found that the formation of these clones is a result of secretion, by cells of the feeder layer, of specific inducers that induce the formation of clones and the differentiation of cells in these clones to macrophages or granulocytes in mice (*8,10,11*) and humans (*12*). After we first detected their presence in culture supernatants (*10,11*), these protein inducers were referred to by a number of names, I shall use the name macrophage and granulocyte inducers (MGI) (Tables 1 and 2). These proteins can be produced and secreted by various normal and malignant cells in culture and in vivo (*1*). Their production can be induced by a variety of compounds (*13–16*), and some cells produce these proteins constitutively (*10,17–20*). Macrophage and granulocyte inducers are a family of proteins that exist in a number of molecular forms that have different biological activities. This cell culture approach has led to the cloning and isolation of growth factors for all the different types of blood cells, including different types of lymphocytes.

NORMAL GROWTH AND DIFFERENTIATION-INDUCING PROTEINS

The family of MGI proteins includes some proteins that induce cell growth (multiplication) and others that induce differentiation. Those that induce growth, which are also required for normal cell viability, we now call MGI-1. These include proteins that induce the formation of macrophage (MGI-1M) (*10,20,21*) or granulocyte clones (MGI-1G) (*10,21,22*) or both (MGI-1GM) (*18,19,23*). The MGI-1 have previously been referred to as mashran g (*24*), colony stimulating factor (CSF) (*25*), or colony stimulating activity (CSA) (*17*) (Table 2). The existence of an antibody that does not react with all forms of MGI-1M or MGI-1G has shown that

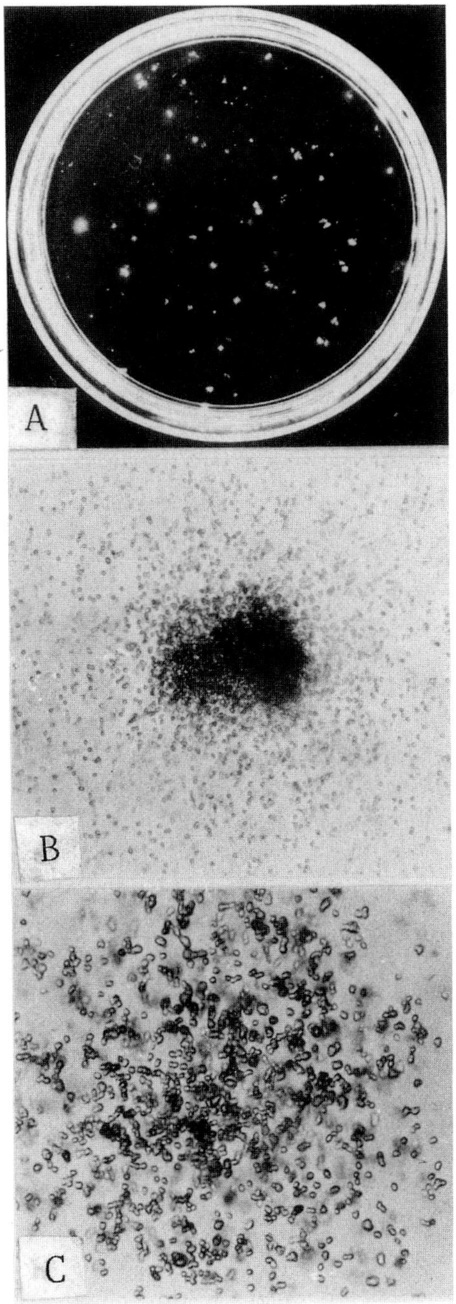

Fig. 1. Clones of macrophages and granulocytes from normal hematopoietic cells incubated with MGI-1. (A) Petri dish with clones (*8*) (B) granulocyte clone (C) macrophage clone (*10*).

there can be different antigenic sites on molecules that belong to the same form of MGI-1 (*21,26*). The other main type of MGI, which we now

TABLE 1
In Vitro Cloning and Clonal Differentiation of Normal Hematopoietic Cells

Result	Source
Cloning and differentiation in liquid medium (mast cells and granulocytes)	*7*
Cloning and differentiation in agar (macrophages)	*8*
Inducer for cloning and differentiation secreted by cells	*8*
Inducer in conditioned medium from cells (for macrophages and granulocytes)	*10,11*
Different inducer for macrophage and granulocyte clones	*10*
Cloning and differentiation of macrophages and granulocytes in methylcellulose	*10*
Confirmation of cloning and differentiation in agar	*9*
Production of inducer for cloning by some leukemic cells	*37*
Cloning and differentiation of human cells	*12,83*
Protein inducer of differentiation that does not induce cloning	*29*

call MGI-2 (*4,21,27*) induces the differentiation of myeloid precursor cells, either leukemic (*28*) or normal (*4,27*), without inducing colony formation. This differentiation-inducing protein (*28,29*) has also been referred to as D factor (*30,31*) and GM-DF (*32*). It has been suggested that there are different forms of MGI-2 for differentiation to macrophages or granulocytes (*27*). The regulation of MGI-1 and -2 appears to be under the control of different genes (*13*). Differentiation-inducing protein MGI-2, but not the growth-inducing protein MGI-1, is a DNA-binding protein (*33*).

These macrophage and granulocyte inducers can be proteins or glycoproteins, depending on the cells in which they are produced, and the presence of carbohydrates does not appear to be necessary for their biological activity (*19*). Their molecular weights are mostly around 23,000 or

TABLE 2
Terminology Used for Proteins that Induce Cloning and Differentiation of Normal Macrophages and Granulocytes

Term	Source
Mashran g	*24*
Colony stimulating factor (CSF)	*25*
Colony stimulating activity (CSA)	*17*
Macrophage and granulocyte inducers (MGI)	*18*
MGI-1 (= mashran g, CSF, CSA) for cloning; MGI-2 (= D factor, DF) for differentiation	*4,21, 67*

multiples of this number (*2,21,34*), and MGI-2 activity is more sensitive to proteolytic enzymes and high temperature than MGI-1 activity (*19*). The MGI-2 have a shorter half-life in serum than MGI-1 (*35*). The ready separability of the different forms of MGI seems to depend on the cells from which they are derived (*21*). Further studies should determine whether tumor cells with the appropriate gene rearrangements and possibly even normal cells under certain conditions, may produce hybrid molecules of different forms of MGI, including hybrid molecules with MGI-1 and MGI-2 activity (*27*).

CONTROL OF GROWTH AND DIFFERENTIATION IN LEUKEMIA

Normal myeloid precursor cells isolated from bone marrow (*36*) require an external source of MGI-1 for cell viability and growth. There are, however, myeloid leukemic cells that no longer require MGI-1 for viability and growth, so that these leukemic cells can then multiply in the absence of MGI-1 (*2,4*). This gives the leukemic cells a growth advantage over the normal cells when there is a limiting amount of MGI-1. Starting with a decreased requirement for MGI-1, this eventually leads to a complete loss of this requirement. Other myeloid leukemic cells constitutively produce their own MGI-1 (*22,37*), and these leukemic cells also have a growth advantage compared to normal cells that require an external source of MGI-1 (Fig. 2). A change in the requirement of MGI-1 for growth, either a partial or complete loss of this requirement or the constitutive production of MGI-1, both give a growth advantage to leukemic cells.

The existence of myeloid leukemic cells that either no longer require MGI-1 for viability and growth or constitutively produce their own MGI-1 raises the question of whether these leukemic cells can still be induced to differentiate to mature cells by the normal differentiation-inducing protein MGI-2. This question has been answered by showing that there are clones of myeloid leukemic cells that no longer require MGI-1 for growth, but can still be induced to differentiate normally to mature macrophages and granulocytes by MGI-2 (Fig. 3) via the normal sequence of gene expression (*2–6*). These mature cells are then no longer malignant in vivo (*26,35,38*). Injection of these myeloid leukemic cells into embryos has shown that after such injection the leukemic cells can participate in hematopoietic differentiation in apparently healthy adult animals (*39,40*).

Injection of MGI-2 into animals, or in vivo induction of MGI-2 by a compound that induces the production of this differentiation-inducing protein, results in an inhibition of leukemic development in animals with such leukemic cells (*26,35*). There are also myeloid leukemic cells that constitutively produce their own MGI-1 and that can be induced to dif-

DIFFERENCES IN GROWTH REQUIREMENT OF NORMAL AND LEUKEMIC CELLS

Type of myeloid cells	Requirement of MGI-1 for growth
Normal	External source
Leukemic	Decrease ⟶ no requirement or Constitutive production

Fig. 2. Differences in MGI-1 requirement for growth in normal and leukemic myeloid cells.

ferentiate by MGI-2. Our results indicate that induction of normal differentiation in myeloid leukemic cells by MGI-2 can be an approach to therapy based on the induction of normal differentiation in malignant cells (*1–3,12,26,28,35,41,42*). There are various forms of MGI-2 that differ in their ability to induce differentiation in different clones of myeloid leukemic cells (*15,26,41,42*).

Leukemic clones that can be induced to differentiate to mature cells by MGI-2 have been found in different strains of mice (*28,32,43–46*). They are referred to as MGI^+D^+ (MGI^+ to indicate that they can be induced to differentiate by MGI-2; D^+ for differentiation to mature cells). The MGI^+D^+ leukemic cells have specific chromosome changes compared to normal cells (*47,48*). These chromosome changes thus seem to involve changes in genes other than those involved in the induction of normal differentiation. There are other clones of myeloid leukemic cells that can also grow without adding MGI-1, but that are either partly (MGI^+D^-) or almost completely (MGI^-D^-) blocked in their ability to be induced to differentiate by MGI-2 (Fig. 4) (*21,44,45,49–52,52a*). These differntiation-defective clones have specific chromosome changes compared to MGI^+D^+ cells (*47,48*).

There are a variety of compounds, other than MGI-2, that can induce differentiation in MGI^+D^+ clones. Not all these compounds are active on the same MGI^+D^+ clone, and they do not all induce the same differentiation-associated properties. The inducers include certain ster-

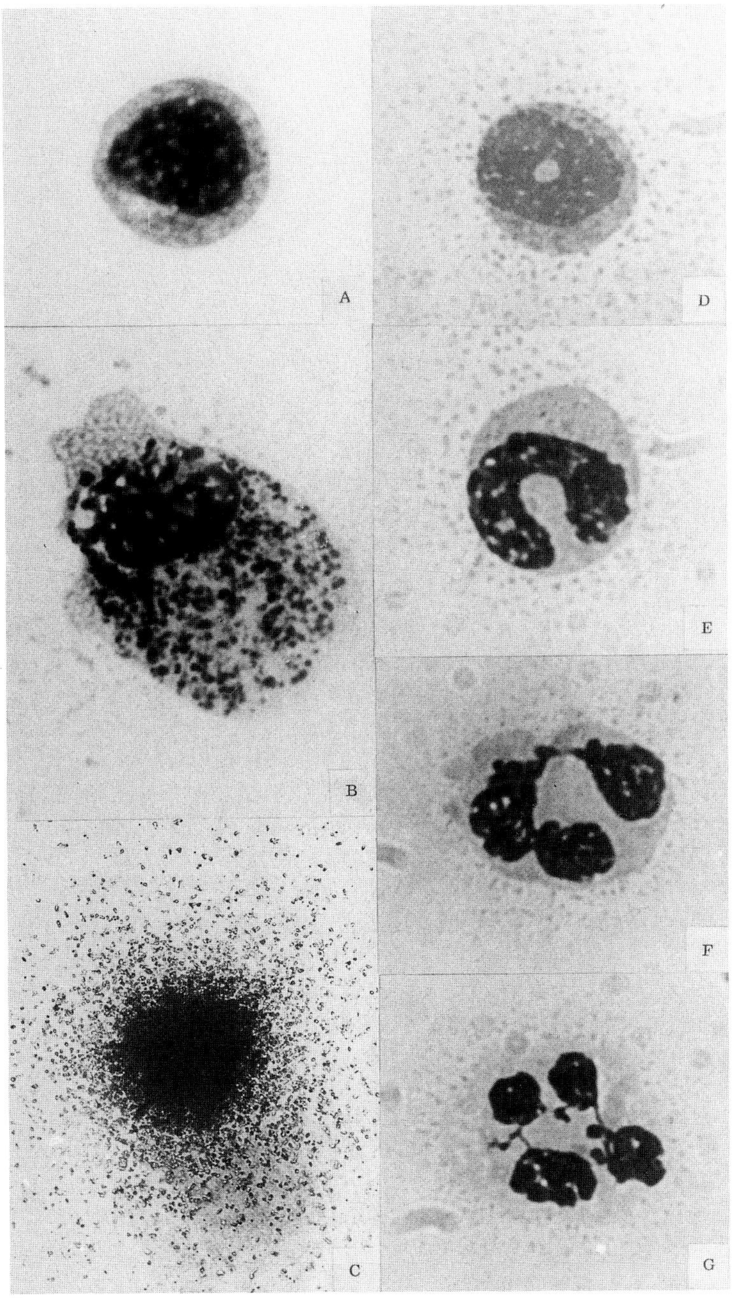

Fig. 3. Differentiation of MGI^+D^+ myeloid leukemic cells to mature macrophages and granulocytes by the normal myeloid differentiation-inducing protein MGI-2. (A) leukemic cells (B) macrophage (C) colony of macrophages (D–G) stages in differentiation to granulocytes (*28*).

oids, lectins, polycyclic hydrocarbons, tumor promoters, lipopolysaccharides, X-irradiation, and compounds used in cancer chemotherapy

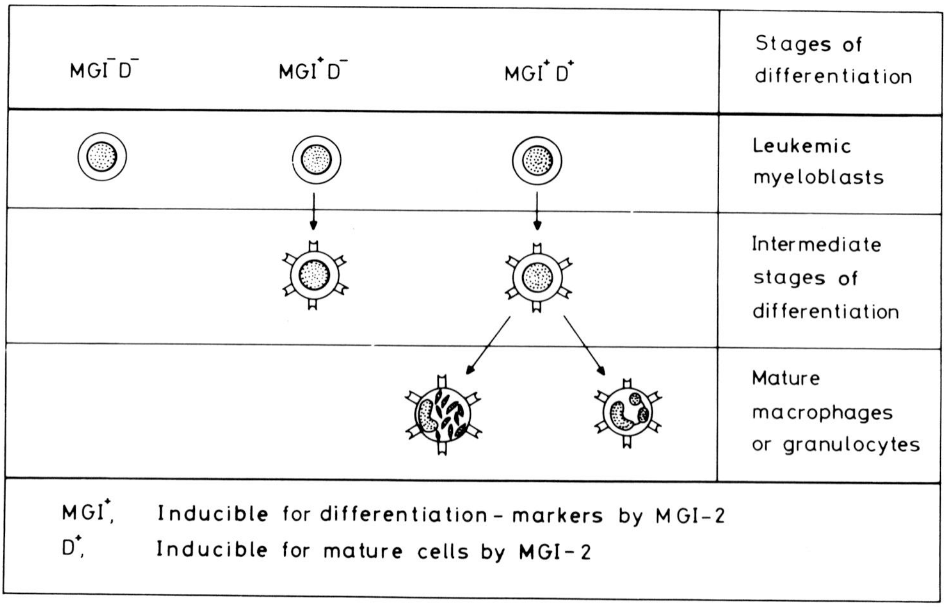

Fig. 4. Classification of different types of myeloid leukemic cell clones according to their ability to be induced to differentiate by MGI-2.

(*2,53*) (Table 3). The existence of clonal differences in the ability of X-irradiation and cancer chemotherapeutic chemicals to induce differentiation may help to explain differences in response to therapy in different individuals (*2*). As a result of these experiments, we have suggested that it may be possible to introduce a form of therapy based on induction of differentiation (*1–3,12,28,35,41,53*). This would include prescreening in culture to select for the most effective compounds, and using these compounds for a low-does chemotherapy protocol aimed at inducing cell differentiation (*53*). Since different myeloid leukemic clones respond differ-

TABLE 3
Compounds used in Cancer Therapy That Can Induce Differentiation in Clones of Myeloid Leukemic Cells at Low Doses

Adriamycin
Cytosine arabinoside
Daunomycin
Hydroxyurea
Methotrexate
Mitomycyin C
Prednisolone
X-irradiation

ently to MGI-2 and other compounds, such differences will also occur in leukemic cells from different patients. Based on these suggestions (*2,3*) some encouraging clinical results have been obtained with the use of low-dose cytosine arabinoside (*54–56*).

ALTERNATIVE PATHWAYS OF DIFFERENTIATION

Some of the compounds that induce differentiation in susceptible clones of MGI^+D^+ leukemic cells, including lipopolysaccharides, nitrosoguanidine and phorbol ester tumor-promoters, such as 12-0-tetradecanoyl-phorbol-13-acetate (TPA), can induce in these clones the production of MGI-2. These compounds thus induce differentiation by inducing in the leukemic cells the endogenous production of the normal differentiation-inducing protein MGI-2 (*13,15,16*). Other compounds such as the steroid dexamethasone, can induce differentiation in MGI^+D^+ clones without inducing MGI-2 (*13*). This steroid induces differentiation by other pathways of gene expression than MGI-2 (*46,57*). The same applies to dimethylsulfoxide (DMSO).

Induction of differentiation in some myeloid leukemic clones requires combined treatment with different compounds (*15,58–61*). In these cases one compound induces changes not induced by the other, so that the combined treatment results in new gene expression. This complementation of gene expression can occur both at the level of mRNA production and translation (*52*). With the appropriate combination of compounds, we have been able to induce all our MGI^-D^- leukemic clones for some differentiation-associated properties (*60,61*). It will be interesting to determine whether the same applies to differentiation of erythroleukemic cells (*62,63*). It is possible that all myeloid leukemic cells no longer susceptible to the normal differentiation-inducing protein MGI-2 by itself, can be induced to differentiate by choosing the appropriate combination of compounds needed to give the required complementation. This can include the use of hormones, such as steroids (*64,65*) or insulin (*60,66*), and different nonphysiological compounds (*2*) with or without MGI-2.

COUPLING OF GROWTH AND DIFFERENTIATION IN NORMAL CELLS

We have developed a simple procedure for isolating normal myeloid precursor cells from the bone marrow (*36*). Incubation of isolated normal myeloid precursors with MGI-1 induces the viability and growth of these normal precursors and results in cell differentiation to macrophages or granulocytes, even without adding the differentiation-inducing protein MGI-2. The incubation of normal myeloid precursors with MGI-1 also results in the induction of MGI-2 (*4,27,67,68*). This induction of MGI-2 can be detected already at 6 h after the addition of MGI-1 (*68*). This induction of MGI-2 by MGI-1 can thus account for the induction of differentiation after adding MGI-1 to the normal cells. The induction of differentiation-inducing protein MGI-2 by growth-inducing protein MGI-1, thus appears to be an effective control mechanism for coupling growth and differentiation in the normal cells.

It has been shown that the receptor for epidermal growth factor has tyrosine-specific protein kinase activity (*69*). This has also been found for receptors for other growth factors, such as insulin (*70*), and presumably also applies to receptors for the myeloid cell growth-inducing proteins MGI-1. The myeloid differentiation-inducing protein MGI-2, but not MGI-1, can bind to cellular DNA (*33*). This shows that growth and differentiation in normal myeloid cells are coupled by induction of a differentiation-inducing DNA-binding protein by a growth-inducing protein. This mechanism for coupling growth and differentiation may also apply to other types of cells. Differences in the time of the switch-on of the differentiation inducer would produce differences in the amount of multiplication before differentiation. The platelet-derived growth factor is structurally related to the simian sarcoma virus oncogene *sis* (*71,72*). It will be interesting to determine whether MGI-1 and -2 are structurally related to any of the known oncogenes.

The multiplication of normal cells is regulated at two control points. The first control is that that requires MGI-1 to produce more cells, which can then differentiate by the MGI-2 induced by MGI-1. The second control is the stopping of cell multiplication that occurs as part of the program of terminal differentiation to mature cells induced by MGI-2. There is thus a coupling of growth and differentiation in normal cells at both these points.

UNCOUPLING OF GROWTH AND DIFFERENTIATION IN LEUKEMIA

As pointed out above, there are $MGI^{+}D^{+}$ clones of myeloid leukemic cells that no longer require MGI-1 for growth, but can still be induced to differentiate normally by MGI-2. These leukemic cells have thus uncoupled the normal requirement for growth from the normal requirement for differentiation. Experiments on the properties of these cells after induction of differentiation by MGI-2 have shown that the normal requirement for MGI-1 for cell viability and growth is restored in the differentiating leukemic cells (*29,67,68*). MGI-1 added to normal myeloid precursors induces the production of MGI-2, so that the cells can then differentiate by the endogenously produced MGI-2. However, in these leukemic cells, MGI-1 did not induce the production of MGI-2, even though, like normal cells, they again required MGI-1 for viability and growth. There was therefore no induction of differentiation after adding MGI-1 (*67,68*). There is another type of leukemic cell that constitutively produces its own MGI-1 and can also show this lack of induction of MGI-2 by MGI-1, so that the cells do not differentiate (*73*). The absence of induction of MGI-2 and MGI-1 therefore uncouples growth and differentiation in these leukemic cells. The lack of requirement of MGI-1 for growth and the absence of induction of the differentiation-inducing protein MGI-2 by the growth-inducing protein MGI-1, are thus mechanisms that uncouple growth and differentiation in MGI^+D^+ leukemic cells (*4,67,68,73*).

In leukemic cells with constitutive production of MGI-1, changes in specific components of the culture medium can result in an autoinduction of differentiation because of the restoration of the induction of MGI-2 by MGI-1, which then restores the normal coupling of growth and differentiation (Fig. 5). These changes in the culture medium include the use of mouse or rat serum instead of horse or calf serum, serum-free medium, and removal of transferrin from serum-free medium (*73*). Autoinduction of differentiation in this type of leukemic cell may also occur under certain conditions in vivo.

The coupling of growth and differentiation in normal cells is regulated at two control points. The uncoupling of growth and differentiation

MGI-1 AND MGI-2 IN GROWTH AND DIFFERENTIAION

Type of myeloid cells	Requirement of MGI-1 for growth	Induction of MGI-2 by MGI-1	Differentiation
Normal	+	Production → of MGI-2	+
Leukemic	+ or −	No production of MGI-2	−
	Constitutive production MGI-1	Production → of MGI-2	+ *

* Autoinduction of differentiation under specific conditions

Fig. 5. Differences in induction of differentiation-inducing protein MGI-2 by growth-inducing protein MGI-1 in normal and leukemic myeloid cells.

is MGI^+D^+ leukemic cells is at the first control point, but the coupling at the second control in normal cells, between the induction of differentiation by MGI-2 and the stopping of multiplication in the mature cells, is maintained. There are differentiation-defective MGI^+D^- leukemic cells, that, like the MGI^+D^+ leukemic cells, no longer require addition of MGI-1 for growth. However, in these cells MGI-2 induces only a partial differentiation, mature cells are not produced, and the cells do not stop multiplying. In addition to uncoupling growth and differentiation at the first control point, MGI^+D^- leukemic cells thus show a second uncoupling between the initiation of differentiation by MGI-2 and the stopping of cell multiplication that occurs as part of the normal program of terminal differentiation. It has been suggested that leukemia originates by uncoupling the first control and that uncoupling of the second control then results in a further evolution of leukemia (*2,4*).

CONSTITUTIVE GENE EXPRESSION IN MALIGNANCY

Since there are leukemic cells that, unlike normal myeloblasts, no longer require MGI-1 for cell viability and growth, the molecular changes required for viability and growth that have to be induced in the normal cells are constitutive in these leukemic cells. This also applies to leukemic cells that constitutively produce their own MGI-1. This suggests that the origin of myeloid leukemia can be a result of a change from an induced to a constitutive expression of genes that control cell viability and growth (*2,4*).

Studies on changes in the synthesis of specific proteins in normal myeloblasts, MGI^+D^+, MGI^+D^-, and MGI^-D^- leukemic clones at different times after adding MGI-1 and -2, using two-dimensional gel electrophoresis (*74*), have directly shown that there have been changes from inducible to constitutive gene expression in the leukemic cells. The results also indicate a relationship between constitutive gene expression and uncoupling of the induction of differentiation by MGI-2 and the stopping of multiplication in the mature cells. The leukemic cells were found to be constitutive for changes in the synthesis of specific proteins that were only induced in the normal cells after treatment with MGI-1. These protein changes, which included the appearance of some proteins and the disappearance of others, were constitutive in all the leukemic clones, derived from different tumors, studied. These have been called C_{leuk}, for constitutive for leukemia. There were other protein changes that were induced by MGI-2 in normal and MGI^+D^+ leukemic cells and were constitutive in MGI^+D^- and MGI^-D^- leukemic cells. There were more of these constitutive changes in MGI^-D^- than in MGI^+D^- leukemic cells. These have been called C_{def}, for constitutive for differentiation defective (Fig. 6) (*74*). These results thus indicate that changes from an induced to a constitutive expression of certain genes is associated with

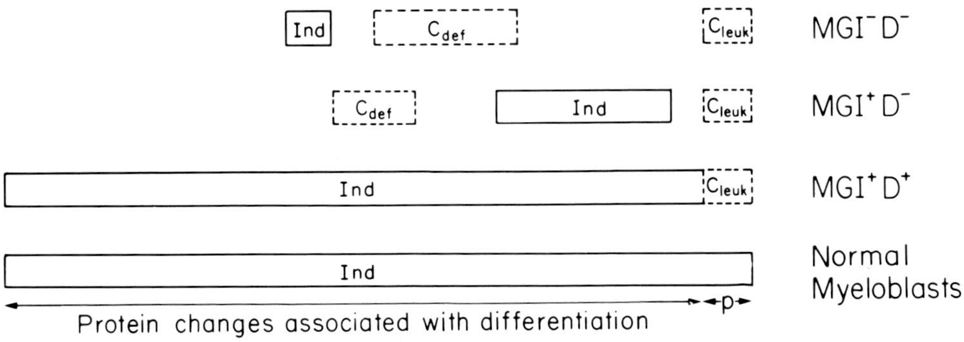

Fig. 6. Schematic summary of protein changes associated with growth (proliferation) and differentiation in normal myeloid precursor cells (myeloblasts) and in different types of myeloid leukemic cells. C_{leuk}; constitutive expression of changes in leukemic compared to normal myeloblasts, C_{def}; constitutive expression of changes in the differentiation defective MGI^+D^- and MGI^-D^- clones compared to MGI^+D^+ clones, Ind; MGI induced changes, and p; protein changes associated with proliferation to produce more cells before differentiation (*74*).

the uncoupling of growth and differentiation, both at the control that requires MGI-1 to produce more cells and at the control of the stopping of cell multiplication that occurs in the formation of mature cells.

The protein changes during the growth and differentiation of normal myeloblasts seem to be induced by MGI-1 and -2 as a series of parallel multiple pathways of gene expression (*74*). It can be assumed that the normal developmental program that couples growth and differentiation in normal cells requires synchronous initiation and progression of these multiple parallel pathways. The presence of constitutive gene expression for some pathways can be expected to produce asynchrony in the coordination required for the normal developmental program. Depending on the pathways involved, this asynchrony could then result in an uncoupling of the controls for growth and differentiation and produce different blocks in the ability to be induced for and to terminate the differentiation process.

We have been able to treat MGI^-D^- leukemic cells so as to induce the reversion of specific C_{def} proteins from the constitutive to the nonconstitutive state. This reversion was then associated with a gain of inductibility by MGI-2 for various differentiation-associated properties. Reversion from the constitutive to the nonconstitutive state in these cells thus restored the synchrony required for induction of differentiation (*61*).

The suggestion therefore derived from these results (*2,4,74*) is that myeloid leukemia originates by a change that produces certain constitutive pathways of gene expression, so that cells no longer require MGI-1 for growth or constitutively produce MGI-1 without inducing MGI-2.

These leukemic cells can, however, still be induced to differentiate normally by MGI-2 added oxogenously or induced in the cells in other ways. The differentiation program induced by MGI-2 can thus proceed normally when it is uncoupled from the growth program induced by MGI-1. This can be followed by constitutive expression of other pathways resulting in the uncoupling of other controls and an asynchrony that interferes with the normal program of terminal differentiation. These second changes then result in the further evolution of leukemia (*4*).

The MGI-1 independent growth of myeloid leukemic cells seems to proceed through stages (*75*), probably starting from a small decrease in MGI-1 requirement to ending, finally, in complete independence of MGI-1. There are presumably also stages in the amount of constitutively produced MGI-1 and the degree of lack of inducibility of MGI-2 and MGI-1. These stages would produce different degrees of asynchrony resulting in differences in the degree of hemotological abnormality, the final stages of asynchrony then resulting in leukemia.

REVERSAL OF MALIGNANCY BY INDUCTION OF DIFFERENTIATION IN VARIOUS TYPES OF TUMORS

These conclusions on the origin and evolution of myeloid leukemia may be applicable to malignant tumors derived from other types of cells whose viability, growth, and differentiation are induced by other physiological inducers. Identification of the physiological inducers of growth and differentiation for different cell types would be a crucial requirement in extending these conclusions to those other tumors. However, even in the absence of such identifications, it appears likely that teratocarcinoma cells (*76,77*) may be comparable to MGI^+D^+ myeloid leukemic cells. The presence of fetal proteins in certain tumors, may also be a result of constitutive gene expression in the tumor of a protein that is induced by the physiological inducer during the developmental program in the normal fetus (*4*). There are probably a variety of tumors in which (A) the original malignancy has a normal differentiation program and the cells are malignant because of uncoupling of the requirement for growth from the requirement for differentiation by changing the gene expression required for growth from inducible to constitutive; and (B) where the further evolution of the tumor results from changes from inducible to constitutive of other pathways of gene expression that produce asynchrony in the normal differentiation program, so that mature nondividing cells are not formed by the physiological inducer of differentiation. However, even these tumors may still be induced to differentiate to form nonmalignant cells by treatment with compounds that can reverse the constitutive to the nonconstitutive state or induce the differentiation program by other pathways. In some tumors, such as sarcomas, reversal of malignancy can be obtained by specific chromosome changes (Fig. 7) (*1,78–82*). But the

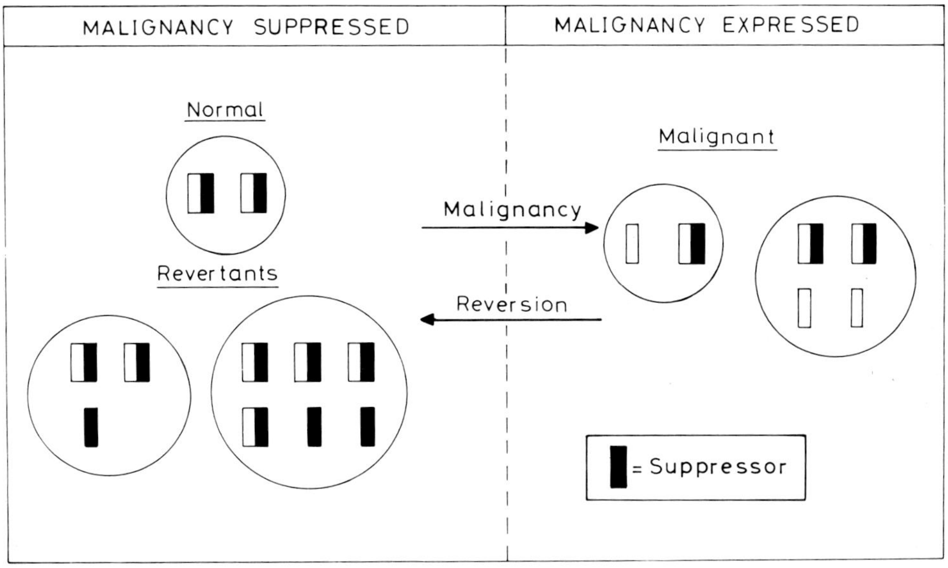

Fig. 7. Gene dosage and the expression and suppression of malignancy. From experiments with normal fibroblasts, sarcoma cells, and their nonmalignant revertants. Genes for □ expression and ■ suppression of malignancy (*79*).

stopping of cell division in mature cells by inducing differentiation induces a reversion of malignancy by bypassing the genetic changes that produce the malignant phenotype.

NEW TYPE OF THERAPY

Our results suggested some novel possibilities for the therapy of myeloid leukemia (*2,3*). The finding of MGI^+D^+ myeloid leukemic cells that can be induced to differentiate normally by MGI-2 suggests the use of MGI-2 injection, stimulation of the production of MGI-2 in vivo, or grafting of MGI-2 producing cells, to induce the normal differentiation of these leukemic cells in patients with myeloid leukemia. This would be an alternative to cytotoxic agents, which kill normal cells as well as tumor cells. There are membrane differences between the cells that differ in their response to MGI-2 that may be useful markers to predict the response of the leukemic cells to MGI-2 in vivo. The MGI^+D^+ leukemic cells can be induced by MGI-2 to again require MGI-1 for cell viability and growth. This suggests that induction of differentiation of the leukemic cells to this stage may also result in the loss of viability and growth of the induced MGI^+D^+ leukemic cells in vivo if there is not enough MGI-1.

The induction of growth of the normal myeloid precursors by MGI-1 and their differentiation to macrophages or granulocytes by MGI-2, also suggests that injected MGI or MGI induced in vivo might be used to restore the normal macrophage and granulocyte population after cytotoxic therapy. MGI therapy may also be useful for treatment of nonmalignant macrophage or granulocyte diseases.

Our results can also help to explain the responses of some, but not all patients, to chemical and irradiation cytotoxic therapy. We have shown that chemicals and irradiations used in therapy can at low doses induce differentiation in clones and myeloid leukemic cells with the appropriate genotype, and that clonal differences in inducibility for differentiation are not necessarily associated with differences in the response of these clones to the cytotoxic effect of these compounds. The growth in vivo of leukemic cells with the appropriate genotype may thus be controlled by the therapeutic agents used not only for their cytotoxic effect, but because they induce differentiation. Differences in the competence of cells to be induced by these agents may thus explain differences in response to therapy in different individuals. The possible induction of MGI-2 by these compounds may also play a role in the therapeutic effects obtained in vivo.

These studies therefore suggest new forms of therapy of leukemia, based on the use of a normal regulatory protein, such as MGI-2, to induce differentiation in malignant cells, and of MGI-1 and -2 to induce a more rapid recovery of the normal cell population after the present forms of therapy. In addition, it is possible to use other compounds that induce MGI-1 or -2 in vivo or can affect mutant malignant cells at differentiation sites that are no longer susceptible to the normal regulator. The malignant cells of each patient should be prescreened for the compounds to which they are susceptible. These possibilities should also be applicable to diseases in other types of cells whose growth and differentiation are controlled by other normal regulators (*2,3*).

REFERENCES

1. Sachs, L. (1974), Regulation of Membrane Changes, Differentiation, and Malignancy in Carcinogenesis. *Harvey Lectures* Academic, NY, Vol. 68, pp. 1–35.
2. Sachs, L. (1978), *Nature* **274,** 535.
3. Sachs, L. (1978), *J. Haematol.* **40,** 509.
4. Sachs, L. (1980), *Proc. Natl. Acad. Sci. USA* **77,** 6152.
5. Sachs, L. (1982), *J. Cell. Physiol. Suppl.* **1,** 151.
6. Sachs, L. (1982), *Cancer Surveys* **1,** 321.
7. Ginsburg, H., and Sachs, L. (1963), *J. Natl. Cancer Inst.* **31,** 1.
8. Pluznik, D. H., and Sachs, L. (1965), *J. Cell. Comp. Physiol.* **66,** 319.
9. Bradley, T. R., and Metcalf, D. (1966), *Aust. J. Exp. Biol. Med. Sci.* **44,** 287.
10. Ichikawa, Y., Pluznik, D. H., and Sachs, L. (1966), *Proc. Natl. Acad. Sci. USA* **56,** 488.
11. Pluznik, D. H., and Sachs, L. (1966), *Exp. Cell Res.* **43,** 553.
12. Paran, M., Sachs, L., Barak, Y., and Resnitzky, P. (1970), *Proc. Natl. Acad. Sci. USA* **67,** 1542.
13. Falk, A., and Sachs, L. (1980), *Int. J. Cancer* **26,** 595.

14. Fibach, E., and Sachs, L. (1975), *J. Cell. Physiol.* **86,** 221.
15. Lotem, J., and Sachs, L. (1979), *Proc. Natl. Acad. Sci. USA* **76,** 5158.
16. Weiss, B., and Sachs, L. (1978), *Proc. Natl. Acad. Sci. USA* **75,** 1374.
17. Austin, P. E., McCulloch, E. A., and Till, J. E. (1971), *J. Cell. Physiol.* **77,** 121.
18. Landau, T., and Sachs, L. (1971), *Proc. Natl. Acad. Sci. USA* **68,** 2540.
19. Lipton, J., and Sachs, L. (1981), *Biochim. Biophys. Acta* **673,** 552.
20. Stanley, E. R., and Heard, P. M. (1977), *J. Biol. Chem.* **252,** 4305.
21. Lotem, J., Lipton, J., and Sachs, L. (1980), *Int. J. Cancer,* **25,** 763.
22. Moore, M. A. S. (1982), *J. Cell. Physiol. Suppl.* **1,** 53.
23. Burgess, A. W., Camakaris, J., and Metcalf, D. (1977), *J. Biol. Chem.* **252,** 1998.
24. Ichikawa, Y., Pluznik, D. H., and Sachs, L. (1967), *Proc. Natl. Acad. Sci. USA* **58,** 1480.
25. Metcalf, D. (1969), *J. Cell. Physiol.* **74,** 323.
26. Lotem, J., and Sachs, L. (1984), *Int. J. Cancer,* **33,** 147.
27. Liebermann, D., Hoffman-Liebermann, B., and Sachs, L. (1982), *Int. J. Cancer* **29,** 159.
28. Fibach, E., Landau, T., and Sachs, L. (1972), *Nature New Biol.* **237,** 276.
29. Fibach, E., and Sachs, L. (1976), *J. Cell. Physiol.* **89,** 259.
30. Maeda, M., Horiuchi, M., Numa, S., and Ichikawa, Y. (1977), *Gann* **68,** 435.
31. Yamamoto, Y., Tomida, M., and Hozumi, M. (1980), *Cancer Res.* **40,** 4804.
32. Burgess, A. W., and Metcalf, D. (1980), *Int. J. Cancer* **26,** 647.
33. Weisinger, G., and Sachs, L. (1983), *EMBO J.* **2,** 2103.
34. Nicola, N. A., Burgess, A. W., and Metcalf, D. (1979), *J. Biol. Chem.* **24,** 5290.
35. Lotem, J., and Sachs, L. (1981), *Int. J. Cancer* **28,** 375.
36. Lotem, J., and Sachs, L. (1977), *J. Cell. Physiol.* **92,** 97.
37. Paran, M., Ichikawa, Y., and Sachs, L. (1968), *J. Cell. Physiol.* **72,** 251.
38. Fibach, E., and Sachs, L. (1974), *J. Cell. Physiol.* **83,** 177.
39. Gootwine, E., Webb, C. G., and Sachs, L. (1982), *Nature* **299,** 63.
40. Webb, C. G., Gootwine, E., and Sachs, L. (1984), *Dev. Biol.* **101,** 221.
41. Lotem, J., and Sachs, L. (1978), *Proc. Natl. Acad. Sci. USA* **75,** 3781.
42. Lotem, J., and Sachs, L. (1983), *Int. J. Cancer* **32,** 781.
43. Ichikawa, Y. (1969), *J. Cell. Physiol.* **74,** 223.
44. Fibach, E., Hayashi, M., and Sachs, L. (1973), *Proc. Natl. Acad. Sci. USA* **70,** 343.
45. Ichikawa, Y., Maeda, N., and Horiuchi, M. (1976), *Int. J. Cancer* **17,** 789.
46. Lotem, J., and Sachs, L. (1977), *Proc. Natl. Acad. Sci. USA* **74,** 5554.
47. Hayashi, M., Fibach, E., and Sachs, L. (1974), *Int. J. Cancer* **14,** 40.
48. Azumi, J., and Sachs, L. (1977), *Proc. Natl. Acad. Sci. USA* **74,** 253.
49. Hoffman-Liebermann, B., and Sachs, L. (1978), *Cell* **14,** 825.
50. Simantov, R., and Sachs, L. (1978), *Proc. Natl. Acad. Sci. USA* **75,** 1805.
51. Simontov, R., Shkolnik, T., and Sachs, L. (1980), *Proc. Natl. Acad. Sci. USA* **77,** 4798.
52. Hoffman-Liebermann, B., Liebermann, D., and Sachs, L. (1981), *Int. J. Cancer* **28,** 615.
52a. Hoffman-Liebermann, B., Liebermann, D., and Sachs, L. (1981), *Dev. Biol.* **81,** 255.
53. Lotem, J., and Sachs, L. (1980), *Int. J. Cancer,* **25,** 561.
54. Baccarani, M., and Tura, S. (1979), *Br. J. Haematol.* **42,** 485.
55. Housset, M., Daniel, M. T., and Degos, L. (1982), *Br. J. Haematol.* **51,** 125.

56. Michalewicz, R., Lotem, J., and Sachs, L. (1984), *Leukemia Res.* **8,** 783.
57. Cohen, L., and Sachs, L. (1981), *Proc. Natl. Acad. Sci. USA* **78,** 353.
58. Krystosek, A., and Sachs, L. (1976), *Cell* **9,** 675.
59. Lotem, J., and Sachs. L. (1978), *Int. J. Cancer* **22,** 214.
60. Symonds, G., and Sachs, L. (1982), *J. Cell. Physiol.* **111,** 9.
61. Symonds, G., and Sachs, L. (1983), *EMBO J.* **2,** 663.
62. Friend, C. (1978), *Harvey Lectures,* Academic, NY, vol. 72, pp, 253–281.
63. Marks, P., and Rifkind, R. A. (1978), *Ann. Rev. Biochem.* **47,** 419.
64. Lotem, J., and Sachs, L. (1974), *Proc. Natl. Acad. Sci. USA* **71,** 3507.
65. Lotem, J., and Sachs, L. (1975), *Int. J. Cancer* **15,** 731.
66. Symonds, G., and Sachs, L. (1982), *Blood* **60,** 208.
67. Lotem, J., and Sachs, L. (1982), *Proc. Natl. Acad. Sci. USA* **79,** 4347.
68. Lotem, J., and Sachs, L. (1983), *Int. J. Cancer* **32,** 127.
69. Ushiro, H., and Cohen, S. (1980), *J. Biol. Chem.* **225,** 8363.
70. Kasuga, M., Frujita-Yamaguchi, Y., Blithe, D. L., and Kahn, C. R. (1983), *Proc. Natl. Acad. Sci. USA* **80,** 2137.
71. Doolittle, R. F., Hunkapiller, M. W., Hood, L. E., Devare, S. G., Robbins, K. C., Aaronson, S. A., and Antoniades, H. N. (1983), *Science* **221,** 275.
72. Waterfield, M. D., Scrace, G. T., Whittle, N., Stroobant, P., Johnsson, A., Wasteson, A., Westermark, B., Heldin, C. H., Huang, J. S., and Deuel, T. F. (1983), *Nature* **304,** 35.
73. Symonds, G., and Sachs, L. (1982), *EMBO J.* **1,** 1343.
74. Liebermann, D., Hoffman-Liebermann, B., and Sachs, L. (1980), *Dev. Biol.* **79,** 46.
75. Collins, S. J., Gallo, R. C., and Gallagher, R. E. (1977), *Nature* **270,** 347.
76. Mintz, B., and Illmensee, K. (1975), *Proc. Natl. Acad. Sci. USA* **72,** 3585.
77. Dewey, M. J., Martin, D. W., Jr., Martin, G. R. and Mintz, B. (1977), *Proc. Natl. Acad. Sci. USA* **74,** 5564.
78. Rabinowitz, Z., and Sachs. L. (1968), *Nature* **220,** 1203.
79. Rabinowitz, Z., and Sachs. L. (1970), *Nature* **225,** 136.
80. Rabinowitz, Z., and Sachs, L. (1970), *Int. J. Cancer* **6,** 388.
81. Hitosumachi, S., Rabinowitz, Z., and Sachs, L. (1971), *Nature* **231,** 511.
82. Yamamoto, T., Rabinowitz, Z., and Sachs, L. (1973), *Nature (New Biol.)* **243,** 247.
83. Pike, B., and Robinson, W. A. (1970), *J. Cell. Physiol.* **76,** 77.

New Strategies in Cancer Chemotherapy

GIANNI BONADONNA

Division of Medical Oncology, Istituto Nazionale Tumori, Milan, Italy

INTRODUCTION

There is unchallenged evidence that during the last quarter of the century progress has been made in developing cancer chemotherapy that is effective against a wide variety of tumors. In particular, specific neoplasms can now be cured by the administration of adequate polydrug regimens. To improve clinical results in various tumors still further, many new strategies have been proposed and applied during the past ten years. They consist mainly of adjuvant chemotherapy regimens for patients at high risk of recurrence, following standard local-regional treatment as well as the delivery of noncross-resistant regimens to overcome the problem of primary cell resistance.

This article will briefly review the main concepts underlying the above-mentioned strategies and will present some of the treatment results achieved utilizing the case series evaluated at the Milan Cancer Institute.

PRINCIPLES OF SYSTEMIC ADJUVANT THERAPY

The concepts underlying systemic adjuvant therapy were introduced into full-scale clinical trials in cancer in the 1970s. These concepts have continued to evolve as second- and third-generation trials based on human disease models rather than on preclinical models. A forerunner of adjuvant therapy was the recognition of possible complementary therapeutic action of various modalities when chemotherapy was first introduced. Subsequently, the demonstration of the successful application of various treatment modalities directed against the early stages of Wilms' tumor led to the concept of combined modality treatment. After many abortive attempts to utilize chemotherapy in the early stages of adult solid tumors in the late 1950s and 1960s, the concept of postsurgical treatment of minimal residual disease emerged. Reflecting an additional evolution are terms such as neo- or protoadjuvant chemotherapy, which refer to the use of systemic therapy prior to a local intervention. It is worth reviewing here the preclinical foundations of adjuvant systemic therapy prior to describing specifically how our concepts have evolved and how these principles have been applied to the management of various tumors, in particular, to breast cancer.

Preclinical Basis

The experimental foundations of adjuvant therapy were derived to a great extent from studies performed by Skipper (*1*), Schabel (*2*), and their co-workers at the Southern Research Institute. Their observations of the survival of mice with transplantable solid tumors of varying age and size, subjected to resection, led them to postulate a great variability in metastatic body burden at the time of surgery. Because of this, empirical considerations have to be utilized in the selection of drugs, treatment schedule, and duration of treatment. Surgical cure rates in solid tumors of mice also show a relationship to the size of the tumor. Of interest is the "break" point in the survival curves, after which no further recurrences were noted. These break points appeared at similar times after surgery, although the percent survival was different in each case, indicating that the biology of the tumor was similar in the various animals, but that the tumor burden accounted for the variability experienced. A corollary to this is that definitive conclusions about therapeutic effectiveness must be witheld until this "break" point is reached, since it is the patients with large tumor burdens who are least likely to benefit from adjuvant therapy and who make up the first portion of the survival curves. As for the selection of drugs and treatment schedules, these must be effective to achieve a net reduction in cell burden by the next cycle; it is apparent that the optimal intensity must generally be identical to the intensity that yields the optimal results in advanced disease.

Martin (*3*) demonstrated that both drug combinations and surgery were essential in making an optimal impact on the survival of mice bearing mammary tumors. However, since the tumor burden is generally lower, if treatments are started in adjuvant situations, the duration of treatment needed to assure eradication is likely to be briefer. The emergence of resistance is also likely to advance in parallel with the progress of the disease, and constitutes a major element in adjuvant therapy failure. However, in accordance with the hypothesis advanced by Goldie and Coldman (*4*), with lower tumor burdens the emergence of resistant cells occurs to a lesser degree.

Immunotherapeutic agents were attractive drugs for the adjuvant setting in view of the unlikelyhood of their having long-term toxicities and of the postulated efficacy of these approaches at low tumor burdens. Experiments in general indicated that agents, such as *Corynebacterium parvum*, acted by antimetastatic effects and only had an effect when given at a specific time prior to surgery in circumstances when surgery alone was clearly beneficial.

Hormones are receiving renewed attention as potential adjuvant treatments. In carcinogen-induced mammary tumors of mice, demonstration that tamoxifen (T) can both delay the appearance of tumors and decrease their overall frequency in a dose-dependent and schedule-dependent manner has been of great interest for its relevance, not only in adjuvant treatment of breast cancer, but because of its possible inhibition of carcinogenic influences. In the animal models, interruption of T may partially eliminate the benefit, since tumors will begin to recur.

The complexities of human disease observation may require additional models for proper interpretation. For example, one may be able to explain late recurrences based on the concept of "dormant" stem cells that have escaped eradication and metastasized. Early recurrences, on the other hand, indicate a failure to eradicate "committed" stem cells expected to be sensitive to chemotherapy, unless their numbers were such to allow emergence of resistance. Strategies underlying the use of various adjuvant therapies are in fact dependent on the intrinsic efficacy of the chemotherapy and the patterns and sites of relapse.

Theoretical Considerations

In the absence of biochemical markers, adjuvant treatment must be empirically chosen, based on the treatment of advanced disease, and applied to situations where staging provides some estimate of risk-to-recurrence (and thereby, tumor burden). Hence, the effective treatment should be applied to patients at risk with equal intensity, but for a shorter duration than in advanced disease. Early treatment at a time of minimum tumor burden is to be preferred, but it should be borne in mind that preoperative treatment may mask staging information. Guidance on how properly to apply adjuvant therapy must be developed from an analysis of comparative experiences with a similarly staged group of patients not receiving treatment. Disease-free survival comparisons will indicate the benefit that might be achieved from adjuvant treatment.

The other side of the coin, however, is that the risks of the adjuvant treatment are experienced by 100% of the population rather than being relegated only to the relapsing population. This relapsing population receives what we term "salvage" therapy. Cost/benefit balance analyses are central to the use of adjuvant vs salvage strategies. When the risks have been rather precisely delineated, and the relapsing population is high and not likely to be cured, such as in breast cancer, the rationale for adjuvant therapy is powerful, and the fulcrum of the balance favors any potential benefit over its cost. On the other hand, in, for example, testicular cancer, the ability to salvage patients with chemotherapy has been well demonstrated. The fulcrum of the balance in this instance shifts toward the cost side. In all these conditions, however, the overriding factor is *efficacy*. When the efficacy or contribution of the adjuvant treatment is likely to be modest, then considerations of the cost to the treated population become most relevant. Such situations often require trials with untreated controls.

Since the treatment given is empirical, its administration has been patterned after certain guiding principles that are derived from advanced disease. However, for the specific purpose of adjuvant treatment, some drugs may be avoided. In particular, drugs that may have long-term sequelae may be excluded. A special situation is the avoidance of drugs that could affect the patient's fertility and gonadal function when equally effective therapeutic alternatives are presumably available. Other empiric strategies have been advanced, based on mathematical models: Late intensification was introduced by Norton and Simon (*5*) and has also been utilized in leukemia studies. It is presently unclear whether such strategies have yielded positive results in adjuvant therapy.

Finally, one must point out other theoretical rationales for adjuvant therapy: (a) a lower tumor burden with presumed lesser resistant fraction; (b) kinetically more favorable circumstances; (c) a less disturbed microenvironment when therapy is started; and (d) a less compromised host.

Example of Breast Cancer

About 12 yr ago, during the period when interest in adjuvant therapy was being revitalized the crucial question to be tested was the ability of available cytotoxic drugs to exert a significant and consistent antitumor effect when administered for prolonged periods of time following standard surgical procedures. The validity of the scientific hypothesis, involving the concept of high incidence of disseminated disease in women with histologically positive axillary lymph nodes and the optimal kinetic conditions of micrometastases (*1,2*), was tested by the first generation of modern adjuvant trials. The correct methodology required the concomitant use of a control group with local-regional therapy alone. Table 1 summarizes the essential results obtained by research teams that followed case series for a minimum of 5 yr (*6*). All seven randomized studies reported a significant increase in the relapse-free survival (RFS) of chemotherapy vs control groups, at least in given subsets with a predominance of premenopausal women. However, in the study arm testing phenylalanine mustard (P) vs control, the Guy's Manchester team apparently failed to confirm the significant superiority of the treated group observed by the National Surgical Adjuvant Breast Project (NSABP). Significant overall survival (OS) advantage has, this far, been reported in given subsets by three research groups. Table 2 summarizes the 10-yr results achieved with cyclophosphamide, methotrexate, and fluorouracil (CMF) chemotherapy in Milan.

TABLE 1
Adjuvant Chemotherapy Studies with a Concomitant Control Group with Local–Regional Treatment Alone in Node Positive Patients[a]

First author	No. Pts	Adjuvant treatment	Median follow up, mo	Essential findings in the treated groups
Fisher	370	P	82	Increased RFS[b] and OS[c] in premenopause, 1–3 nodes
Bonadonna	386	CMF	92	Increased RFS and OS in premenopause
Senn	118	LMF + BCG	82	Increased RFS in postmenopause. No OS difference.
Wheeler	227	CVFM	60	Increased RFS (mainly in postmenopause). No OS difference.
Rubens	312	P	48	No significant difference after P.
	316	CMF	48	CMF superior in premenopause. No OS difference.
Morrison	512	AVCMF	45	Increased RFS in premenopause. No OS difference.
Smith	322	CMF vs CMF + RT	42	Increased RFS in pre- and postmenopause after RT + CMF and CMF vs postoperative RT alone. Increased OS after RT + CMF vs RT.

[a]Adapted from G. Bonadonna and P. Valagussa (*6*).
[b]RFS; relapse-free survival.
[c]OS; overall survival.

TABLE 2
Essential Findings of First CMF Study; 10-yr Results

	CTR, %	CMF, %	*P*
RFS, total	31.4	43.4	<0.001
nodes 1–3	37.7	50.8	0.002
>3	20.3	27.9	0.05
Pre	31.4	48.3	0.0005
Post	32.2	38.2	0.32
Survival, total	47.3	55.2	0.10
nodes 1–3	53.4	60.5	0.22
>3	33.5	43.8	0.22
Pre	44.8	59.0	<0.02
Post	50.1	52.1	0.89

The important issue is the durability of disease-free status following primary combined treatment. In the first Milan adjuvant trial (i.e., control vs CMF), although there was a progressive decline over time in the RFS for both treatment arms, there was little difference in the comparative proportion of patients continuously disease-free at each interval. This would suggest that the percent difference of women benefitting from surgery vs surgery plus chemotherapy was already evident by the third year postmastectomy, and the time of clinical manifestation of relapse was probably relative only to the growth characteristics of micrometastases. In other words, the fraction of patients bearing micrometastatic colonies sensitive to CMF chemotherapy could be predicted with sufficient accuracy at the 3-yr analysis. Local-regional recurrence was reduced in the chemotherapy group vs control group, thereby eliminating this manifestation as an indication of the routine use of radiotherapy.

The above-mentioned findings, implying the concept that the maximum cell kill of sensitive tumor cells occurs within the first few treatment cycles, are supported by the results of the second Milan adjuvant trial, testing six vs twelve cycles of CMF. At 8 yr after mastectomy, the actuarial analysis of RFS failed to detect a significant difference between the two treatment groups (*6*). The 3–5 yr results of other research institutions utilizing Leukeran, methotrexate, and fluorouracil (LMF) or Adriamycin plus cyclophosphamide confirmed the lack of significant therapeutic advantage by prolonged treatment. Also, the overall 5-yr RFS with CMF administered for 24 cycles, as reported by investigators of Memorial Hospital, was practically the same as that observed in Milan. Details concerning the prognostic relevance of axillary lymph nodes, menopausal and receptors status, and dose levels are discussed elsewhere (*6*).

Earlier studies failed to demonstrate a 5-yr overall survival advantage with adjuvant castration alone, and castration plus long-term prednisone-prolonged survival only in the small subset of premenopausal women who were older than 45 yr of age. During the past few years, increasing emphasis has been placed on the use of adjuvant endocrine therapy, either alone or combined with chemotherapy. The main reasons this avenue of clinical research was stimulated were the possibility of selecting patients through the receptor status, the heterogeneity of breast cancer cells, and, last but not least, the lack of severe toxicity associated with additive hormonal manipulations. In the vast majority of ongoing clinical trials, the antiestrogen compound at the daily dose of 20–30 mg for 12 or 24 mo, represented the drug of choice. At present, only the 2–3-yr results are available (*6*). However, as occurred with adjuvant chemotherapy, great hopes, as well as controversies are raised. The largest series of patients in which postoperative T administered in conjunction with chemotherapy was that studied by the NSABP group (*7*). The 3-yr analysis showed that chemohormonal therapy phenylalanine mustard, fluorouracil, and tauroxifen (PFT) yielded markedly superior RFS only in women 50 yr of age or older. The effectiveness of combined therapy was related to the level of estrogen (ER) and progesterone (PR) receptors, and women with levels of 10 fmol or more displayed the greatest benefit. However, the trial provided discordant results in the subcategories of 50–59 and 60–70 yr of age. In the first subgroup, survival following PFT was inferior compared to combination chemotherapy when tumor PR was low, whereas in the older age group, PFT was superior to phenylalanine mustard plus fluorouracil (PF) regardless of hormone receptor levels. In patients under 50 yr of age, both RFS and OS were adversely affected by PFT when PR levels were under 10 fmol or the effects of PF and PFT were similar when either receptor or both were high. The recent 4-yr actuarial analysis of the Eastern Cooperative Oncology Group (ECOG) showed that in premenopausal women, 52% of whom had ER + tumors, there was no difference in terms of RFS and OS among the three groups given CMF, CMF plus prednisone (CMFP), and CMF plus prednisone and tamoxifen (CMFPT) for 1 yr. In particular, the addition of T to CMFP was not detrimental. In postmenopausal women less than 65 yr of age, there was no survival difference; however, CMFP and CMFPT gave superior results to the control group with ER − tumors (*6*).

Preliminary favorable results were recently reported following use of adjuvant T alone in five randomized case series using different forms of local-regional modality as a concomitant control group (*6*). Not all patients were assayed for hormonal receptors, and when the observed findings were related to ER, some investigators were not able to show selective benefit according to receptors status. The two most important trials are those from Canada and Great Britain. The Canadian study was analyzed at a median followup of 3.1 yr; the total RFS was significantly improved in 1–3 and >3 modal subsets for the T-treated group (30 mg/d for 2 yr) compared to the controls. Survival was similar in the two treatment groups even in the presence of receptor positivity. In the British trial, life-table analysis showed a significant reduction in breast cancer recurrence (20.5% for the control group vs 14.2% for the T group; $P = 0.01$) and a prolongation of the disease-free interval by 9 mo. At an average period of followup of 35.1 mo, an overall survival difference was reported favoring T-treated patients ($P = 0.013$). Finally, positive initial results limited to the RFS are currently being observed by the Royal Marsden Hospital, where the investigators are utilizing adjuvant postoperative aminogluthetimide vs control.

Considering the above reported, and at times contradictory, findings, the lack of prolonged followup for most of the ongoing studies and the fact that not all trials have included a homogeneous, and therefore comparable, type of patient (i.e., women with node-positive or -negative; hormonal receptor positive, negative, or undetermined, T administered for 12 or 24 mo), it appears advisable at this time to wait a few more years before drawing meaningful conclusions on the role of surgical adjuvant endocrine therapy.

In conclusion, current evidence indicates that adjuvant chemotherapy for primary breast cancer is still in a transitional phase of clinical research because much remains to be elucidated in terms of optimal drug combinations, intervals, and duration of treatment. For these reasons and because of the inconsistency of some published results, not every form of systemic adjuvant therapy can be presented as a standard treatment approach. However, it is undeniable that constant progress has been made during the past decade, particularly in subsets with a limited number of positive nodes, and this was achieved at the expense of inevitable, but acceptable, drug toxicity. What may be considered an overall limited success should be translated into the total number of the high-risk women who today can remain continuously disease-free for five or more years following adjuvant treatments of relatively short duration.

All those who have expected more successful results from the initial clinical trials will certainly be disappointed. But physicians should be aware that the magnitude of the problems encountered during the past decade was indeed much greater than we had originally thought. To overcome all the obstacles related to tumor cell heterogeneity, simple treatments have proved not to be enough. Therefore, other clinical trials utilizing more complex and pharmacologically suitable drug regimens are now required if we are to further improve current results. These efforts will take many years. While waiting for the development of such optimal therapies, we would reiterate our opinion that treatment regimens that have already resulted in sustained improvement at the expense of moderate-to-minimal toxicity and have been tested through randomized studies can be widely and judiciously applied in node-positive patients, and that these patients may be expected to benefit from such therapy in terms of prolonged disease-free survival. It is highly advisable that adjuvant therapy be selected, performed, and monitored by trained medical oncologists. They will be less tempted than other clinicians to prescribe cytotoxic drugs on incorrect dose schedules or to reject adjuvant therapy because its actual cure rate has not yet been established. A patient objectively informed about current potential benefits vs toxicity will also help physicians in making the final decision. Adjuvant therapy administered to node-negative women or treatments involving preoperative chemotherapy must for the present be restricted to research centers.

We strongly recommend that physicians persist in entering patients into clinical trials because there is no other reliable method to generate the data needed to convert unsupported opinions to hard facts. However, investigators should become more aware that unless fixed rules on uniform patient selection, stratification, and mode of therapy, as well as reporting technique, are settled on an international basis, randomization, *per se,* will not be enough to ensure scientific progress. Some of the present results do appear inconsistent because the mosaic of current trials lacks coordination. Because breast cancer is itself a mosaic, repetitive trials based on different eligibility criteria or utilizing modified drug regimens will often provide answers that will not be consistent with the initial hypothesis (*6*).

NEW CONCEPTS IN DRUG RESISTANCE

Many research physicians are currently following with increasing interest the development of new concepts, as well as the interpretation of clinical findings, related to drug resistance and cancer chemotherapy. The hypothesis that chemotherapy failure could often be attributed to drug resistance actually represents the application of an old concept to a very common problem. Cell resistance to anticancer drugs was generally described in the past as the result of acquired resistance, i.e., a process related to specific adaptive mechanisms that become operative within individual cells under the influence of a particular drug. However, rigorous experimental studies carried out on a variety of neoplastic cell lines (as well as microbial systems) indicated that the basis for cell resistance involves other mechanisms. "All biological systems have an inherent probability or rate of undergoing phenotypic variations as a consequence of random changes in the genetic material (*4*)." This new concept, originating in the pioneering studies of Luria and Delbruck (*8*) in bacterial systems, was recently reexamined by Goldie and Coldman and adapted to the area of cancer chemotherapy. Their assumption was that "the proportion as well as the absolute number of resistant cells will increase with time and the fraction of resistant cells within tumor colonies of the same size will vary depending on whether mutation occurs as an early or late event (*4*)." Therefore, we should expect considerable fluctuation in the size of the fraction of resistant tumor cells within a given group of patients despite their being of the same histological subtype, having similar clinical or pathologic stages, and carrying roughly equivalent tumor burden.

Since spontaneous and permanent mutation to phenotypic drug resistance could occur as an intrinsic property of genetically unstable, rapidly growing tumor-cells, Goldie and Coldman developed a mathematical model that related curability by anticancer drugs to the time of appearance of single, double, or multidrug-resistant cell lines (*4*). The frequently observed variability of response to cancer chemotherapy following the administration of similar (if not identical) drug regimens can find, in part, a strong theoretical rationale in the spontaneous mutation to multiple levels of drug resistance. In fact, if the first mutant cell appears early in the growth of a given clone, it will have the time to generate a large number of progeny; this cell, by reaching a certain critical size, will further mutate to double and then to multidrug resistance, thus rendering the tumor incurable. By affecting only a limited fraction of sensitive neoplastic cells, chemotherapy will in this case produce only a small and transient diminution in the total cell population. In contrast, if the first mutant cell has appeared much later in the expansion of the clone, the ratio sensitive to resistant tumor cells will be high, and the same drug(s) will induce a comparatively higher complete remission rate. However, even in this case, the total neoplastic cell population will not be eradicated, since a cure will occur only in the absence of resistant mutants following exposure to an effective drug regimen of appropriate sequence, intensity, and duration (*1,4*).

Thus, since a major cause of chemotherapy failure is the selection and overgrowth of permanent-drug-resistant neoplastic-cell populations, both the tumor cell burden and the mix of drug-sensitive and -resistant cells represent crucial prognostic variables for patients undergoing chemotherapy. The lack of improved clinical benefits after maintenance chemotherapy in malignant lymphomas, as in other neoplastic diseases, once complete remission has been properly assessed (*9*), and following long-term adjuvant chemotherapy (*6*) lend support to the above-mentioned concepts. The failure of temporarily effective chemotherapy, commonly observed in all neoplasms, is associated with the overgrowth of tumor stem cells resistant to one or more, but not necessarily all, of the individual drugs in the combination. This concept is in line with the clinical observation that malignant lymphomas often respond to retreatment with the same drug combination [e.g., mechlorethamide, vincristine, procarbazine, and prednisone (MOPP) in Hodgkin's disease (HD) and cyclophosphamide, vincristine, and prednisone in follicular lymphomas] if the relapse occurs following at least 12 mo from initial complete remission (CR).

Since tumor cell heterogeneity and spontaneous drug resistance before drug exposure in tumors reaching a critical size constitutes an important barrier to a cure, even in diseases exquisitely sensitive to many anticancer drugs, treatment strategies utilizing effective non-cross resistant combinations delivered in the appropriate sequence could yield superior results compared to a single polydrug regimen. Starting with HD, alternating noncross-resistant chemotherapy was first successfully applied on the basis of clinical findings (*9*), and is currently being studied in other neoplastic diseases, following the above mentioned concepts.

COMBINATION CHEMOTHERAPY IN MALIGNANT LYMPHOMAS

During the past few years, the problem of curability of advanced malignant lymphomas [HD and non-Hodgkin's lymphomas (NHL)] has been interpreted in the light of new concepts in drug resistance. In fact, although impressive clinical results were recently reported as a consequence of MOPP chemotherapy for HD and in high-grade NHL utilizing various drug combinations, relapse following CR still occurred in a considerable fraction of responders. This is particularly true in the presence of certain prognostic variables. Furthermore, if the duration of first remission is less than 12 mo, retreatment with the same drug combination is often ineffective, and switching to a noncross-resistant regimen is required to achieve a second CR (*9*). Today, reexamination of old findings can easily attribute the lack of a curative effect, following a single- or multidrug regimen, to the ratio of drug-resistant to sensitive neoplastic cells, either present at the start of chemotherapy or surviving at cessation of treatment and at relapse. The initial empirical, but successful, application in HD of alternating noncross-resistant regimens lends support at the clinical level to mathematical hypothesis (*4*) and experimental laboratory observations (*1*).

ALTERNATING DRUG COMBINATIONS IN HODGKIN'S DISEASE

As previously mentioned and summarized in a recent publication (*9*), the first idea—to design a prospective controlled clinical trial with two noncross-resistant regimens delivered in alternating fashion for patients previously untreated with chemotherapy—originated in the clinical observations that: (a) patients resistant to MOPP chemotherapy showed either progressive disease during the first 2–4 cycles or disease recurrence within a few months (often less than 6 mo) following an initial complete response; and (b) MOPP and Adriamycin, bleomycin, vinblastine, and dacarbazine (ABVD) appeared to be equally effective noncross-resistant combinations, and the median time to CR with either regimen was the third cycle. Details concerning trial design, patient characteristics, and preliminary 5-yr results were reported in several publications during 1982 (*9*). A few more patients are presently evaluable and the essential comparative data are summarized in Table 3.

TABLE 3
Alternating Chemotherapy vs MOPP in Stage IV Comparative 7-yr Overall Results[a]

	MOPP, %	MOPP/ABVD, %	P
Complete remission	74.4	88.9	0.14
Freedom from progression	35.1	68.1	0.001
Relapse-free survival	44.4	76.8	0.004
Survival of CR_s	73.2	90.3	0.038
Total survival	61.1	82.3	0.043

[a]Adapted from G. Bonadonna et al. (*9*).

The updated results confirm the previous findings in terms of disease progression during chemotherapy, CR rate in various prognostic subgroups, and survival results. Not all comparative findings reach statistical significance because in given subgroups the number of patients is small; however, the general trend in favor of alternating chemotherapy remains quite evident. In particular, in the analysis of total survival the difference between MOPP and MOPP/ABVD became evident at the 7-yr analysis.

Our current study (*9–11*) attempts to meet more closely the stringent requirements needed for the initial assumption to work. We are now testing, through a prospective randomized study activated in July 1982, the efficacy of cyclical MOPP/ABVD through different sequences. The control arm consists of MOPP/ABVD alternating after each full monthly cycle of either chemotherapy; on the contrary, in the experimental arm half-cycle of MOPP (mechlorethamine and vincristine on d 1, procarbazine and prednisone from 1–7) it is alternated within a month period with half cycle of ABVD given on d 15. Theoretically, this new alternating sequence therapy, by testing more appropriately the Goldie and Coldman hypothesis, should be able to further improve the five-year RFS achieved in our previous study. This clinical trial will require a large series of patients and is being performed in high-risk groups such as those with stage IA-IIA in bulky disease, and stages IIB, III (A and B), and IV (A and B). Folowing CR, low-dose radiotherapy will be delivered only to the area(s) with bulky disease at the start of chemotherapy. Otherwise, no further treatment will be administered once CR is achieved by alternating drug therapy. The CCNU, etoposide, and prednimustine (CEP) chemotherapy, with or without irradiation, will be reserved for patients not achieving CR following six cycles of either alternating sequence chemotherapy because, in our experience, its administration in 40 patients resistant to MOPP and ABVD induced remission in 60% (CR 35%, PR 25%) for a median duration in excess of 17 mo (*12*).

Following our initial reports on ABVD, alone or alternated with MOPP, other research groups have utilized two noncross-resistant regimens in tandem. The first group was that of Memorial Hospital, who alternated MOPP and ABVD chemotherapy in 91 patients with stages IIB, III, and IV. This prospective nonrandomized study also included low-dose radiotherapy (2000 rads) to the sites of originally bulky disease. In untreated patients, the CR rate was 88% and the 4-yr RFS was 84%, a result very similar to that achieved in Milan. The same institution is currently evaluating two (MOPP/ABVD) vs three-drug combinations (CCNU/Alkeran/vindesine, or CAD/MOPP/ABV) (*13*). In the study arm, dacarbazine was dropped from ABVD to reduce vomiting, whereas in both arms, low-dose (2000 rads) radiotherapy was delivered to initially involved regions.

The other research group was the Yale Group, which treated with MOPP/ABVD and low-dose irradiation, in a nonrandomized fashion, a high-risk subgroup with HD, characterized by patients more than 40 yr of age and/or with stage-IV disease with multiple sites of involvement. In the above-mentioned sub-groups, the survival was 87% compared to 58% for those treated with mechlorethamine, vincristine, vinblastine, procarbazine, and prednisone (MVVPP) and radiation therapy in previous years (*14*). The Southwest Oncology Group (*15*) has recently reported that MOPP–bleomycin, adriomycin, prednisone (MOPP–BAP) (mechlorethamine and vincristine on d 1, Adriamycin on d 8, procarbazine and prednisone from d 2 through 7, as well as from 9 through 12) produced significantly better CR rates, as well as significantly superior five-year RFS and survival, than did MOPP–bleomycin. The results of this study clearly emphasize the therapeutic effectiveness of Adriamycin within a polydrug regimen for HD. More recently, Klimo and Connors (*16*) utilized an alternating sequence very similar to that currently being tested in our institute. In fact, mechlorethamine, vincristine, procarbazine, and prednisone are administered on d 1 and Adriamycin, bleomycin, and vinblastine on d 8 of each monthly cycle for a total of eight cycles. Dacarbazine was deleted to reduce vomiting. In a limited series of patients with advanced or recurrent HD, the CR rate was impressive (untreated patients 10/11; following first relapse, 8/8; following second relapse 4/4). After a median treatment followup of 12 mo, only 2/23 patients had relapsed (2/4 with multiple relapses following prior treatments).

CONCLUSION

Despite advances in the chemotherapy of several malignancies, new approaches are required, at least in given histopathologic and stage subgroups.

Phenotypic heterogeneity appears to be a major stumbling block in the cure of cancer with chemotherapy. This statement applies to the treatment of advanced disease as well as in an adjuvant situation. The first step toward overcoming the important clinical problem of drug resistance was made by utilizing various single polydrug regimens. Alternating chemotherapies, such as MOPP/ABVD for HD, and to a large extent also methotrexate, Adriamycin, cyclophosphamide, etoposide /MOPP (Pro-MACE/MOPP) in high grade NHL, was found more effective in eradicating the singly and doubly resistant lymphoma stem cells and in preventing the emergence of double- or multidrug resistant neoplastic cells during treatment. The dominant variables affecting the cure rate of neoplasms treated with chemotherapy are numerous, and with this number of critical variables, retrospective and prospective computer simulations might be useful in reducing the current empiricism in planning alternating combinations and deducing how new anticancer drugs can be used more effectively.

REFERENCES

1. Skipper, H. E., and Schabel, F. M., Jr. (1984), *Cancer Treat. Rep.* **68,** 43.
2. Schabel, F. M., Jr. (1977), *Cancer* **39,** 2875.
3. Martin, D. S., Fugman, R. A., Stolfi, R. L., and Hayworth, P. E. (1975), *Cancer Chemother. Rep.* **5,** 89.
4. Goldie, J. H. (1983), *Breast Cancer Res. Treat.* **3,** 129.
5. Norton, L., and Simon, R. (1977), *Cancer Treat. Rep.* **61,** 1307.
6. Bonadonna, G., and Valagussa, P. (1985), *J. Clin. Oncol.* **3,** 253.
7. Fisher, B., Redmond, C., Brown, A., Wickerham, D. L., Woemark, N., Allegre, J., Escher, G., Lippman, M., Saveov, E., Wittliff, J., Fisher, E. R., Plotkin, D., Bowmon, D., Wolter, J., Bornstein, R., Desser, R., Frelick, R., and cooperating Investigators (1983) *J. Clin. Oncol.* **1,** 227.
8. Luria, S. E., and Delbruck, M. (1943), *Genetics* **28,** 491.
9. Bonadonna, G. (1982), *Cancer Res.* **42,** 4309.
10. Bonadonna, G., Santoro, A., Valagussa, P., Viviani, S., Zucali, R., Bonfante, V., and Banfi, A. (1984), *Current Status of the Milan Trials for Hodgkin's Disease in Adults. Second International Conference on Malignant Lymphomas,* Lugano, Switzerland, 1985.
11. Bonadonna, G., Viviani S., Bonfante V., Valagussa, P., and Santoro A. (1984), *Proc. Amer. Soc. Clin. Oncol.* **3,** 254.
12. Bonadonna, G., Viviani, S., Valagussa, P., Bonfante, V., and Santoro, A., 1985, *Sem. Oncol.* **12**(Suppl 2): 23.
13. Strauss, D. J., Myers, J., Koziner, B., Lee, B. J., Nisce, L., and Clarkson, B. D. 1982, *Proc. Am. Soc. Clin. Oncol.* **1,** 160.
14. Kapp, D. S., Prosnitz, L. R., Farber, L. R., Bertino, J. R., Cadman, E. C., and Fischer, D. B. *Proc. Am. Soc. Clin. Oncol.* **1,** 166.
15. Jones, S. E., and Dixon, D. O., *Cancer* (in press).
16. Klimo, P., and Connors, J. M. (1983) *Proc. Am. Soc. Clin. Oncol.* **2,** 208.

Summary and Prospects

William F. Raub

On Behalf of the NIH, Bethesda, MD

Through the presentations recorded in this volume, a succession of distinguished investigators illuminate some of the frontiers of biomedical knowledge. They help us to share the excitement of their quests, stimulate us to reflect upon the known and potential value of their findings, and call to mind something Alfred Nobel wanted us never to forget—that science offers apparently limitless power with which we can affect our lives and our environment. We could honor his memory in no better way than to use that power for the maximum good that we can imagine.

This volume illustrates in detail some of the latest confluences of biomedicine and physical science. Example after example has been given of how uncommonly creative individuals have gained new insights into life processes in health or disease by crossing those imaginary boundaries that we use to partition the knowledge base of science. Few would deny the necessity for this disciplinary organization of science; indeed, pedagogues, archivists, and administrators would be rendered almost helpless without it. But one lesson seems clear. If we do not continually resist the tendencies to compartmenalize concepts and sequester methods, we risk doing more to postpone seminal discovery than to foster it.

As manifested in the previous pages, the fruits of multidisciplinary endeavor are nowhere more evident than in the recent history of molecular and cell biology. Scientists, administrators, and newspersons alike, in attempting to describe the many rapid-breaking developments, have nearly worn out the terms "new biology" and "biological revolution." Yet who can dismiss their repeated references as either excessive or exaggerated? The fact is that we are experiencing the greatest upheaval that the life sciences have ever undergone. And few who are touched by it can remain dispassionate about the prospects for improved human health and well-being, especially when one recognizes that much of this payoff will occur within our lifetime.

Even an over-the-shoulder glance down the paths of biomedical inquiry leading to today will discern the footprints of physics, chemistry, mathematics, and engineering. Modern biology and medicine abound with accouterments derived from these disciplines. For example, there are radioisotopes and scintillation counters for dissecting biochemical events ever more finely; spectrophotometers and other devices for measuring the optical surrogates of chemical and biological processes; light and electron microscopes for observing the fine structure of normal and pathological tissue; mass spectrometers, nuclear magnetic resonance spectrometers, and X-ray and electron diffractometers for elucidating the structure of biologically important molecules; imaging devices, such as gamma cameras and tomographic scanners, for the noninvasive study of physiological and biochemical events in human subjects; macro- and microelectrodes of several descriptions for studying the excitatory phenomena that are associated with cellular functions, especially the workings of those cells that constitute the central nervous system; and—last, but not least—computers for virtually every form of information handling from process control to data-base management to mathematical modeling. The day is long past when one could escape the rigors of the "hard sciences" by opting for a career in biology.

Multidisciplinary achievements generally seem to come about as a consequence of the day-to-day strategic choices of individual scientists in their laboratory or clinical milieu. In some cases, investigators succumb to siren songs emanating from disciplines outside their previous experience and, in so doing, go on to solve biological problems. In other cases investigators look to other domains of knowledge only after they have grown frustrated with the apparent inadequacies of their own. Only in exceptional instances does the integration of ideas from different disciplines seem to be traceable to detailed planning by or specific direction from some central authority or funding source.

The National Institutes of Health (NIH) has long recognized the tremendous value of investigator initiative in the planning and direction of research. We are convinced that fostering such decentralized decision-making is the key to improving human health—our principal mission. Therefore, we expect the scientists in our own laboratories and clinics to take the initiative in choosing topics for study; and, though we monitor their progress carefully with the aid of peer reviewers, we try not to constrain their options any more than necessary. Similarly, we expect applicants for grants to present and justify prospective projects of their own selection, and, in awarding funds to those applicants judged by their peers to be most deserving, we accord the recipients considerable discretion with respect to the direction and/or pace of their work. Virtually every review of the NIH has endorsed its policy of modulating the flow of biomedical research gently along a broad front rather than directing the tide into narrow channels.

But even if, as we claim, it is imperative that the NIH foster investigator initiative in the form of discrete research projects, that does not mean this approach is sufficient. At any given time some important and apparently tractable research problems in biomedicine are likely to be overlooked or left untouched; hindsight is all too good at bringing them into bold relief. Although history shows that seemingly uncoordinated, atomistic, asynchronous assaults on the unknown have been extraordinarily productive of new biomedical knowledge, few observers, if any, will assert that there is no room for improvement.

What are the plausible complements to investigator-initiated research projects? Some scientists and lay persons call for more attempts at corporate planning, including the specification of targets, milestones, and timetables. Others are wary of efforts to program discovery—particularly in the basic research arena—but believe that large groups of scientists (e.g., more than ten) are more likely to solve important problems than the two- or three-person teams that carry out the bulk of current activities. The NIH uses both these approaches selectively, especially where the need for a multidisciplinary orientation is obvious. For example, some new health care procedures cannot be evaluated unequivocally without carefully planned and tightly controlled multicenter clinical trials; deliberate juxtaposition of laboratory and clinical research activity frequently is seen as the means of choice for achieving resonance between the test tube and the bedside. Nevertheless, the NIH repeatedly places the discrete, investigator-initiated project at center stage, trusting in the lessons of history and choosing humility over hubris.

Multidisciplinary research presents many real or apparent barriers right from the outset, even when investigator initiative is working superbly. Scientists trained in different disciplines frequently have difficulty melding their respective fields of expertise, vocabularies, and traditions. Scientists working in fields that are conceptually far apart generally are separated geographically as well and thereby tend to miss the planned and chance human encounters that trigger new approaches to old problems. And, given the inevitable and time-honored pressures to meet the standards and expectations of their peers, scientists have considerable incentive to work within established paradigms and to stay in close touch with their disciplinary origins. Fortunately, every generation of scientists includes some whom these barriers cannot deter for long.

Less obvious, perhaps, as a restraining influence on multidisciplinary projects is the tendency of life scientists to put biological processes into service as experimental tools. In some instances, such as the classical LD_{50} measurement, the bioassay is crude to the point of embarassment; one is hard-pressed to believe that contemporary science could not produce something better if the search were sufficiently broad and vigorous. By contrast, some bits of biotechnology are so ingenious that they leave current physical science methodology in their wake; three prominent examples are the Ames test, hybridomas, and recombinant DNA. Ironically, such uniquely biological approaches to life science problems can be so successful in meeting their immediate purposes that they temporarily quench enthusiasm for seeking alternatives through other disciplines.

The Ames test (*1*) employs bacterial mutants that are unable to grow if the culture medium lacks the amino acid histidine. When millions of these bacteria are simultaneously subjected to some type of potentially mutagenic insult (e.g., chemical or radiation), a few bacteria will undergo mutation such that they subsequently begin to synthesize histidine and grow into visible colonies. The test therefore yields an unequivocal, rapid, and inexpensive indication of induced DNA change and for many purposes is far superior to mutagenesis assays that require intact animals. To learn as much or more at comparable cost with physicochemical methods alone—e.g., isolating and comparatively characterizing megabase stretches of DNA under various test conditions—is as yet out of reach.

A similar analysis obtains with respect to hybridomas (*2*). Fusion of antibody-secreting plasma cells and myeloma cells yields hybrids that not only produce monoclonal antibodies, but also are long-lived in culture and form tumors in animals. The chemical products of these living factories generally bind only with a particular antigen, thereby enabling scientists to home in on molecular targets in vivo with a degree of specificity previously unattainable. If hybridomas had not been invented, achievement of the equivalent capability through conventional means would not come quickly and, in all likelihood, would require ambitious and wide-ranging efforts to separate, purify, and characterize myriad molecules (e.g., receptors on cell surfaces), as well as to design, synthesize, and produce large quantities of high-specificity reagent molecules.

Recombinant DNA (*3, 4*) rounds out the illustrations. Using current techniques for augmenting the genetic makeup of bacteria or other cells, scientists are well along toward being able to elicit virtually any gene product, such as human proteins, from cells in culture. Previous attempts to produce complex biomolecules by physicochemical means—e.g., laboratory synthesis or large-scale extractions from cellular material—frequently proved to be either unsuccessful or prohibitively expensive. The technological alternatives to recombinant DNA are likely to suffer by comparison for the forseeable future.

These illustrations of biological ingenuity are in no sense intended to downplay the importance of multidisciplinary initiatives. Active research in the areas mentioned must and will continue. However, the examples emphasized the propensity of biology to offer some of the means to its own elucidation. The unity of science remains for us to exploit as best we can, synthesizing concepts and transplanting methods from field to field, as necessary, to serve our ends. To this continuing commitment we need add only our recognition that, in dealing with nature, the straight path is not always the shortest.

REFERENCES

1. Ames, B. N., McCann, J., and Yamaski, E. (1975), *Mutation Res.* **31,** 347.
2. Kohler, G., and Milstein, C. (1975), *Nature* **256,** 495.
3. Jackson, D. A., Symons, R. H., and Berg, P. (1972), *Proc. Natl. Acad. Sci. USA* **69,** 2904.
4. Cohen, S. N., Chang, A. C. Y., Boyer, H. W., and Helling, R. B. (1973), *Proc. Natl. Acad. Sci., USA,* **70,** 3240.

Author Index

B

Baldwin, J. P., 35
Baltimore, David, 17
Bonadonna, Gianni, 243
Bradbury, E. M., 35

C

Calvin, Melvin, 189
Cooper, Leon N, 103
Cormack, A. M., 151

G

Glaser, Donald, 145

H

Halvorson, Harlyn O., 171
Hubel, David H., 91

L

Lauterbur, Paul, 211

N

Nicolini, Claudio, 67

P

Pauling, Linus, 3
Prigogine, Ilya, 217

R

Raub, William F., 257

S

Sachs, Leo, 225

T

Temin, Howard, M., 9

Subject Index

A

A-MuLV, 22
Acid, aspartic, 173
  phosphatase, 182
Acremonium, 174
Adriamycin, 248
Alcohol, 196
  terpenoid, 203
Alu, sequences, 14
Antibiotic, 174
Antibodies, 4, 31
  monoclonal, 260
Antigens, 4
ATPase, 183
ART, multiplicative, 213
Aspergillus, 172
Association, 114
Autoinduction, 235
Axons, 106

B

Bacillus subtilis, 178
*Bam*HI, 176, 183
Basis, preclinical, 244
Benard, 220
Biology, 223
Biotechnology, 171
Blobs, 91, 94, 100
Bloch, 211
Blood, cells, 226
  squid, 104
Brain, 148
  human, 104

C

Calotropis procera, 201
Cancer, 17
  breast, 246
  testicular, 245
Capacity, heat, 72
Carboxylase, -ribulose-1,
    5-biphosphate, 185
Cells, blob, 100
  cortical, 93
  differentiation, 225
  dormant stem, 245
  HeLa, 77
  hematopoietic, 227, 228
  leukemic, 235
    $MGI^{-}D^{-}$, 237
  nonblob, 100
  resistant, 244
CEP, 254
Chemotaxis, 223
Chemotherapy, cancer, 243
Chirality, 221
Chromatin, 35
  active, 64
  fibers, 59
  higher-order, 48
  H1-depleted, 62
  hydrated, 55
  inactive, 62
  shearing of, 68
Chromatosome, 48, 49
Chromosome, 78
  histone-depleted, 71
  scaffold, 77
Cleavage, peptidase, 182
Climate, 221
Clones, $MGI^{+}D^{+}$, 230
Cloning, 176, 226
CMF, 248
Coccidiosis, 186
Coil, *rf,* 213
Colicin, 177
Colliculus, 91
Color, 148
Columns, 100
Complementariness, 5

Compound, isomeric C_5, 203
Computers, 107
Conditions, rearing, 138
Connections, synaptic, 110
Connectivity, binocular, 137
Contrast, 58
Convection, 220
Conversions, coal, 190
Copaifera multijuga, 205, 207
Core, hydrophobic, 47
  particles, 39, 41, 42
Cortex, striate, 91, 95, 98
  visual, 105, 122
    experimental, 122
Cyclase, pyrophosphate, 207
Cyclosporin A, 186
Cytotoxic, 240

D

D_2O, 37, 40
Dehydration, 59
Deletion, 26
Dendrite, 106
Dependence, ionic strength, 68
Deuterium, 40
Diagram, 133
Dictyostelium discoideum, 223
Differentiation, 233
  clonal, 226
Diffraction, 35
  neutron, 55
  neutron-fiber, 57
  patterns, X-rays, 57
  peak, 61
Dimethylallyl-pyrophosphate, 203
Dioxide, carbon, 189
Disease, Hodgkin's, 253
Distribution transverse-radial, 55
Disturbances, electrical, 105
  chemical, 105
DNA, 9, 41, 67, 70
  c, 14
  euchromatic, 82
  halo of free loops of, 78
  heterochromatic, 82
  length, 72
  loops of -, 71
  recombinant, 174
  replication, 77
  strand brakes, 81
  twisted, 70
Dominance, ocular, 139
Drosophila, 12, 71, 78
Drug, combination, 244
  resistance, 251
Dynamics, classical, 219
  thermo-, 219

E

*Eco*RI, 176
Electrode, 98
Elements, movable genetic, 12
Engineering, genetic, 171, 207
  protein, 184
Enthalpy, 72
Entropy, 219
Envelope, nuclear, 76, 77
  oocyte nuclear, 77
Enzymes, 175
  isolated, 184
  kinetics, 223
  restriction, 175
Equilibrium, 219
  non-, 220
Escherichia coli, 29, 172
Ethidium Bromide, 69
Euphorbia lathyris, 198, 199
  tirucalli, 199
Evolution, 9
Eye, 91

F

Fermentation, galactose, 172
Fiber, rope-like, 81
Fibrosome, 83
Filament, 10 nm chromatin, 50
  30 nm chromatin, 53
Fossil, non-, 192
Fuels, production, 199
  renewable, 189
Function, internal-structure, 43
  nonlinear, 128
  radial-distribution, 43
  scattering, 43
  shape, 43
Fusion, 26

G

Galactosidase, β, 172
Gaps, 93
Gemma, 214
Gene, 84, 173, 207
  abl, 21

ampicillin-resistant, 178
dosage, 239
erb A, 28
expression, 236
gag, 15
high-order structure,
histone, 83
insertion, 177
interferon, 182
invertase structural, 180
myb, 28
raf, 28
splicing, 180
tetracycline-resistant, 176
transplant, 207
Geniculate, 91
Genomes, 71
Genotype, 240
Germ-line, 14
Globulin, 14
Gradients, 213
Granulocytes, 227
Growth, control, 229
normal, 226
uncoupling, 234
Guinier, plots, 40
region, 53
Gyration, radius, 41
of chromatin, 50

H

Helix,
α-, 181
-coil, 75
Hepatitis B, 13
virus (pararetrovirus), 14
Histones, 41, 48, 70, 71
acetylation, 64
H1, 36, 48, 68
H2A, 36
H2B, 36
H3, 36
H5, 48
HMG, 64
non-, 70
Hormones, 245
Hologram, 108
Humidity, relative, 57
Hybridization, 83
Hydrocarbons, 190
Hydrogen, chloride, 5
bond, 5
Hyperphosphorylation, H1, 62

I

Images, T_1, 213
Imaging, magnetic resonance, 159
NMR, 211, 214
Immunotoxins, 185
Information, 104
Inhomogeneities, 213, 223
Instability, 128
Interferon, human, 182
Invertebrates, 109
Ions, 71
Isomers, D, 222
L, 222
Isoprenoid, 203
Isotopes, 159

J

Junction, 111
synaptic, 107, 111

K

K^+, 104
Kinase, 29

L

Lactamase, β-, 178
Laws, Newton's, 217
Learning, 115
neuron, 103
Leukemia, differentiation in, 229, 234
murine, 19
Levels, Zeeman energy, 211
Library, D, 30
V, 30
J, 30
Life, 217
LMF, 248
Logic, 112
Loops, 48
LTR, 13
Lymphocytes, PHA-stimulated human, 77
Lymphoid, -B, 20
Lymphomas, malignant, 253

M

*Mbo*I, 176

Macaques, 94
Macrophages, 227
  mature, 231
Magnetization, nuclear, 212
Malignancy, 225, 236
  reversal, 238
Mappings, distributed, 110
Marrow, bone, 229
Materials, 189
Matrix nuclear, 80
Mechanisms, visual, 145
Membrane, 105
Memory, 103
  distributed, 108
  long-term, 112
  short-term, 112
  storage, 109
Microbes, 171
Microscopy, fluorescence, 78
  freeze-etching electron, 80
Model, 110
  mathematical, 119
    drapery-like, 73
    rope-like, 73
    Mark II ice-cube, 100
Modification, network, 115
  passive, 117
MOPP, 252, 255
  -BAP, 255
  Motion, 148
  Myeloid, 229

N

Na, 104
Network, neural, 105, 110
Neurons, 130
  network, 103
Newton, 217
Nuclease, micrococcal, 67
Nuclei, 69
Nucleofilament, 68
Nucleosome, 35, 39, 42, 71
  sub-, 39
  Supercoil, 55
Nucleus, 77
  cell, 57

O

Oblate, 45
Oil, 189
  crude, 202
Oncogenes, 17
Operon, *lac*, 178
Organization, brain, 103
Oxidase, cytochrome, 94, 96

P

pBR322, 178, 179
pC194, 179
pH, low, 184
Paradox, C, 84
Particles, ribosomal, 77
Penicillium, 174
PET, 157
Phase, 133
  G1, 77
  G2, 62
  meta, 62
  S-G2, 78
*PHO*81, 183
Physics, 217, 223
Plants, 189
  hydrocarbon-producing, 197
Plasmids, 177
Polymerase, RNA, 83
Polypeptides, 77, 180
Pores, 67
  nuclear, 77
Positrons, 157
Problem, binocular, 136
  monocular, 134
Process, stochastic, 121
Processing, image, 77
Profiles, thermal denaturation, 68
  DNA, 68
  chromatin, 68
Promoter, 178
Protein, 237
  Abelson, 19
  differentiation-inducing, 226
  HMG, 36
  maltose-binding, 181
  MGI, 226
  MGI-1, 229
  MGI-2, 229
  MGI$^{+}_{+}$ D$^{-}_{+}$, 230
  MGI D, 230, 239
  nonhistone, 83
  secretory, 181
Psychophysics, 100

R

Radioastronomy, 162
Radon, 151
Rays, X, 36, 55
Reaction, Belousov-Zhabotinski, 221
Receptor, hormonal, 250
  lambda-phage, 181
Recognition, 113
Recollection, 113
Regions, A–T, 83
Regulation, enzyme, 172
Relapsing, 245
Repeats, inverted, 13
Replicon, hybrid, 179
Residue, phosphate, 29
Resonance, nuclear magnetic, 159
Results, mathematical, 130
Retina 145, 148
Retroviruses, 9
Rings, 57
RNA, 77
  m-, 83, 84, 180
  viruses, 9

S

Saccharomyces cerevisiae, 179, 182
*Sau*3A, 176
Scaffold, 78
Scanning, 151
Scatter, neutron, 35, 41
Scattering, differential light, 72
  length, 37
Secretion, protein, 180
Sedimentation, 70
Selectivity, 130
Sequence, 84
Sesquiterpenes, 207
Site, primer binding, 11
Sites, active, 184
Skeleton, 71
Spacing, Bragg, 55
Specificity, biological, 3
Spectrometer, Varian A-60, 213
Stain, cytochrome oxidase, 97
Staphococcus aureus, 178
Stereo, 148
Strand, β-extended, 181
Structure, high-order gene, 82
  nuclear, 67
  quaternary, 67, 73
  quinternary, 73, 75
Stuhrmann, 41
Subunits, chicken-erythrocyte chromatin, 40
*SUC*2, 180
Sugar, cane, 195
Supercoil, 54, 55
Superhelix, density, 70
Survival, 245
Synapses, cortical modification, 125
System, central nervous, 107
  immune, 26

T

Terminal regions, N-, 62
  C-, 62
Terpenes, 204
Test, Ames, 260
Tetrafilament, 300-Å, 72
Therapy, 239
  systemic adjuvant, 243
Thermodynamics, 219
Thinking, logical, 103
Thymus, calf, 55
Time, relaxation, T_1, 211
  relaxation T_2, 212
Titration, fluometric, 69
Tomography, Computer Assisted, 152
  Positron Emission, 157
Tract, polypurine, 11
Transcription, 83
  reverse, 12
Transformation, 176
Transition, salt-induced, 53
Transposons, 12
Trees, hydrocarbon-producing, 203
Tubulin, 14
Tumors, 212
  carcinogen-induced, 245
Turns, negative superhelical, 71, 72
Tyrosine, 29

U

Ubiquitination, 36
Urchins, sea, 83

V

v-abl, insert, 24
  sequences, 26

Van der Waals, 5
Vector, *Hind*III cloning, 178
  host-, 177
Viruses, DNA, 9
  endogeneous, 9
  exogeneous, 9
  RNA, 9
  Rous sarcoma, 18
  Abelson, 18
Viscosity, 79
Vision, 145
  color, 91
Volume, nuclear, 80

Y

Yeast, 179